About IFPRI

The International Food Policy Research Institute (IFPRI), a CGIAR research center established in 1975, provides research-based policy solutions to sustainably reduce poverty and end hunger and malnutrition. IFPRI's strategic research aims to foster a climate-resilient and sustainable food supply; promote healthy diets and nutrition for all; build inclusive and efficient markets, trade systems, and food industries; transform agricultural and rural economies; and strengthen institutions and governance. Gender is integrated in all the Institute's work. Partnerships, communications, capacity strengthening, and data and knowledge management are essential components to translate IFPRI's research from action to impact. The Institute's regional and country programs play a critical role in responding to demand for food policy research and in delivering holistic support for country-led development. IFPRI collaborates with partners around the world.

www.ifpri.org

About CGIAR

CGIAR is a global research partnership for a food-secure future. CGIAR science is dedicated to reducing poverty, enhancing food and nutrition security, and improving natural resources and ecosystem services. Its research is carried out by 15 CGIAR Centers in close collaboration with hundreds of partners, including national and regional research institutes, civil society organizations, academia, development organizations, and the private sector.

www.cgiar.org

GLOBAL FOOD POLICY REPORT

2022

CLIMATE CHANGE & FOOD SYSTEMS

A Peer-Reviewed Publication

Copyright © 2022 International Food Policy Research Institute (IFPRI).

This publication is licensed for use under a Creative Commons Attribution 4.0 International License (CC BY 4.0). Subject to attribution, you are free to share, copy, and redistribute the material in any medium or format, and adapt, remix, transform, and build upon the material for any purpose, even commercially.

THIRD-PARTY CONTENT: The International Food Policy Research Institute does not necessarily own each component of the content contained within the work. The International Food Policy Research Institute therefore does not warrant that the use of any third-party-owned individual component or part contained in the work will not infringe on the rights of those third parties. The risk of claims resulting from such infringement rests solely with you. If you wish to re-use a component of the work, it is your responsibility to determine whether permission is needed for that re-use and to obtain permission from the copyright owner. Examples of components can include, but are not limited to, tables, figures, or images.

The boundaries and names shown and the designations used on the maps in this publication do not imply official endorsement or acceptance by the International Food Policy Research Institute (IFPRI).

RECOMMENDED CITATION: International Food Policy Research Institute. 2022. *2022 Global Food Policy Report: Climate Change and Food Systems*. Washington, DC: International Food Policy Research Institute. https://doi.org/10.2499/9780896294257

This is a peer-reviewed publication. Any opinions expressed herein are those of the authors and are not necessarily representative of or endorsed by the International Food Policy Research Institute (IFPRI).

International Food Policy Research Institute
1201 Eye Street, NW
Washington, DC 20005-3915 USA
www.ifpri.org

ISBN: 978-0-89629-425-7
ISSN: 2329-2873
DOI: https://doi.org/10.2499/9780896294257

Photo credits
COVER: Prashanth Vishwanathan/IWMI.
CHAPTER IMAGES: p. 16 atravellens/Shutterstock.com; p. 28 PradeepGaurs/Shutterstock.com; p. 38 Tiwuk Suwantini/Shutterstock.com; p. 47 Sven Torfinn/Panos Pictures; p. 58 Mikkel Ostergaard/Panos Pictures; p. 81 D.Cz./Shutterstock.com; p. 90 Sven Torfinn/Panos Pictures; p. 107 C. de Bode/CGIAR; p. 146 Jordi Ruiz Cirera/Panos Pictures; p. 161 PradeepGaurs/Shutterstock.com

BOOK LAYOUT: Lee Dixon
EDITORIAL MANAGER: Pamela Stedman-Edwards

Contents

FOREWORD ... 3

ACKNOWLEDGMENTS ... 5

CHAPTER 1 **Climate Change and Food Systems:** Transforming Food Systems for Adaptation, Mitigation, and Resilience ... 6
Johan Swinnen, Channing Arndt, and Rob Vos

CHAPTER 2 **Repurposing Agricultural Support:** Creating Food Systems Incentives to Address Climate Change ... 16
Rob Vos, Will Martin, and Danielle Resnick

CHAPTER 3 **Trade and Climate Change:** The Role of Reforms in Ensuring Food Security and Sustainability ... 28
Joseph W. Glauber

CHAPTER 4 **Research for the Future:** Investments for Efficiency, Sustainability, and Equity ... 38
Gert-Jan Stads, Keith Wiebe, Alejandro Nin-Pratt, Timothy B. Sulser, Rui Benfica, Fasil Reda, and Ravi Khetarpal

CHAPTER 5 **Climate Finance:** Funding Sustainable Food Systems Transformation ... 48
Eugenio Díaz-Bonilla and Ruben Echeverría

CHAPTER 6 **Social Protection:** Designing Adaptive Systems to Build Resilience to Climate Change ... 58
Daniel O. Gilligan, Stephen Devereux, and Janna Tenzing

CHAPTER 7 **Landscape Governance:** Engaging Stakeholders to Confront Climate Change ... 64
Ruth Meinzen-Dick, Wei Zhang, Hagar ElDidi, and Pratiti Priyadarshini

CHAPTER 8 **Nutrition and Climate Change:** Shifting to Sustainable Healthy Diets ... 72
Marie T. Ruel and Jessica Fanzo

CHAPTER 9 **Rural Clean Energy Access:** Accelerating Climate Resilience ... 82
Claudia Ringler, Alebachew Azezew Belete, Steven Matome Mathetsa, and Stefan Uhlenbrook

CHAPTER 10 **Bio-innovations:** Genome-Edited Crops for Climate-Smart Food Systems ... 90
José Falck-Zepeda, Patricia Biermayr-Jenzano, Maria Mercedes Roca, Ediner Fuentes-Campos, and Enoch Mutebi Kikulwe

CHAPTER 11 **Food Value Chains:** Increasing Productivity, Sustainability, and Resilience to Climate Change ... 100
Alan de Brauw and Grazia Pacillo

CHAPTER 12 **Digital Innovations:** Using Data and Technology for Sustainable Food Systems ... 106
Jawoo Koo, Berber Kramer, Simon Langan, Aniruddha Ghosh, Andrea Gardeazabal Monsalue, and Tobias Lunt

REGIONAL DEVELOPMENTS ... 114

Africa ... 116
Jemimah Njuki, Samuel Benin, Wim Marivoet, John Ulimwengu, and Caroline Mwongera

Middle East and North Africa ... 120
Clemens Breisinger, Amgad Elmahdi, Yumna Kassim, and Nicostrato Perez

Central Asia ... 124
Kamiljon Akramov, Kahramon Djumaboev, and Roman Romashkin

South Asia ... 128
Aditi Mukherji, Avinash Kishore, and Shahidur Rashid

East and Southeast Asia ... 133
Kevin Chen and Yue Zhan

Latin America and the Caribbean ... 137
Eugenio Díaz-Bonilla, Carolina Navarrete-Frias, and Valeria Piñeiro

PROJECTIONS FROM IFPRI'S IMPACT MODEL: CLIMATE CHANGE AND FOOD SYSTEMS ... 146

NOTES ... 162

Foreword

This year's *Global Food Policy Report* on food systems transformation and climate change echoes the somber warning issued by recent IPCC reports: as we continue to degrade the environment and push beyond our planetary boundaries, we are entering a "Code Red for Humanity." Food systems are inseparably linked to this unprecedented crisis, which threatens the food security, nutrition, and health of billions of people. Our food systems are not only severely impacted by climate change, requiring an urgent focus on adaption, but also play a role in causing about one-third of global greenhouse gas emissions, with two-thirds of that resulting from agriculture, forestry, and other land use. Investing in food systems transformation is a key piece of the climate change puzzle, yet it is vastly underfunded, with only a small part of climate finance directed toward this goal.

In 2021, as the COVID-19 pandemic continued to trigger health and economic crises around the world, the international community came together to recognize the centrality of food systems for meeting development and sustainability goals. The first-ever United Nations Food Systems Summit advanced food systems to the top of the global policy agenda, and the UNFCCC COP26 commenced plans to truly incorporate agriculture into COP27 in 2022. But these developments fall far short of what is urgently needed: a wide range of investments in climate-positive research, development, policies, and programs rooted in food systems.

The 2022 *Global Food Policy Report* highlights a range of evidence-based policies and innovations that should be prioritized and implemented now to tackle adaptation and mitigation in our food systems. Drawing on research from IFPRI and other CGIAR centers, it offers lessons that can help us better achieve food security, nutrition, and sustainability through climate-positive financing, innovation, and governance.

Going forward, research on transforming food systems to deal with climate change will remain at the heart of the 2030 CGIAR Research and Innovation Strategy. This strategy guides science and innovation initiatives at IFPRI and One CGIAR to advance the transformation of food systems, as well as land and water systems, in a climate crisis. The COVID-19 pandemic as well as the current upheaval of global food markets, caused by a series of conflicts around the world, have made this research strategy even more essential and urgent.

We hope that the 2022 *Global Food Policy Report* will support transformation by contributing to global policy discussions and to the many national and local policy discussions and reforms that will be essential to food systems transformation. We look forward to engaging and working together with many partners around the world to contribute to this transformation and thus to a better future.

JOHAN SWINNEN
Director General, IFPRI
Global Director, Systems Transformation, CGIAR

Acknowledgments

The *2022 Global Food Policy Report* was prepared under the overall leadership of Johan Swinnen, Channing Arndt, and Rob Vos and a core team comprising Charlotte Hebebrand, Pamela Stedman-Edwards, Jamed Falik, and Sivan Yosef.

Text and data contributions were made by Kamiljon Akramov, Alebachew Azezew Belete, Rui Benfica, Samuel Benin, Patricia Biermayr-Jenzano, Clemens Breisinger, Kevin Chen, Alan de Brauw, Stephen Devereux, Eugenio Díaz-Bonilla, Kahramon Djumaboev, Ruben Echeverría, Hagar ElDidi, Amgad Elmahdi, José Falck Zepeda, Jessica Fanzo, Ediner Fuentes-Campo, Andrea Gardeazabal Monsalue, Aniruddha Ghosh, Daniel O. Gilligan, Joseph W. Glauber, Ahmed Kamaly, Yumna Kassim, Ravi Khetarpal, Avinash Kishore, Jawoo Koo, Berber Kramer, Simon Langen, Tobias Lunt, Will Martin, Steven Matome Mathetsa, Wim Marivoet, Brian McNamara, Ruth Meinzen-Dick, Enoch Mutebi Kikulwe, Aditi Mukherji , Caroline Mwongera, Carolina Navarrete-Frias, Alejandro Nin-Pratt, Jemimah Njuki, Grazia Pacillo, Nicostrato Perez, Valeria Piñeiro, Pratiti Priyadarshini, Shahidur Rashid, Fasil Reda, Danielle Resnick, Claudia Ringler, Maria Mercedes Roca, Roman Romashkin, Marie Ruel, Gert-Jan Stads, Timothy B. Sulser, Janna Tenzing, Stefan Uhlenbrook, John Ulimwengu, Rob Vos, Abdul Wajid Rana, Keith Wiebe, Sivan Yosef, Yue Zhan, and Wei Zhang.

Production of the report was led by Pamela Stedman-Edwards. Lee Dixon was responsible for design and layout. Editorial assistance was provided by Claire Davis and Gillian Hollerich.

We would like to thank IFPRI's donors, including all those who supported IFPRI's research through their contributions for the CGIAR Fund: https://cgiar.org/funders/.

CHAPTER 1

Climate Change and Food Systems
Transforming Food Systems for Adaptation, Mitigation, and Resilience

JOHAN SWINNEN, CHANNING ARNDT, AND ROB VOS

Johan Swinnen is director general, International Food Policy Research Institute (IFPRI), and global director, CGIAR Systems Transformation Group, Washington, DC. **Channing Arndt** is division director, Environment and Production Technology Division, IFPRI, Washington, DC. **Rob Vos** is division director, Markets, Trade, and Institutions Division, IFPRI, Washington, DC.

KEY MESSAGES

- Climate change is a growing threat to our food systems, with impacts becoming increasingly evident. Rising temperatures, changing precipitation patterns, and extreme weather events, among other effects, are already reducing agricultural yields and disrupting food supply chains. By 2050, climate change is expected to put millions of people at risk of hunger, malnutrition, and poverty.

- Aspirations for food systems are extremely high. Global summits in 2021 highlighted the central role of food systems transformation in the world's response to climate change as well as meeting multiple other development goals. Action to address climate change is underway but must be hastened by accelerating innovation, reforming policies, resetting market incentives, and increasing financing.

- Adaptation is urgent, but feasible for food systems. Food production, distribution, and consumption practices must all be adapted to climate change to better support rural livelihoods and provide healthy diets for all, even as population and income growth increase the demand for food.

- Food systems contribute substantially to greenhouse gas emissions and must play a role in mitigation through changes in agricultural practices and land use, more efficient value chains, and reduced food loss and waste.

- Many promising innovations and policy approaches show potential to address climate change in food systems while also increasing productivity, improving diets, and advancing inclusion of vulnerable groups. These range from new crop varieties, clean energy sources, and digital technologies to trade reforms,

landscape governance, and social protection programs. All of these will require substantial increases in funding for R&D and other investments in sustainable food systems transformation.

- Food systems policies that create better market incentives, strengthen regulation and institutions, and fund R&D for climate-resilient technologies and practices are needed to catalyze and accelerate climate action.

Climate change is a growing threat to our food systems, with grim implications for food and nutrition security, livelihoods, and overall well-being, especially for poor and vulnerable people around the world. The imperative for urgent action on climate change — both to achieve the major emissions reductions needed to limit global warming and to increase adaptive capacity and resilience of food systems — is drawing global attention.

The impacts of global warming are becoming increasingly evident. Higher temperatures, changing precipitation patterns, sea level rise, and growing frequency and intensity of extreme weather events such as droughts, floods, extreme heat, and cyclones are already reducing agricultural productivity, disrupting food supply chains, and displacing communities.[1] At the same time, food systems are estimated to contribute more than a third of the global greenhouse gas (GHG) emissions responsible for climate change,[2] placing food production at the center of attention as both a contributor to global warming and a critical sector for mounting an adaptive response to climate change.

Looking forward, modeling scenarios, created by researchers at the International Food Policy Research Institute (IFPRI) together with other CGIAR colleagues, indicate that rising temperatures will negatively impact agricultural yields, driving up prices and resulting in increased hunger, especially in Africa (see IMPACT data in this report for details).[3] The goal of ending hunger will remain elusive even by 2050, especially considering the additonal impacts of extreme weather events, local shocks, and global crises, such as COVID-19 and the current war in Ukraine, that will push many more people into poverty and hunger. Thus, beyond its direct impacts on production, climate change will create cascading effects on livelihoods and sustainability through interconnections among economic, environmental, social, and political spheres.

Even in the absence of climate change, food systems face enormous challenges and demands. Hunger and malnutrition are rising, and over 3 billion people currently cannot afford a healthy diet.[4] Food systems are the world's largest "employer," but for many, particularly women, youth, and other vulnerable groups, agriculture-based livelihoods are precarious. In addition, food systems are major contributors to environmental degradation beyond GHG emissions, including deterioration of water resources and loss of habitat and biodiversity, which compromise environmental services that support food production.

Yet global aspirations for food systems are extremely high. As was made clear at the 2021 UN Food Systems Summit (UNFSS), food systems must play a central role in achieving multiple, pressing sustainable development and climate goals, from the

BOX 1 INTERNATIONAL SPOTLIGHT ON FOOD SYSTEMS AND CLIMATE ACTION

Key international events over the past year have cemented the centrality of food systems transformation in the climate change and Sustainable Development Goals (SDG) agendas. Global calls for "building back better" after COVID-19 include a push for more sustainable, healthy, and equitable food systems. The chorus of voices for change suggests that now may be the moment.

The United Nations convened its first-ever Food Systems Summit (UNFSS) in September 2021, marking an important shift from prior World Food events (1992, 1996, 2002). By moving to a food systems view that encompasses the production, processing, transport, and consumption of food, the UNFSS highlighted the role of global food systems in achieving the SDGs by 2030. The close links among food security and nutrition goals, climate goals, and many of the other SDGs point to the "the need to confront the realities of balancing food production with climate action, affordable food with healthy diets, and stable food supplies with fair and open trade."[a] Likewise, the December 2021 Tokyo Nutrition for Growth Summit highlighted the links between food systems and nutrition and climate change.[b] At the close of the UNFSS, the UN Secretary General outlined the need for concrete follow-up at the national level, as countries prepare pathways to transform food systems and achieve their climate commitments.[c]

The UNFCCC COP26 held in November 2021 stressed that much more action is required to meet commitments to net zero emissions, and countries were asked to strengthen current targets. In the realm of agriculture and land use (AFOLU), 137 countries pledged to halt and reverse forest and land degradation by 2030, and over 100 countries pledged to reduce methane emissions, including those from the agriculture sector.[d] The Koronivia Joint Work on Agriculture, an important workstream of the UNFCCC, highlighted the key role of soil and nutrient management practices and livestock management systems, and signaled that a draft decision on agriculture will be released in 2022. While such promises are encouraging, previous commitments have not been met. The 2009 pledge to provide US$100 billion per year from 2020 through 2025 for climate change adaptation in developing countries, for example, has been postponed for several years.

The 2021 commitments will require concrete follow-up by national governments to ensure real change. Significant shifts in public and private investment will be essential, an issue that was discussed at both UNFCCC and UNFSS (see Chapter 5, Box 1). The 2022 UN Conference on Biodiversity and the World Trade Organization ministerial conference, also planned for 2022, will provide further opportunities to advance global climate and food systems action.

local to the global level. This message was strongly reinforced at last year's UNFCCC COP26 in Glasgow (Box 1). These aspirations envision food systems that are far more nature-positive, deliver improved and more resilient livelihoods, empower disadvantaged groups, and produce a healthy mix of foods at affordable prices. Food systems are called on to accomplish all these goals in the context of a rapidly changing climate and while making a substantial contribution to achieving net zero GHG emissions by 2050.

Action to address climate change has begun, but it urgently needs to be hastened by accelerating innovation, reforming policies, resetting market incentives, and increasing financing for sustainable food systems transformation. This year's *Global Food Policy Report* sets out a broad range of opportunities for accelerated action that should be considered by international and domestic forums for policy and investment decision-making.

This first chapter has two purposes. First, we ground this report in current realities related to the climate change adaptation and mitigation demands upon food systems. These issues have been highlighted and explored in recent global forums and publications such as the UNFSS, COP26, and the Sixth Assessment Report of the IPCC, so we discuss them only briefly here. The second half of this chapter summarizes the major findings and recommendations presented in this report.

ADAPTATION AND MITIGATION IN FOOD SYSTEMS

ADAPTATION IS URGENT, BUT FEASIBLE FOR FOOD SYSTEMS

The world remains far from achieving the emissions reductions needed to constrain warming to 1.5°C above pre-industrial levels. While this goal remains feasible, the longer we wait, the more difficult achieving it will become. For each tenth of a degree that the global average temperature rises above 1.5°C, human and environmental costs are expected to escalate at increasing rates. For the foreseeable future, climate change will continue to disrupt food systems with greater frequency and severity, unless action is taken now.

Food production, distribution, and consumption practices must be adapted to climate change in order to support rural livelihoods and provide healthy diets for all, even as population and income growth increase the demand for food. At the farm level, adaptation must address changing growing conditions, water scarcity, droughts and floods, increased risks of destructive weather events, and related risks of disease and pests. Along value chains, storage and logistics will also be affected by climate change, and price volatility will increase, with implications for processors, traders, and consumers as well as farmers. Climate change will put increasing pressure on scarce resources, which can increase the risk of conflicts. Such conflicts can affect entire value chains and are a major driver of increases in global food insecurity and hunger.[5] Recent experience with the COVID-19 pandemic has shown us how disruptions in one part of a value chain can have wide-reaching impacts. But it has also shown us how food systems, including public and private sector actors, can respond and adapt quickly to severe shocks.

A number of promising innovations show potential to support adaptation and build resilience while also increasing productivity. New crop varieties can better withstand climate shocks as well as improve yields. Solar energy can be used to improve product storage as weather conditions worsen, and also contribute to mitigation. Digital technology can expand access to knowledge and services in rural areas, allowing producers to adapt practices to local conditions and improve market access. Many climate-smart innovations, such as no-till farming, agroforestry, and landscape management, will also support mitigation by sequestering carbon or reducing emissions. However, technical innovations will never reach their full potential without the right enabling environments, including policy incentives and governance approaches that promote climate-positive change and inclusion of all food systems actors. Policies and institutions at the local, national, and international levels need to incentivize the development and adoption of new technologies and practices and ensure adequate finance. They must recognize potential trade-offs — in terms of yields and efficiency — between sustainable systems and existing or other modern farming practices and between sectors, such as water and energy.

Policies must also facilitate coordinated action across stakeholders and ensure equitable systemic transformation for all.

FOOD SYSTEMS HAVE A VITAL ROLE TO PLAY IN MITIGATION

Despite international commitments to reduce GHG emissions, total net anthropogenic GHG emissions continue to rise. Stepped-up efforts to reduce GHG emissions are required of developed economies, and progress by developing and emerging economies is also necessary, but it is important to bear in mind that the least developed countries accounted for only 3.3 percent of global GHG emissions in 2019.[6] Absolute emissions from fossil fuels in some developed countries (Western Europe, North America, Australia, and New Zealand) have been trending downward, largely as a result of policy support and technological advances, including development of clean energy sources.[7] For now, developed countries are leading the way in these innovations, but as these and other technologies mature, they must be swiftly adopted in developing country markets.[8]

Food systems account for as much as 34 percent of total greenhouse gas (GHG) emissions stemming from agriculture and land use, storage, transport, packaging, processing, retail, and consumption.[9] Continued technological progress in the energy and transport sectors can reduce fossil-fuel use and emissions throughout food systems, including in irrigation, processing, transport, cold storage, and waste recycling, where emissions are currently increasing. But two-thirds of food system GHG emissions – or about 21 percent of total emissions from all sources – are from agriculture, forestry, and other land use (AFOLU) (Figure 1).[10]

AFOLU can deliver substantial emissions reductions and carbon sequestration.[11] It is the *only* economic sector with serious potential to become a net emissions sink – pulling more GHGs out of the atmosphere than it emits – through creation and protection of carbon sinks such as forests.[12] Given that some sectors (energy, industry, transport, buildings) will not reach net zero emissions by 2050, AFOLU must reach negative emissions to achieve the topline goal of COP26: *Secure global net zero emissions by mid-century.* Viewed in this way, AFOLU must achieve significantly larger total emissions reductions than other sectors.

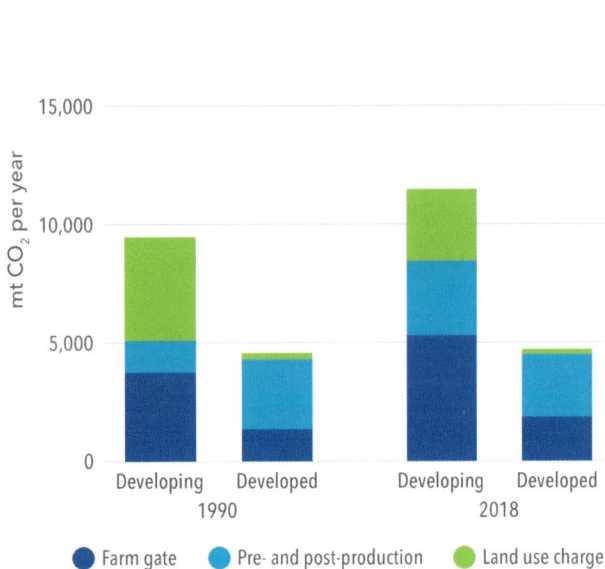

FIGURE 1 Greenhouse gas emissions from food systems, 1990 and 2018

Source: Constructed using data from F.N. Tubiello, K. Karl, A. Flammini, et al., "Pre- and Post-production Processes along Supply Chains Increasingly Dominate GHG Emissions from Agri-food Systems Globally and in Most Countries," *Earth Systems Science Data Discussion* [preprint 2021].

Yet realizing this potential requires addressing substantial barriers, such as insufficient institutional and financial support, uncertainty regarding long-term increases in sequestration, risks of carbon sequestration reversal, and our ability to measure and verify sequestration.[13]

Globally, land use change and management accounts for almost half of total CO_2 emissions from AFOLU (Figure 2).[14] Net agricultural land expansion is concentrated in the developing world,[15] and between 2003 and 2019, cropland expanded by about 9 percent globally, principally due to agricultural expansion in Africa and South America. Conversely, land use change in the United States is providing a net sink, offsetting about 12 percent of total US emissions (and more than all US emissions from agriculture).[16] For developed countries, the top priority should be measures that will turn their landscapes into larger net sinks for emissions. For developing countries, the priority should be fostering agricultural practices that both raise productivity and turn the tide on AFOLU emissions.

FIGURE 2 Global AFOLU emissions shares by source

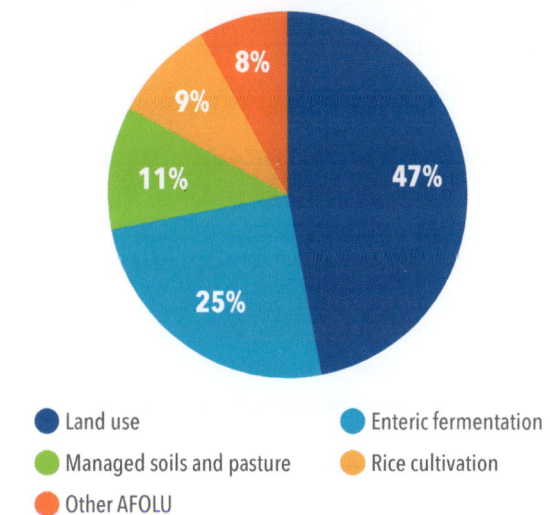

- Land use: 47%
- Enteric fermentation: 25%
- Managed soils and pasture: 11%
- Rice cultivation: 9%
- Other AFOLU: 8%

Source: Constructed using data from W.F. Lamb, et al., "A Review of Trends and Drivers of Greenhouse Gas Emissions by Sector from 1990 to 2018," *Environmental Research Letters* 16, 7 (2021): 073005.

Note: AFOLU = agriculture, forestry, and other land use.

Significant GHG mitigation can also be achieved in AFOLU by reducing nitrous oxide emissions from fertilizers and methane emissions from paddy rice and enteric fermentation (from cattle and other ruminant digestion) as well as decreasing emissions intensity within sustainable production systems and reducing food loss and waste. From the demand side, shifting food consumption toward healthy diets has also been found to have substantial potential for emissions reduction.[17] Combined, these efforts could move the world toward net zero emissions.

POLICY RECOMMENDATIONS

Promising policy responses for adaptation, mitigation, and resilience to climate change are explored in this *Global Food Policy Report*. Eleven thematic chapters and a regional section, covering six major developing regions, examine policy options and opportunities for change. These are arranged in three broad groupings: 1) Global frameworks for policies and incentives (Chapters 2–5), 2) Inclusion and diversity, livelihoods, and resilience (Chapters 6–8), and 3) Sustainable production and consumption (Chapters 9–12). Here we review broad findings and recommendations from these chapters that can support climate change responses in the short term and build resilience and capacity for the future.

R&D FOR CLIMATE-RESILIENT, RESOURCE-EFFICIENT, AND SUSTAINABLE INNOVATIONS IN FOOD SYSTEMS. A promising portfolio of technology innovations could accelerate sustainable food systems transformation. Many of these innovations have proven potential to both raise productivity and reduce GHG emissions intensity in agrifood production. Irrigation technologies, such as drip irrigation and solar power pumps, can both improve yields and reduce emissions (Chapter 9). New genome-editing technologies, such as CRISPR, have proven capable of rapidly developing crop and animal varieties suited for climate change adaptation and mitigation (Chapter 10). Improved cold chain technologies, powered by solar energy, and new drying methods are increasing food quality and availability and reducing food loss and waste, particularly for perishable nutritious foods (Chapter 11). Digital innovations are revolutionizing production, markets, and delivery throughout food systems, with great potential for improving productivity and quality and reducing natural resource use and food loss and waste. This broad array of innovations in data use stretches from precision agriculture, improved weather forecasting, and use of robotics to blockchain-based product quality and sustainability traceability and to e-logistics and e-commerce for enhanced value chain efficiency (Chapters 11 and 12).

Development and adoption of such "disruptive" innovations requires investment in R&D. A review of evidence on the benefits of past investments in R&D for innovation in agrifood systems finds that the benefit-cost ratio of such investments can be at least 10 to 1, contributing much more to reducing poverty and hunger than other development investments (Chapter 4).[18] Looking forward, an investment in R&D equivalent to just 1 percent of agricultural output could produce a sustained increase of 30 percent in food production (Chapter 2).[19] In addition, investment in development and adaptation of such "green" innovations for use in low- and middle-income countries (LMICs) could help reduce AFOLU emissions by some 40 to 50 percent.[20] Despite these substantial benefits, food systems R&D is notably underfunded, especially in LMICs (Chapter 4), with only a tiny fraction of

> **Appropriate design of policies, institutions and governance systems at all scales can contribute to land-related adaptation and mitigation while facilitating the pursuit of climate-adaptive development pathways (*high confidence*)**
>
> – International Panel on Climate Change 2019

agricultural innovation investment targeting environmental outcomes. Based on this evidence, this report recommends that:

- Public investments in R&D for productivity-increasing and emissions-reducing innovations should be doubled from current levels, with at least $15 billion of the increase for innovations benefiting food systems in LMICs.

- R&D investment should focus on innovations for sustainable intensification in LMICs, both on and beyond the farm.

- Global and regional mechanisms for knowledge sharing, such as the CGIAR system, should be enhanced and strengthened to facilitate technology diffusion that benefits countries with limited domestic research capacity (Chapter 4).

- Governments should create stronger enabling environments to attract private sector investment for agrifood innovations and to spur adoption of improved technologies and practices, including resetting distortionary market incentives created by agricultural support and trade regulations (Chapter 2, 3, 5) and improving regulation for safe adoption and market acceptance of new technologies (Chapter 10).

HOLISTIC, INCLUSIVE GOVERNANCE AND MANAGEMENT OF WATER, LAND, FORESTS, AND ENERGY RESOURCES. Improved efficiency of natural resource use will be essential to increasing productivity while reducing environmental degradation and GHG emissions. The close links among water, energy, land use, and food systems require integrated policy responses to climate change in order to prevent undesirable trade-offs among development goals. Use of modern energy technologies, for example, is essential for raising productivity – including for pumping of irrigation water and storage, transport, and processing of food products – but fossil-fuel use contributes to GHG emissions. Expanding access to "clean" energy sources, including solar power, in the agrifood sector is therefore critical, but brings its own risks of increased exploitation of water and other farm inputs (Chapter 9).

Governance through integrated landscape approaches has potential to achieve sustainable use of land, water, forest, and energy resources (Chapter 7), including long-term productivity and greater sequestration of CO_2. However, integrated landscape management is complex; it requires effective stakeholder engagement, inclusive governance, adequate coordination among local, regional, and cross-border natural resource management, and compromise among diverging economic and political interests (Chapter 7). To promote integrated landscape approaches and sustainable resource use, this report recommends that:

- Agricultural, food, and climate change policies should explicitly consider landscape dimensions and provide incentives for integrated landscape management through local governance, including development of multistakeholder platforms that can build support for collective action on climate change.

- Land tenure and access rights to other natural resources for farmers, rural households, and communities should be strengthened to motivate investments in sustainability and participation in landscape governance (Chapter 7).

- Governments should promote adoption of clean energy sources in agrifood systems through an enabling environment and appropriate financial incentives for the use of wind and solar power and decentralized electricity grids (Chapter 9)

- Identification of productive-use locations that could jointly support energy, water, and food security can be used to attract investments that increase productivity and sustainability (Chapter 9).

PROMOTING HEALTHY DIETS AND INCREASED SUSTAINABILITY OF FOOD PRODUCTION. Globally, undernourishment and micronutrient deficiency continue to rise even as overweight and obesity are becoming more prevalent. Both forms of malnutrition now affect about a quarter of the world's population, with some people suffering from both, and poor diets are among the largest global health risks.[21] Making healthy diets affordable and influencing consumers to make healthier choices is key to overcoming these global nutrition challenges and can be well aligned with addressing climate change (Chapter 8). The ecological footprint of healthy diets – those without excessive consumption of highly processed foods and red meats – has been found to be much lower than that of prevailing diets across the world, but especially those in advanced countries.[22] Changing dietary habits is not easy, however. Key policy directions include the following:

- All countries should adopt national food-based dietary guidelines. These can be a key policy instrument to translate global evidence on healthy and sustainable diets into practical, culturally appropriate, and context- and population-specific dietary recommendations (Chapter 8).

- Innovation policies should prioritize R… nutrient-rich foods (including fruits and… to make healthy diets more affordable… Targeted consumer subsidies and remo… taxes on healthy foods will also help to l… costs of healthy diets for low-income households (Chapter 8).

- Consumers can be encouraged to make healthy, sustainable food choices through changes in the food environment, including use of food standards, labeling, and certifications that warn of unhealthy foods and signal the nutritional value and environmental footprint of foods (Chapter 8).

IMPROVING VALUE CHAIN EFFICIENCY, FACILITATING TRADE, AND REDUCING FOOD LOSS. Climate change impacts – including shifts in crop production, rising temperatures, changing humidity levels, and more frequent extreme weather – will affect whole value chains through which agricultural products are traded, aggregated, processed, and sold to consumers. Incentives for producers and other value chain actors will be altered as climate change reduces the effectiveness of some inputs, increases risks, and impacts transaction costs. International trade can play a key role in softening these impacts – by reducing price volatility and providing access to food for countries that have suffered a drop in production. Trade and investments in climate-smart practices all along value chains can support adaptation through increased efficiency of resource use and reductions in food loss. Climate-smart practices in value chains can also support mitigation. A large share of agrifood sector GHG emissions (35 percent) is generated beyond the farmgate, largely by energy use in long supply chains and food waste and loss (Figure 1).[23] Policy priorities include:

- While efforts to reduce transport-related GHGs should be continued, free and open trade should be an integral part of climate-smart agricultural and food policies. Trade allows countries to obtain nutritious foods at the lowest cost and can be a key tool for adaptation in the face of weather-related shocks. Globally, trade can also promote more efficient use of natural resources and thus help reduce GHG emissions from agrifood production (Chapter 3).

- Investments along value chains for efficient and safe storage and transport of food crops and products, including low-emissions cold chains for perishable products and other measures to prevent spoilage and safety hazards, can improve access to healthy diets and reduce food waste and loss (Chapter 11).

- Increasing consumer demand for sustainably produced foods, for example through certification programs, can create incentives for changing practices along entire value chains (Chapter 11).

INCLUSION AND SOCIAL PROTECTION. Poor rural populations who depend on agrifood systems for food and livelihoods are among the most vulnerable to the impacts of climate change. These groups remain underserved in many ways, including access to markets, finance, and knowledge, as well as infrastructure, energy, and natural resources, and hence have limited capacity to benefit from innovations in food systems. The precariousness of their livelihoods leaves them likely to lose income, experience hunger and malnutrition, and suffer long-term impacts on well-being when climate-related shocks occur. Policy reforms must both improve access to services, markets, and financial tools, particularly for women and other vulnerable groups, and ensure a secure safety net for all vulnerable populations. Social protection systems now cover about 2 billion people around the world, and while their potentially important role in addressing climate change challenges has been recognized for more than a decade, they now need to be expanded to do so more effectively. Increased resilience, empowerment, and agency among the disadvantaged can support multiple social, economic, and environmental benefits.

- Climate-positive food systems transformation will require development of context-appropriate institutions and in "soft" infrastructure inclusive of rural and urban food system actors, including equal access to digital climate services, innovative insurance tools, advisory services and actionable information, and financial services to support increased productivity and sustainability (Chapters 9 to 12).

- Women's participation, along with that of other vulnerable groups, should be strengthened across resource governance, including in clean energy systems, water systems, landscapes, crop development, and digital innovations (Chapters 7, 9, 10).

- Social protection programs can provide a safety net for vulnerable groups and support sustainable food systems transformation, including the transition to more climate-resilient crops and to off-farm and urban employment (Chapter 6).

- Expanding "adaptive" social protection programs that comprise traditional social assistance, humanitarian responses, and disaster relief, and that are integrated with complementary climate investments targeted to the poor, can immediately reduce the impact of shocks and support inclusion in food systems transformation (Chapter 6).

- Improved real-time monitoring of food crisis risks is needed to take early and preventative action to protect vulnerable populations in contexts affected by conflict, natural resource scarcity, and exposure to climate shocks (Chapter 6).

REORIENTING FINANCIAL FLOWS AND ATTRACTING NEW FINANCE. The future of food systems depends on access to sufficient funding to facilitate a shift to sustainable production and consumption and to better livelihoods. Current financial flows – including agricultural support, international development funds, and private investment – are at best insufficient and at worst counterproductive to climate-resilient development. They often support unsustainable and unhealthy production while undervaluing environmental impacts. As much as $350 billion per year will be needed to meet climate-related goals in food systems (Chapter 5), much of which could be "reoriented" from existing sources. In addition, many smallholders and small and medium enterprises lack access to finance needed to transform their production practices and to weather climate shocks. Moving forward, investment for environmental, social, economic, and nutrition goals could be increased in several ways.

- Reform of existing counterproductive incentives created by current agricultural, trade, and investment policies can mobilize both public and private finance for climate-positive food systems transformation and reorient funds toward climate finance (Chapters 2, 3, and 5).

- Public support to agriculture, totaling an estimated $620 billion per year worldwide, should be repurposed toward R&D for green innovations and incentives to producers to adopt and invest in climate-smart technologies and practices. Such innovations should focus on increasing productivity, reducing emissions, and enhancing resilience in food production (Chapter 2).

- International development funds should be clearly targeted to meeting climate and sustainability goals, and used to leverage or crowd-in private funds from global capital markets (Chapter 5).

- Reorientation of consumer demand – through better information, food environments, and fiscal tools – will also create incentives for producers to adopt and invest in sustainable and climate-resilient practices (Chapters 5 and 11).

- Innovative mechanisms for tapping additional resources, such as publicly guaranteed "green bonds" or climate-change transparency requirements for banks and investors, should be explored to ensure climate finance needs will be met (Chapter 5).

Achieving these reforms and ensuring widespread adoption of innovations will be politically challenging. Policy solutions, including incentives, institutions, and financing, will need to be tailored to a wide range of contexts and to balance environmental, nutritional, economic, and social goals. Regions, countries, and local landscapes will need to establish their own priorities, address potential trade-offs, build constituencies for reform, and ensure that reforms generate widespread benefits. The Regional Developments section of this report discusses the varied context of the large global regions, and points to some promising opportunities.

Investing in policies and innovations to support sustainable food systems will contribute significantly to global economic prosperity, poverty reduction, food security, and healthy diets as well as to planetary health.[24] Reaching the ambitious goals set for food systems will require inclusive, holistic approaches that consider all components of our food systems, from local to global and from farmer to consumer. Charting optimal pathways for sustainable food systems transformation will be challenging, but we must step up our efforts now to ensure our global future.

CHAPTER 2

Repurposing Agricultural Support
Creating Food Systems Incentives to Address Climate Change

ROB VOS, WILL MARTIN, AND DANIELLE RESNICK

Rob Vos is division director and **Will Martin** is a senior research fellow, Markets, Trade, and Institutions Division, International Food Policy Research Institute (IFPRI), Washington, DC. **Danielle Resnick** is a nonresident fellow, IFPRI, and a fellow with the Brookings Institution, Washington, DC.

KEY MESSAGES

- Agricultural support policies transfer around US$620 billion per year to the farm sector worldwide.

- Support policies based on subsidies and trade barriers are highly distortive to markets and are also regressive, as most support is provided to larger farmers. On balance, the incentives this support creates appear to increase greenhouse gas emissions that contribute to climate change.

- Support provided to the farm sector in the form of subsidies involves budget allocations that can be reallocated. Better outcomes could be achieved if even a small portion of agricultural subsidies were repurposed into investments in R&D dedicated to productivity-enhancing and emissions-reducing technologies. Repurposing would create multiple wins — mitigating global climate change, reducing poverty, increasing food security, and improving nutrition.

- Because current support policies are often politically popular and serve well-organized interests, reform will be difficult, especially since repurposing would need to be internationally coordinated for greatest effect.

- Successful reform will require detailed analysis of winners and losers and thoughtful strategies. Creating constituencies for reform at the national level and in international forums will be essential to build political consensus for concerted global action.

Agricultural support policies provide enormous transfers of resources to farmers – about US$620 billion per year worldwide in 2018-2020 – and enjoy strong political support in both developed and developing countries. Some agricultural support policies, such as input subsidies, have boosted global food production, particularly of staple crops, thereby reducing hunger and poverty. Yet, there are serious concerns about their impacts on achieving sustainable, healthy, and inclusive food systems. Redirecting, or "repurposing," agricultural subsidies toward investments that support both increased production and greater sustainability – such as agricultural research and development (R&D) and rural infrastructure – has the potential for win-win-win gains for people, planet, and prosperity.

CURRENT IMPACTS OF AGRICULTURAL SUPPORT AND OPTIONS FOR REFORM

Current agricultural support goes largely to agricultural producers, primarily in forms that affect market prices and distort incentives for producers and consumers (Box 1). Support coupled to output or input use increases output, thus increasing greenhouse gas (GHG) emissions from agricultural production and land conversion for agriculture. Support provided through trade barriers, however, may reduce global emissions by reducing demand for output. The strong focus of many agricultural support policies on promoting staple crops has improved access to basic calories but has done much less to improve dietary diversity. Moreover, impacts of the support are often regressive – benefiting wealthier commercial farmers, while denying poorer farmers access to markets – and, when provided through trade protection, raise the cost of food and harm poor consumers.

Government support to agriculture is often justified by perceived needs to protect farm incomes, ensure food availability, and promote agricultural productivity. Of the $620 billion total, individual producers received about $540 billion in "positive" support[1] per year (2018-2020) through market price support and subsidies. However, its efficiency in delivering benefits to farmers is low, estimated at 35 percent,[2] with the remainder either shared with consumers or dissipated as economic waste. Many interventions create trade

> **BOX 1 CURRENT AGRICULTURAL SUPPORT**
>
> Current agricultural support (provided by 54 countries for which comparable data are available) amounts to US$616 billion per year, net of taxes on agriculture (2018–2020).[a] Of this, positive direct support to farmers amounts to $540 billion per year, but some farm activities (often exports) are also taxed, at $104 billion globally per year. Thus, net direct farm support averaged $446 billion per year in 2018–2020 (Figure A).
>
> **BOX FIGURE A** Agricultural producer support by main types of support, 2018–2020 (billions of US$ per year)
>
>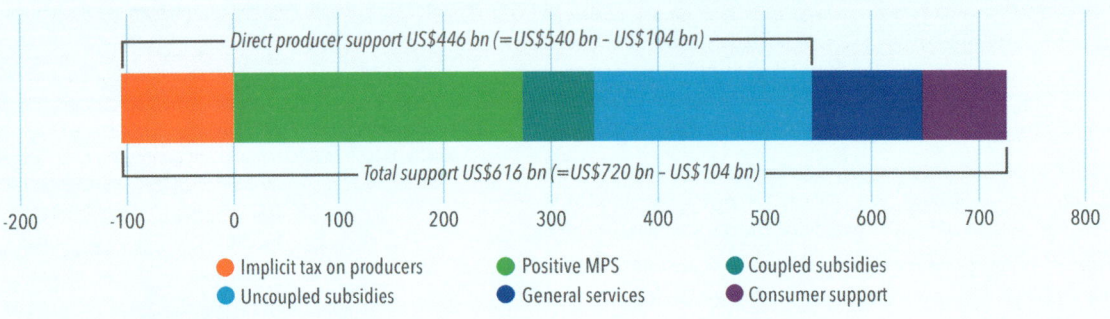
>
> **Source:** OECD, *Agricultural Policy Monitoring and Evaluation 2021* (Paris: 2021).
>
> The "positive" direct support to farmers includes trade measures and market price support, valued at $272 billion per year in total. This support generally does not entail use of government budget resources. Rather, it involves implicit transfers from consumers to producers by creating a price gap between domestic market prices and border prices for specific agricultural commodities. Border measures can take the form of import licenses, tariffs, tariff rate quotas, or export bans that raise domestic prices, benefiting the farm sector. Some emerging and developing countries, including Argentina, India, Indonesia, Kazakhstan, Russia, and Viet Nam, implicitly tax producers of certain agricultural commodities through export taxes or export restrictions, which depresses the domestic price of these products. This "negative" market price support amounted to $104 billion per year, as mentioned above.
>
> Support measures requiring fiscal expenditures amounted to $448 billion per year. These include direct transfers to producers and consumers such as farm output or input subsidies, consumer food subsidies, and spending on public goods in support of agricultural development. Three-quarters of this support ($268 billion) goes directly to farmers: $66 billion in the form of subsidies directly coupled to levels of production and/or to input use and $202 billion in less directly coupled payments to farmers, such as payments to land. Only a limited portion of budgetary support is for R&D and agricultural innovation systems, infrastructure, and other general services for the sector, with only 4 percent of total support allocated specifically to R&D. In 2018–2020, direct support to consumers in the form of food subsidies amounted to 11 percent of total positive support (or $78 billion per year globally).

conflicts between countries and very few help reduce greenhouse gas (GHG) emissions, despite the threat of devastating climate change impacts on agriculture, especially in tropical zones. Only a small share of total support is invested in public goods, including R&D and rural infrastructure, although both the private and social returns of such investments are estimated to be very high (see Chapter 4).

The need for reforms is now well recognized,[3] and the urgency of reducing GHG emissions and adapting to climate change has added impetus to the calls for reform. However, recent studies — discussed below — have shown that simply eliminating all existing support would not greatly reduce emissions, but would depress farm incomes, increase poverty, and raise the cost of healthy diets.[4] Public discourse thus has shifted to how existing support might be repurposed to create better incentives for producers and consumers. The 2021 United Nations Food Systems Summit (UNFSS) called for such repurposing as part of a just rural transition to sustainable food systems.[5]

GLOBAL SCENARIO ANALYSIS: REMOVING ALL SUPPORT

In a series of recent studies, IFPRI estimated the impact of a complete withdrawal of current agricultural support on GHG emissions, farm output, poverty, food security, and diets using its global model, MIRAGRODEP.[6] A first, perhaps surprising, result is that current measures have only a small influence on the overall (global) volume of agricultural production (Figure 1), although they do have important impacts in individual countries. Second, at the global level, withdrawals of domestic subsidies and border measures have offsetting impacts on production and emissions. Removing subsidies *reduces* both global food output and emissions, but removing border protection, which acts as a tax on demand, slightly *increases* global output and emissions in protecting countries. The combination of removing both subsidies and border support slightly *reduces* global output and GHG emissions from agriculture (Figure 1), lowers farm output, and raises the cost of healthy diets. Thus, simply abolishing all support would not be a game-changer and would involve trade-offs between environmental, economic, and social objectives.

FIGURE 1 Global implications of repurposing domestic support

Source: M. Gautam, D. Laborde, A. Mamun, W. Martin, V. Piñeiro, and R. Vos, *Repurposing Agricultural Policies and Support: Options to Transform Agriculture and Food Systems for Better Health of People, Economies and the Planet*, Technical Report (Washington, DC: World Bank and IFPRI, 2022).

Note: Green bars indicate movement toward societal goals; orange/red bars indicate movement away from societal goals.

The impacts of removing all agricultural subsidies differ substantially between rich and poor countries (Table 1). The drop in farm income per worker would be four times larger in developed countries than in developing countries. Farm employment would decline in developed countries but increase in developing countries, where higher world prices would induce a supply and employment response. However, poverty in developing countries would increase, as higher food prices push more people below the poverty line. GHG emissions would fall by over 6 percent in developed countries, but worldwide they would fall by only 1.5 percent because agricultural production would shift to developing countries.

Clearly, agricultural policy reform must be carefully thought through to achieve the drastic reductions in GHG emissions that are needed to avert disastrous climate change impacts. Given the multiple goals that food systems are now called upon to address, how can the substantial resources that support agriculture be repurposed in ways that simultaneously provide strong incentives to reduce GHG emissions, improve food system efficiency and farm productivity, and help combat poverty, hunger, and malnutrition?

GLOBAL SCENARIO ANALYSIS OF REPURPOSING SUPPORT

Existing subsidies can be repurposed in ways that would make significant progress toward achieving both global climate and food security goals. Additional model-based analysis conducted by IFPRI and the World Bank[7] indicates that investing an additional 1 percent of agricultural output value in R&D for technologies that both increase the efficiency of production and reduce emission intensities — such as modified diets for ruminants and alternate wetting and drying for rice — complemented by incentives to farmers for the adoption of those technologies could achieve greater gains with fewer trade-offs than simply eliminating subsidies.

This scenario assumes an internationally concerted strategy in which all countries shift resources from current market-distorting subsidies toward more spending on R&D that reduces emissions and, by raising productivity, creates incentives for farmers to adopt the improved technologies. The scenario results are promising: global welfare and food output increase; food prices fall, making food and healthy diets more affordable for many people; and poverty rates fall worldwide (Figure 1). Global GHG emissions from agriculture and land use change would drop by about 40 percent, both because of the direct reduction in emissions from crop production and because higher productivity reduces the need for agricultural land. Farm incomes would fall with the removal of subsidies, although returns to farm labor would rise if policy reform were combined with rural development policies to facilitate a benign movement of labor out of agriculture.

TABLE 1 Impacts of abolishing all agricultural subsidies by country group (percent change)

	World	Developed	Developing
Macroeconomic			
National real income	0.05	0.05	0.04
Farm Sector			
Real farm income per worker	-4.51	-11.36	-2.70
World prices	2.93	2.93	2.93
Production volume - crops	-1.31	-2.56	-1.02
Production volume - livestock	-0.49	-1.10	-0.07
Social			
Farm employment	-0.53	0.25	-0.60
2040 poverty at PPP $3.20	0.05	-0.01	0.06
Nutrition/Diets			
Dairy consumption per capita	-0.42	-0.49	-0.37
Veg & fruit consumption per capita	-0.48	-0.54	-0.45
Healthy diet food prices	1.70	2.17	1.44
Climate			
Emissions from production, % of ALU	-0.59	-1.52	-0.38
Emissions from land-use change, % of ALU	-0.89	-4.52	-0.07
Total emissions, % of ALU	-1.48	-6.04	-0.44
Nature			
Agricultural land	-0.06	-0.15	-0.01

Source: M. Gautam, D. Laborde, A. Mamun, W. Martin, V. Piñeiro, and R. Vos, *Repurposing Agricultural Policies and Support: Options to Transform Agriculture and Food Systems for Better Health of People, Economies and the Planet*, Technical Report (Washington, DC: World Bank and IFPRI, 2022).

Note: ALU = agriculture and land use.

The reduction in GHG emissions could be increased further through complementary policies not considered in this scenario analysis. These could include measures — such as nutrition education, food standards, and taxation — that influence food demand and dietary choices to reduce excess consumption of unhealthy or emissions-intensive food products (see Chapter 8).

POLITICAL CONSTRAINTS TO POLICY REFORM

Reallocation of agricultural support to R&D focused on productivity-enhancing and emissions-reducing technologies could produce better outcomes for food security and nutrition and for the natural environment, especially if carried out in an internationally coordinated manner. However, even the best reform agenda will inevitably face considerable political hurdles.

Political economy studies typically focus on three factors that shape policy outcomes: interests, institutions, and ideas.[8] Interests matter because different groups have different goals and different abilities to organize to promote those goals. Institutions, such as electoral processes, land tenure systems, and international organizations, matter because they filter whose interests prevail in policy processes and shape the scope of potential decisions. Ideas matter because they influence both the goals of stakeholders and the policies used to achieve them. Four case studies of agricultural support policies presented here provide insight into the interplay among interests, ideas, and institutions in generating support for particular interventions and in achieving reforms in the face of often-formidable obstacles (Boxes 2–5).

THE POWER OF INTERESTS AND IDEAS. In recent years, agricultural policy in India has pushed domestic food prices below world levels, helping consumers but hurting many farmers (Box 2).[9] This is consistent with a common global pattern: despite large agricultural labor forces, many lower-income countries tax agriculture, while rich countries with few farmers generously subsidize the sector.[10] In wealthier countries, smaller numbers of farmers tend to be more effective in lobbying for their interests than urban consumers, who spend less of their incomes on food than consumers in poorer countries. In addition to pressures from special interest groups, policymakers often design policies based on broad perceptions of what is considered best for agriculture and food security. Indian policymakers have designed policies based on the "idea-driven" goal of national self-sufficiency in staple foods, seeing self-sufficiency as synonymous with food security. In this regard, India's input subsidy schemes (especially for fertilizer and electricity) and its active R&D program have helped maintain the supply of staple foods, and hence self-sufficiency, despite relatively low farm prices. Input subsidies garner strong political support from both farmers and input providers, interest groups who are perhaps less aware of the losses they suffer from the low food prices (see South Asia in the regional section). Where goals are in conflict, the interest in the self-sufficiency goal seems to be strong enough to override the goal of relatively low food prices — with India's protection on import-competing commodities generally positive, and that for exportables generally negative.[11]

Another important goal of India's agricultural policy is price stabilization for key staple crops. The country's government has pursued this through a combination of trade measures and public stockholding.[12] Purchases for price stabilization and subsidized food distribution to the poor are influenced by minimum support prices. When the Indian government attempted to reform agricultural marketing arrangements in 2020, it encountered intense resistance from different interest groups, including farmers, commission agents, and state governments, and eventually dropped the reforms.[13]

INSTITUTIONAL COMMITMENTS. The evolution of agricultural policies in China highlights a combination of shifting social interests, institutional factors, and ideas (Box 3). During the 1980s, the government taxed agriculture to provide low-cost food for urban consumers. In the 2000s, government policies shifted dramatically toward support for agriculture. This policy change was, in part, a response to the widening urban-rural income gap associated with rapid economic growth and constraints on outmigration from rural areas. In addition, China's commitments at the World Trade Organization (WTO) had an important institutional influence by limiting the growth of agricultural trade

BOX 2 FARM SUPPORT AND OBSTACLES TO REFORM IN INDIA

Bharat Ramaswami
Professor of Economics, Ashoka University, Haryana, India

India's agricultural policies support farmers through input subsidies (fertilizer, electricity, and hence groundwater) and consumers through low food prices. Key staples, namely rice and wheat, receive substantial price support, and subsidized food distribution schemes rely on public procurement that likewise benefits farmers. Electorally driven credit subsidies (created by forgiving formal sector loans) are also sizable. However, agriculture does not provide a viable livelihood for most Indian farmers, with 86 percent of farms working less than 2 hectares and mostly growing staple foods.[a]

Expansion of India's farm support is a perennial election theme. Its enormous political traction persists despite the fact that price subsidies have perpetuated the bias toward staple crops and hampered structural transformation and growth. At the same time, interventionist agricultural trade policies generally protect consumers and implicitly tax producers.

Subsidies have also contributed to environmental degradation and greenhouse gas (GHG) emissions; most notably, methane emissions from rice cultivation are sustained through rice price supports and electricity subsidies. Depletion of valuable water resources is also aggravated by support policies, both directly (through electricity subsidies that promote groundwater withdrawal) and indirectly (through output subsidies that promote overproduction of water-intensive rice). However, the policy debate on environmental damage stemming from agriculture has focused on air pollution, as crop-residue burning is a major contributor to poor air quality in northern India. Crop-residue burning is a common practice in the paddy-wheat crop rotation sustained by support prices.[b] Reduction of GHG emissions has yet to receive similar attention.

Resolving trade-offs between supporting livelihoods and food security, on the one hand, and environmental sustainability on the other, is a challenge in India. Current agricultural subsidies amount to about 2 percent of GDP, but account for about 20 percent of farm income.[c] Any repurposing of support, including toward R&D and promotion of climate-smart policies, could thus cause hardship for poor farmers.

In the past decade, successive Indian governments have experimented with reforms. Historically, open-ended procurement of rice and wheat has been the primary mechanism to provide price support to farmers. The system is logistically demanding, however, and leaves the government with unwanted stocks. For other crops, policymakers have favored price deficiency payments, which are easier to administer despite being expensive and reproducing some of the market distortions of the procurement system. Policymakers increasingly see direct (uncoupled) transfers as an alternative to these distortionary subsidies. Progress has been made in financial systems to facilitate such payments, but gaps remain in reaching all farmers, in part because of poor land records and insufficient digital connectivity.

Agricultural policy reform would serve India's national interests and potentially make an important global contribution to climate change mitigation, but it lacks political ownership. In addition, the country's federal structure gives state governments considerable influence over agricultural policies. Ignoring these constitutional constraints on federal authority has proved costly. In 2020, the central government pushed through reforms to liberalize agricultural markets, but was forced to reverse course when it encountered strong opposition from state governments, commission agents, and farmers, culminating in a year of demonstrations involving more than 5,000 protests.[d] These political interests would be less constraining in an economy that offered plentiful jobs in the nonfarm sector; as things stand, reforms may have to be designed through consensus and then carried out incrementally to gain the necessary political support.

---

protection, leading policymakers to support farm incomes through lump-sum payments to farmers. More recently, China has also increased public investments in R&D and innovations in technologies and practices that raise agricultural productivity, reduce GHG emissions, and enhance carbon sequestration – a combination of approaches that helps both achieve self-sufficiency goals and meet commitments under international agreements on climate change (including the Paris Climate Accord). This exemplifies how

BOX 3 CHINA'S EFFORTS TO REFORM AND "GREEN" AGRICULTURE

Jikun Huang
Professor of Agricultural Economics, China Center for Agricultural Policy, Peking University, Beijing

China's agricultural performance has been impressive, averaging 4.5 percent annual sectoral growth and 7 percent annual growth in farm incomes since the 1980s, while substantially diversifying production. Yet, many challenges remain.[a] The rural–urban income gap has widened, and agricultural expansion has come at the cost of natural resource degradation and high greenhouse gas (GHG) emissions.

Achieving self-sufficiency in staple foods and stability of domestic food prices are policy priorities in China. The Chinese government implicitly taxed agriculture until the early 1990s by keeping urban food prices low. This policy was reversed in the mid-1990s as concerns grew about the expanding rural–urban income gap and urban consumers became less concerned about food prices. The government allowed domestic prices to rise above world market prices and began providing direct payments to farmers – thus shifting from taxation of producers to protection of domestic production. As a result, the nominal rate of protection (NRP) in agriculture increased from –50 percent in 1981 to around +13 percent in recent years, with direct payments adding 5 percentage points (as reflected in the nominal rate of assistance, NRA; see Figure B).

The reversal in China's agricultural policies might have been even greater if it had not been limited by the country's commitment to multilateral trading rules. For instance, protection of domestic rice production would likely have been higher if not for China's commitment to a tariff binding (cap) of 65 percent at the World Trade Organization (WTO). While the country's policymakers remain committed to ensuring grain self-sufficiency, they managed to do so without raising protection for rice, unlike other high-growth economies in the region.

To further support farm incomes, in 2004, the Chinese government introduced a direct payment scheme largely decoupled from agricultural production and increased support through crop procurement schemes. Despite the huge fiscal cost, these reforms had only a modest effect on average farm incomes, and benefits from procurement were unequally shared. As a result, the government phased out public procurement of all commodities, except for rice, wheat, and cotton, and converted all farm subsidies to lump-sum income transfers to farmers in 2015.

Environmental concerns and international commitments to reducing GHG emissions led the Chinese government to enhance its Store Grains (Food) in Land (SGiL) and Store Grains (Food) in Technology (SGiT) programs to raise productivity, enhance food security, and promote sustainable production. The program enlargement, introduced in 2015, included large-scale investments in "high-standard farmland," defined as land with a high degree of resilience to impacts of droughts and floods, water-saving production practices, high yields, and soil improvement. Through the SGiT, public expenditure on agricultural R&D was raised to RMB 26 billion (about US$4.1 billion), overtaking US spending and making China the world's largest public investor in agricultural R&D.[b] The additional R&D is primarily focused on biotechnology and digital technology.

In 2016, the Chinese government also introduced a special project to reduce fertilizer and pesticide use and a subsidy program to promote the use of organic fertilizers. In 2018, Technical Guidelines on Green Agricultural Development were issued, promoting low-carbon and circular-economy technologies to raise productivity, reduce GHG emissions, and enhance carbon sequestration. This strategy is part of China's effort to comply with its commitments under the Paris Accord to reduce GHG emissions by 2030 and achieve carbon neutrality by 2060.

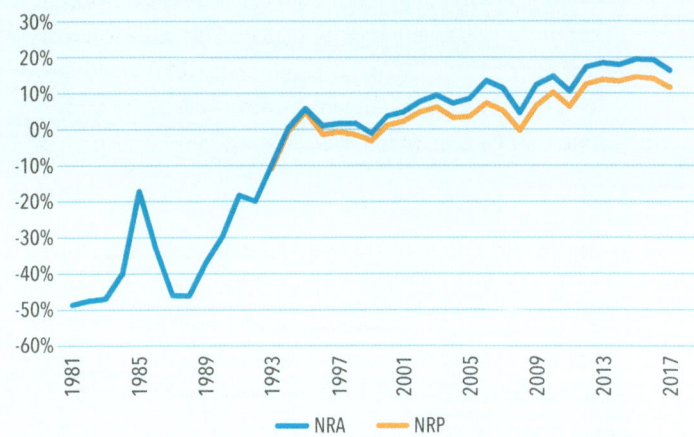

BOX FIGURE B China's support to agriculture, 1981–2017

Source: Data compiled from J.K. Huang and G.L. Yang, "Understanding Recent Challenges and New Food Policy in China," *Global Food Security* 12 (2017): 119–126, and OECD, *Agricultural Policy Monitoring and Evaluation 2021* (Paris: 2021).

Note: Nominal rate of protection (NRP) is calculated as support from border protection divided by the value of agricultural production at world prices. The nominal rate of assistance (NRA) is calculated as support from all sources divided by value of agricultural production at world prices.

institutions (international agreements, in this case) can drive national policy reform.

SHIFTING INTERESTS. European policy reforms between the 1980s and the early 2000s illustrate how even policies that are rooted in long-held ideas, like food self-sufficiency, and heavily supported by powerful interest groups can be fundamentally changed (Box 4). Europe's original Common Agricultural Policy (CAP) maintained high support prices for farmers. But in the

---

BOX 4 THE POLITICAL ECONOMY OF EU AGRICULTURAL POLICY REFORM

Johan Swinnen
Global Director, CGIAR Systems Transformation Science Group, and Director General, International Food Policy Research Institute, Washington, DC

When the European Union (EU)'s Common Agricultural Policy (CAP) was designed in the 1960s, it featured administratively determined market price support, with an important role for import barriers. Farm organizations had lobbied strongly for this system to protect them against internal and external competition. The policy found support in widely felt concerns about food security – typically identified with food self-sufficiency – given the challenges of accessing food in many parts of Europe during and after World War II.

High support prices ignited a strong supply response and turned the EU into a major commodity exporter by the 1980s. The farm support required export subsidies, provoking the ire of other agricultural exporters, particularly the United States, which responded with its own program of export subsidies. As world agricultural prices fell to unprecedented lows during the mid-1980s, pressures from other countries increased as did budgetary pressures, with rising costs of export subsidies and storage. The unsold stocks accumulated in embarrassing "butter mountains" and "wine lakes."

Agricultural exporters pushed hard for reform of global agricultural trade during the Uruguay Round of global trade negotiations (1986–1993). Given European desire to contribute to the Uruguay Round and the internal problems with the price support system, important CAP reforms were introduced in 1963.[a] Reforms reduced support prices and replaced them with direct payments to farmers.

The prospect of accession of 10 Eastern European countries with large agriculture sectors to the EU in the 2000s caused much concern. Expectations were that, unless the CAP was further reformed, their accession would lead to exploding budgets, a massive inflow of cheaper Eastern agricultural products, and a conflict with WTO agreements. Food safety and animal welfare crises in the 1990s compounded the pressure for reform. In addition, reform was made easier by institutional changes, as decisions no longer required unanimous agreement of EU member states, removing the veto power of those most opposed. This resulted in the 2003 reform that decoupled farm subsidies from production decisions, while maintaining the overall level of farm support and allowing the gradual integration of the Eastern European countries in the CAP.[b]

Environmental goals have been gradually integrated into Europe's agricultural policies. Subsequent incremental reforms over the past 30 years have introduced environmental policies and shifted more of the budget to such measures.[c] However, the global food price spikes in 2008–2011 provided arguments for those lobbying against environmental measures that restricted input use and production – weakening pro-environment reforms.[d]

Current reforms aim to build a Farm-to-Fork strategy as part of an EU-wide Green Deal that is designed to make Europe the first climate-neutral continent by 2030.[e] The reforms include payments to farmers conditional on reduced use of pesticides and fertilizers, a shift to organic farming practices, and adoption of new technologies that reduce GHG emissions from agriculture. Development and adoption of new, lower-emission technologies will reduce emissions from both production and land use change. A trade-off is that the reduction in fertilizer use and shift to organic farming practices could reduce productivity and thus create pressure to expand agricultural land, be it in the EU or elsewhere, potentially leading to increased global GHG emissions from land use change or a shift to regions with higher emission intensities.[f] As a decade earlier, high food prices in global markets in 2022 trigger the same political economy reactions, reinforcing lobbying pressure from farmers and agribusiness against environmental policies that would reduce productivity and thus the EU's potential to produce food.

face of rising budget costs for farm and export subsidies and pressures from trading partners to reduce export subsidies, the European Commission initiated a major overhaul of the CAP in the 1990s. During the 2000s, new ideas – namely, growing environmental concerns – drove further reforms to European agricultural policies. Incremental changes to reduce agriculture's environmental footprint were introduced from 2003, and more extensive reforms are underway as part of Europe's effort to become the first climate-neutral continent.[14] The reform proposals include payments to farmers conditional on their adoption of more sustainable practices.

COORDINATING TRADE-OFFS ACROSS GOALS. The United States' biofuel program targets three goals – energy self-sufficiency, farm income support, and emission reductions (Box 5). Combining these goals helped build political support over the years for the biofuel program from farmers, investors in biofuel production, and environmental advocates. Much of the support is provided by mandates, which face less budget scrutiny because they have no direct fiscal cost, and which may result in continued production even when it is uneconomic. However, to allay concerns that the use of food grains for biofuel would reduce food availability, the program set targets for expanding biofuel production from nonfood feedstocks. The efficiency of the technology using such inputs for biofuel production is still unproven, explaining in part why the targets have not even remotely been achieved. This failure sends a cautionary note about setting environmental targets without allocating R&D resources to help achieve them.

COLLECTIVE ACTION CHALLENGES. While these four experiences provide lessons for policy reform, they do not address all the types of policy challenges facing national policymakers. For example, collective action problems associated with management of land and water resources require strong institutions, such as water user groups (see Chapter 7). The weakness of such institutions in many parts of sub-Saharan Africa appears to explain the poor coverage and performance of surface irrigation schemes in the region.[15]

Collective action problems also contribute to underinvestment in agricultural R&D globally. Poor countries often underinvest in R&D because constituents cannot see tangible benefits from these investments in the short term.[16] Small countries have less incentive to invest in R&D because they receive only a small share of the benefits from research findings of broad applicability, and hope to benefit from spillovers from other countries' investments and innovations (see Chapter 4).[17] The CGIAR international research system was developed to address these collective action problems. However, an international system also requires strong national agricultural research systems that can adapt improved technologies and practices to local conditions and can promote their adoption. Creating incentives for developing country governments to allocate more resources to national R&D systems remains a challenge. An interesting funding model is the collective agreement among producer organizations in Côte d'Ivoire to provide funding for reinvestment in the Inter-Professional Fund for Agricultural Research and Extension for services to all agricultural sectors.[18]

A GLOBAL REPURPOSING AGENDA

Repurposing agricultural support clearly holds great promise for generating more sustainable, resilient, inclusive, and equitable food systems. Existing government agricultural support budgets offer a potential source of public finance for innovations and incentives to producers and consumers. Currently, only an eighth of total government support to agriculture is invested in R&D, inspection and control systems, and rural infrastructure – all areas where the private sector tends to under-provide – while three-quarters is allocated to individual producing firms, many of which are commercial and large-scale operations, thus reinforcing inequality. Hence, a strategy to mobilize both public and private finance for food systems transformation should include repurposing of the agricultural support that contributes to solving serious environmental, food security, and equity problems.

Current beneficiaries will undoubtedly resist policy reforms, while those who might gain from reforms are likely to be uncertain about the benefits or insufficiently organized to mobilize for change. Consequently, most policy reforms emerge from development of policy instruments that improve the

> **BOX 5 THE POLITICAL ECONOMY OF BIOFUEL POLICIES IN THE UNITED STATES**
>
> Biofuel policies in the United States are an energy and agricultural strategy with important environmental dimensions. Biofuel policies were first introduced in the 1970s, with the goal of replacing expensive petroleum-based fuels and lead-based additives then used to improve engine performance. They were also supported by interest groups – first farmers and then ethanol producers. As concerns about global greenhouse gas (GHG) emissions increased, biofuels were increasingly justified on environmental grounds.[a]
>
> Support for biofuels was initially provided by a subsidy in the form of a tax credit.[b] Production of ethanol tripled between 2000 and 2007, thanks to the combination of a fixed subsidy and a sharp rise in the price of oil. Reforms in 2005 and 2007 introduced a mandate for the use of biofuels, with targets rising from 13 billion gallons in 2010 to 36 billion gallons in 2022. This policy was enormously popular with ethanol distillers and blenders, who otherwise would face substantial uncertainty about profitability and throughput; however, the mandate makes the demand for feedstock unresponsive to price changes, hence likely increasing the volatility of grain prices.
>
> Because of concerns that transferring large shares of grain output to production of biofuels would raise food prices,[c] the mandate required only a 25 percent increase in conventional biofuels and targeted a twentyfold increase in advanced biofuels, mainly from vegetation unsuitable for human consumption. However, at the time, there was no established technology to achieve this increase, nor have substantial advances been made yet, with the result that advanced biofuel output has increased only sixfold.[d]
>
> Another drawback to promoting ethanol for environmental purposes is that while bioethanol use may decrease fossil-fuel emissions, its production increases emissions through the land use change required to grow bioenergy crops.[e] Considering only the land use change entailed within the United States, recent estimates suggest that US ethanol has a higher GHG intensity than oil-based gasoline.[f]
>
> Several lessons can be drawn from this experience. One is that environmental goals, and particularly mitigation of climate change, may provide important pressure for change. A second is that it may be helpful to build coalitions, including among interests with different but potentially compatible goals – such as energy self-sufficiency and farm income support – to achieve rapid, widely supported reform. However, no single instrument such as biofuel policy can hope to achieve multiple goals, so additional policy instruments are needed.[g] Finally, simply mandating a goal, such as a major expansion of output using new technologies, is unlikely to be successful unless it is backed by investments in targeted R&D.

balance between gains and losses – such as the EU's provision of financial support to farmers who engage in forest conservation and organic practices – or identifying windows of opportunity for change.[19] Windows of opportunity for national reforms may come from international agreements, including the WTO and Paris Climate Accord. Such agreements could also provide an opportunity for developing an internationally concerted repurposing agenda.

The case for such an agenda is easily made. Climate change is an existential threat to food systems globally and the repurposing scenarios analyzed in this chapter clearly show that international cooperation for repurposing achieves superior outcomes on all environmental, economic, and social dimensions for all countries compared with current non-cooperative agricultural support policies. Nonetheless, getting to a common approach will not be easy. This is so because some key tools for emissions reduction – such as carbon taxes and transferable emissions quotas – work less well in agriculture than in sectors dominated by energy-use emissions. For instance, it is difficult to monitor and tax emissions from livestock or rice production. Thus, a carbon tax would create little incentive to change production techniques. Regulatory approaches, such as mandating reduced use of chemical fertilizer or targeting levels of organic farm production, may be ineffective in lowering emissions if they reduce yields (as the evidence suggests) and thus increase the agricultural land footprint, and hence, emissions from land use change.

A detailed analysis of societal gains in the short and long run and of likely winners and losers could help to build support for repurposing. Reallocation of resources to R&D focused on raising productivity and reducing emissions is expected to produce major societal gains, including benefits for those farmers who benefit from current support. However, the gains from innovation in sustainable production methods may be perceived as uncertain, and adoption may come at a cost to producers in the short run. Compensatory payments to losers and to offset adoption costs for producers could help win political support. Importantly, appropriate regulations, such as mandates on the use of renewable energy or limits on the conversion of land for farming, may be essential to overcome the resistance of some agricultural producers to more environmentally sustainable reforms.

Shifting resistance that is tied to ideas, such as the notion that self-sufficiency should be prioritized, may require policy analysis to overcome misperceptions about the impacts of particular policies and reframing of reform benefits in new ways to secure political support. It may require identifying policy options that minimize the cost of a goal that cannot be changed — for instance, replacing a goal of zero imports or exports of any staple with a broader goal of net food self-sufficiency.

Lastly, there are interactive and mutually reinforcing dynamics between the domestic and global policy arenas. Creating constituencies for reform at the domestic level is essential to achieving global action. To spur domestic action and overcome resistance, an even-handed global diffusion of technologies and financial resources is needed to let all countries reap the benefits of agricultural policy reform. Given that climate change and environmental sustainability transcend borders and that national policies have strong international spillover effects, international coordination is essential. However, reaching a common understanding of the benefits of acting together (and the cost of failure) will not be easy. Intense dialogue, informed by continuous and credible assessments of the gains to be obtained and trade-offs to be reckoned with, will be essential to smart repurposing of agricultural support.

CHAPTER 3

Trade and Climate Change
The Role of Reforms in Ensuring Food Security and Sustainability

JOSEPH W. GLAUBER

Joseph W. Glauber is a senior research fellow, Markets, Trade, and Institutions Division, International Food Policy Research Institute, Washington, DC.

KEY MESSAGES

- Climate change is projected to cause significant regional shifts in agricultural production, potentially reduce productivity, and increase the volatility of crop and livestock production.

- Trade allows countries to obtain nutritious foods at the lowest possible cost, and so will be a key component in any strategy to help countries to feed and nourish their populations. Trade can also promote more efficient use of natural resources and potentially reduce greenhouse gas (GHG) emissions.

- Food imports make up a growing share of low- and middle-income country food consumption.

- Government implementation of mitigation and adaptation policies may effectively help address climate change, but concerns arise if those policies run counter to international trade rules. In particular, proposed measures such as carbon border adjustment measures and "climate-smart" agricultural policies could directly conflict with World Trade Organization (WTO) trade rules if they distort production and trade.

- Climate-smart policies such as increasing agricultural productivity and reducing emission intensities through investments in R&D are minimally trade-distorting and one of the most effective ways to address climate change.

- Free and open trade should be seen as integral to any climate-smart agriculture strategy because, globally, it can lead to a more efficient use of resources and can help reduce GHG emissions from global agricultural production.

- To facilitate trade and help to meet global goals for resilience and mitigation, countries should avoid policies and strategies that distort trade, and should pursue further liberalization of agrifood trade through reductions in tariff and nontariff barriers, trade-distorting domestic support, and export subsidies and restrictions.

Climate change poses a major threat to the ambitious global commitments to ending hunger and all forms of malnutrition by 2030, set out in the Agenda for Sustainable Development.[1] Climate change is projected to cause significant regional shifts in agricultural production, potentially reduce productivity, and increase the volatility of crop and livestock production. Reducing malnutrition and hunger in this context will require a concerted effort to help producers adapt to adverse climate outcomes, adopt climate-smart agricultural practices, and mitigate the substantial contribution of agriculture to climate change.

International trade allows countries to obtain nutritious food at the lowest possible cost, and so will be a key component in any strategy to help countries feed and nourish their populations.[2] Trade can also promote more efficient use of natural resources and potentially reduce greenhouse gas (GHG) emissions. Over the past 25 years, imported foods have provided an increasingly large share of calories and nutrients consumed globally, as population growth has outstripped domestic supplies in some regions and productivity gains have allowed other regions to become major exporters. Those trends are expected to continue and arguably will accelerate as climate change affects global and regional agricultural productivity and volatility.[3] However, pressures from changing climate conditions and growing populations mean business-as-usual policies will not be enough. To further facilitate access to affordable, nutritious foods, progress is needed in reducing and repurposing trade-distorting support (see Chapter 2), improving market access, and addressing new issues, such as guarding against export disruptions and bans.

Yet, trade draws criticism for its impacts on nutrition. Growth in trade has included more trade in food products that are considered "less healthful" and are often blamed for rising obesity and poor nutrition.[4] To address the adverse health effects of overconsumption of processed foods, some have advocated using trade policy measures such as tariffs, import bans, or labeling to limit imports.[5] This ignores the fact that nutrition is most directly a consumption issue, and measures that target consumption directly, such as consumption taxes, are almost always better than trade measures for addressing these problems.[6]

Similarly, trade has been criticized on environmental grounds. For example, recent proposals by

> **BOX 1 CARBON BORDER ADJUSTMENT MEASURES**
>
> Carbon border adjustment measures (CBAMs) would levy taxes on imports to account for the level of carbon emissions embodied in the imported good. When applied only to imports, CBAMs may violate WTO national treatment principles if foreign and domestic producers are treated differently. Unequal treatment can distort prices of importables relative to exportables, create competitiveness concerns in export industries, generate economic waste, and likely create highly divisive trade conflicts and deterioration in the terms of trade for developing countries. A recent proposal has been made for a CBAM that would apply equal levies on domestic production and on imports by symmetrically rebating the carbon tax on exports in the manner of a VAT export rebate.[a] Such an approach could shift the base for carbon taxation from output to consumption and intermediate input use, and thus potentially lower the cost of achieving reductions in emissions.

the European Union, Canada, and by some proponents in the United States would implement carbon border adjustment measures (CBAMs) to help reduce GHG emissions (Box 1). These would require that carbon-intensive imports incur either carbon charges or a carbon-based tariff.[7] CBAMs aim to deter carbon "leakage" – for example, if firms shift carbon-intensive production to countries that do not tax GHG emissions (or tax at a low rate) and then export the goods to countries that regulate those emissions, CBAMs could level the playing field and help reduce emissions.

Other proposals would repurpose domestic support to promote climate-smart or nutrition-smart agricultural practices that would reduce GHG emissions, sequester carbon, or promote production of nutritious foods. Depending on how such policies are implemented, these measures could distort trade and potentially conflict with existing trade rules under the World Trade Organization (WTO). This concern has led some to question whether WTO trading rules should be modified to allow policies to take health- or climate-related outcomes into account.[8]

GROWING IMPORTANCE OF AGRIFOOD IMPORTS

Trade in agricultural products has more than tripled in value and almost doubled in volume since 2000 (Figure 1). This growth has been driven, in part, by increased demand for red meat, dairy, and poultry products, particularly in developing countries, and by increases in nonfood uses of cereals, particularly for biofuels.[9] Remarkably, despite the disruption of global supply chains caused by the COVID-19 pandemic, agricultural trade volume declined by only 2 percent in 2020, while trade volumes in energy-related goods fell by 3 percent and manufacturing goods by 6 percent.[10]

Import penetration (that is, imports as a percent of global consumption) has increased for many agricultural products in recent decades (Table 1). In 2020/21, wheat imports accounted for about one-quarter of total global consumption, up from 17 percent in 1995/96. For rice, the import share more than doubled, from 4 percent of consumption in 1995/95 to 9 percent in 2020/21, and soybean imports rose from 25 percent of total consumption in 1995/96 to 45 percent today. Among meat products, only chicken imports have remained flat as a share of consumption; however, import volumes increased substantially, as global chicken consumption more than doubled over the past 25 years.[11]

Import penetration rates for dairy products such as cheese and butter are down substantially from 1995/96 levels. This reflects the removal of distortionary subsidies by large dairy exporters like the European Union and United States in response to new WTO disciplines on export subsidies. (The United States and European Union had relied on export subsidies to dispose of surplus dairy commodities that built up because of their domestic support policies.)[12] In contrast, whole and skimmed milk powder imports have risen in recent years, as use of imported protein concentrates has become increasingly common

FIGURE 1 Growth of world trade, 2000–2020

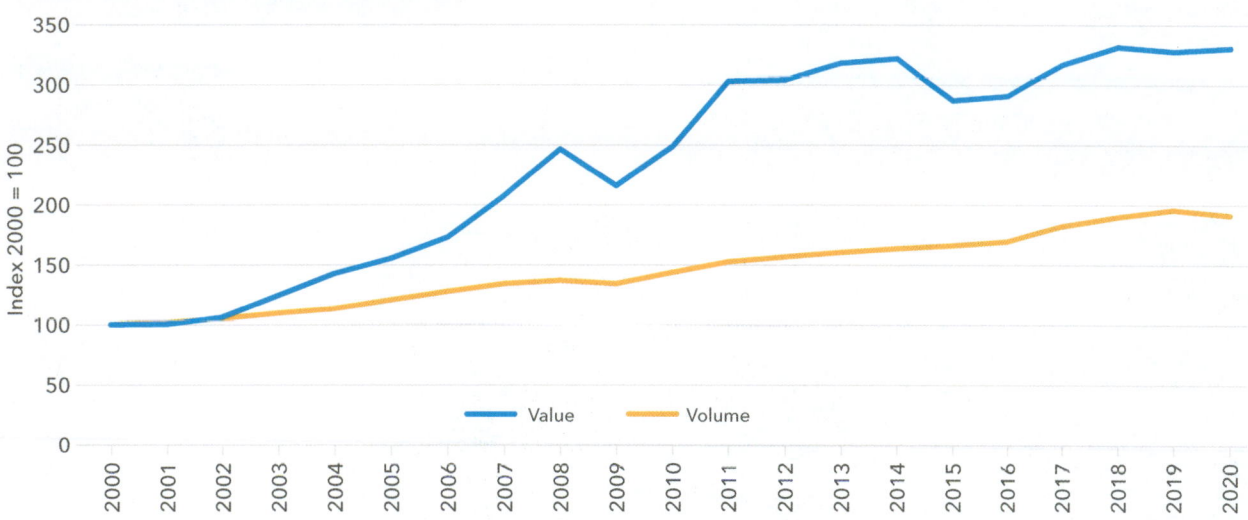

Source: Data from World Trade Organization, *World Trade Statistical Review 2021* (Geneva: 2021).

in food manufacturing and for reconstitution into liquid milk.

At the regional and country levels, cereal imports have grown as a share of consumption (import dependency) since 1999–2001 in many areas where food insecurity is high, such as Africa south of the Sahara (Table 2). Population growth, income growth, urbanization, and changes in consumer preferences have increased demand for wheat, maize, and rice, often outpacing productivity gains for locally produced cereals and driving up imports. Cereal imports in North Africa, the Caribbean, and West Asia now account for half or more of their total cereal consumption. While these imports address food needs, the carbon footprint of some imports, such as rice and livestock products, can be large (see Chapter 2). For example, rice imports for Africa south of the Sahara

TABLE 1 Global import penetration by commodity (imports as percent of consumption)

COMMODITY	1995/96	2000/01	2005/06	2010/11	2015/16	2020/21
Maize	12	11	12	11	13	16
Rice	4	6	7	8	9	9
Wheat	17	18	18	22	24	25
Soybeans	25	31	30	36	42	45
Vegetable oil	34	33	39	41	41	40
Sugar	29	31	32	31	32	31
Beef and veal	10	11	12	12	13	17
Chicken	10	8	9	10	10	10
Swine	4	4	5	6	7	11
Butter	18	6	6	3	3	5
Cheese	21	7	8	7	6	8
Milk powder	47	29	25	27	28	36

Source: Data from US Department of Agriculture, Foreign Agricultural Service, PSD Online.

have grown by about 4 percent annually since 2000, with deleterious impacts on GHG emissions.

Exporting regions show negative import dependencies. Some major cereal-producing regions such as South America and Europe (including Russia and Ukraine) have seen large increases in exports relative to domestic consumption since 1995/96, while cereal exports from other areas including North America, though large, have declined relative to the growth in domestic consumption.[13]

Trading patterns have shifted over the past 25 years as well. Developing countries have become increasingly important suppliers and consumers in world markets, and now account for about 40 percent of world food trade (Figure 2). South-South trade (not shown) alone accounts for over 20 percent of world food trade.[14]

Trade in processed agricultural products has grown along with agricultural trade, more than tripling between 2000 and 2012 before leveling off (Figure 3). Processed products also increased as a share of total agricultural exports, from 42 percent in 2000 to over 46 percent by 2007. The subsequent drop in this share reflects a relative drop in prices, rather than a drop in total volume of processed-product trade. Since 2013, the share by value has recovered, approaching 46 percent again by 2018. Table 3 provides a breakdown of the processed-product trade. In terms of (un)healthy diets, it is important to note the high rates of growth in snack foods (annual growth rate of 8.2 percent), fats and oils (9.0 percent per year), and syrups and sweeteners (8.1 percent per year).

TABLE 2 Cereal import dependency by region (percent of cereal consumption from imports)

REGION	1999–2001	2011–2013	2016–2018
Eastern Africa	13.2	14.8	16.2
Middle Africa	33.9	37.1	31.5
Northern Africa	47.9	47.5	52.4
Southern Africa	9.5	10.4	20.4
Western Africa	17.8	22.6	23.9
North America	−36.0	−23.8	−27.5
Europe	−6.0	−16.1	−29.3
Central America	36.3	33.9	36.7
Caribbean	77.1	74.2	70.3
South America	−4.0	−25.2	−27.4
Central Asia	−21.0	−18.7	−24.1
Eastern Asia	9.7	10.5	9.3
Southern Asia	4.0	−2.3	1.0
Southeastern Asia	4.4	5.8	6.2
Western Asia	39.1	49.3	49.2
Australia and New Zealand	−181.6	−187.6	−154.0

Source: Data from FAOSTAT, https://www.fao.org/faostat/en/#data

FIGURE 2 Developing countries' share of world food trade, 1995–2019

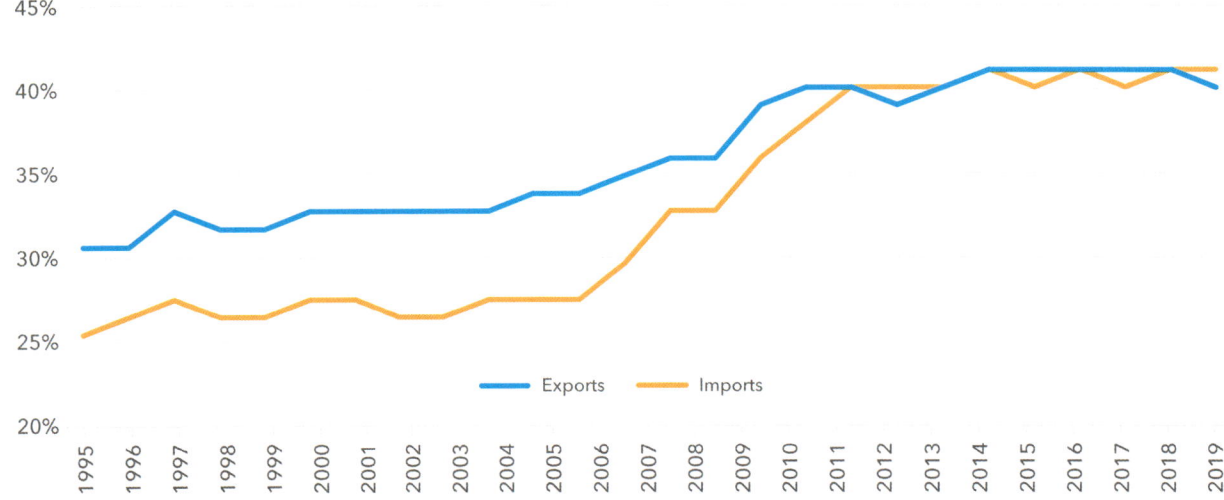

Source: Data from World Trade Organization, *World Trade Statistical Review 2021* (Geneva: 2021).

FIGURE 3 Growth in global trade in processed agricultural products, 2000–2018

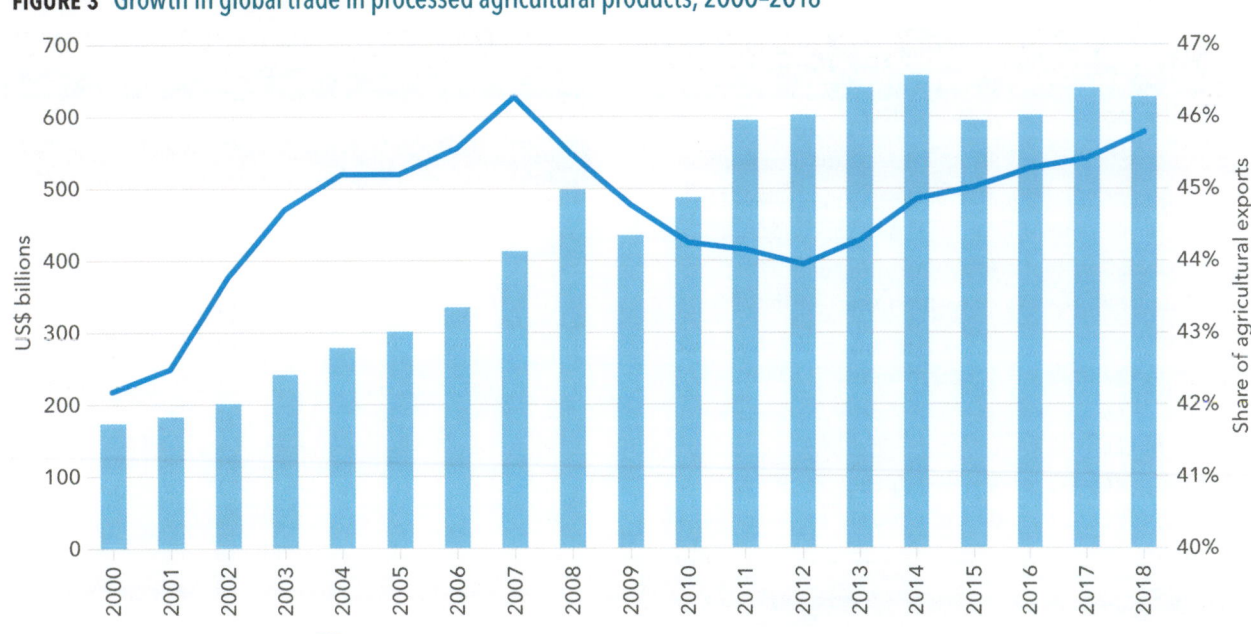

Source: Data from UNCTAD Stat 2021.

TABLE 3 Trade in processed food products (US$ billions)

ITEM	2000	2005	2010	2015	2018
Processed Agricultural Food Total	172.6	299.2	484.2	590.5	623.6
Alcoholic Beverages	29.0	45.8	63.9	74.7	85.7
Food Preparations & Ingredients	17.9	32.3	51.3	69.2	76.5
Processed/Prepared Dairy Products	23.5	38.4	58.3	64.0	76.3
Non-Alcoholic Beverages	16.1	28.0	44.6	58.0	59.5
Chocolate and Confectionery	13.0	23.3	38.2	47.3	48.5
Snack Foods	10.5	19.9	30.0	39.6	43.9
Fats & Oils	17.6	35.7	75.4	80.6	82.4
Processed Vegetables & Pulses	11.9	19.0	29.9	36.8	31.0
Prepared/Preserved Meats	8.5	15.6	24.0	27.8	28.7
Pasta & Processed Cereals	4.7	8.7	13.5	16.3	16.2
Canned, Dried & Frozen Fruit	5.4	8.3	13.0	18.8	17.2
Condiments, Sauces, Jams & Jellies	3.8	6.9	11.2	14.5	14.9
Dog & Cat Food	3.4	5.6	8.8	11.5	13.4
Spices	2.8	3.2	6.5	10.8	6.8
Baby Food	1.6	2.8	5.7	9.0	11.9
Syrups & Sweeteners	1.8	3.5	6.8	8.0	7.2
Soups	1.0	1.5	2.1	2.5	2.4
Processed Egg Products	0.3	0.5	0.8	1.2	1.2

Source: Data from UNCTAD 2021.

IMPEDIMENTS TO TRADE

The WTO Agreement on Agriculture (AoA), which entered into force in 1995, provided disciplines in the areas of market access, domestic support, and export competition.[15] Yet, despite those reforms and the subsequent growth in agricultural trade, significant trade barriers remain in the form of: 1) tariff and nontariff border measures that affect market access (including export restrictions); 2) domestic subsidies that distort production and trade; and 3) export subsidies, including export credits and other concessional sales.

MARKET ACCESS

Agricultural tariff rates remain high for certain product groups. For WTO members, the simple average bound tariff rate (that is, the maximum permitted rate) for agricultural products is over 50 percent.[16] However, applied (actual) tariff rates are generally much lower, in part because countries aim to keep food prices low.[17] Table 4 shows the average applied most-favored-nation (MFN) tariff rates for selected food product groups in the United States, European Union, Brazil, China, and India. In general, applied tariffs are higher for dairy, meat, and sugar products and lower for bulk products like cereals and oilseeds. As most poor families in developing countries are net purchasers of food,[18] lower tariffs and increased market access would likely improve the nutritional status of both rural and urban food-insecure households.[19] In addition, increased market access through lower border measures could reduce GHG emissions, according to recent analysis, if meat and dairy production shift to more GHG-efficient regions (that is, where GHG emissions per unit of meat or milk output are lower).[20]

In addition to explicit tariffs on goods, there are numerous nontariff barriers (NTBs), such as sanitary and phytosanitary restrictions, regulations on product packaging, and more recently, public and private standards that impose labeling and product certification requirements.[21] For example, some countries have required labeling to address nutritional and environmental concerns. While standards and other NTBs can serve useful roles, such as providing information or ensuring product integrity, they can also impose costs on food manufacturers that could raise consumer prices. And if applied in a discriminatory fashion (for example, on imports but not on domestic production), they risk violating WTO rules.

Moreover, using tariffs on foreign supplies of products considered unhealthy or environmentally damaging may decrease imports. But by providing implicit protection to domestic suppliers, such tariffs may increase domestic production, thus defeating the aim of the policy. An excise tax targeted at domestic sales of the good is preferable, because it should reduce all consumption of that good, not just imports, and is consistent with WTO rules.[22] Measures such as quantitative restrictions and trade standards can have more direct effects on food availability but need to be based on scientifically valid safety concerns and be nondiscriminatory and consistent with WTO rules.

Concerns about discriminatory practices also arise around the CBAMs being discussed in the European Union, Canada, the United States, and elsewhere.[23] Currently, there is little agreement on standards for CBAMs (for example, measuring GHG emission coefficients based on production practices). Further, a proliferation of different sets of standards among major trading partners risks creating a "spaghetti

TABLE 4 Average applied most-favored-nation tariff rate, selected food product groups (percent ad valorem equivalent)

MEMBER	Animal products	Dairy	Fruits & vegetables	Cereals & products	Oilseeds & products	Sugar	Fish
USA	2.3	18.4	4.6	3.1	7.2	13.8	0.7
EU	15.6	37.1	10.6	13.7	5.3	24.5	11.6
Brazil	8.3	18.3	9.7	10.7	7.9	16.5	10.3
China	13.2	12.3	12.2	19.5	10.9	28.7	7.2
India	30.8	35.7	30.2	32.9	33.9	50.9	29.9

Source: Data from *World Tariff Profiles 2021* (WTO, ITC, and UNCTAD).

FIGURE 4 Producer support as percentage of the value of agricultural production, 1986–2020

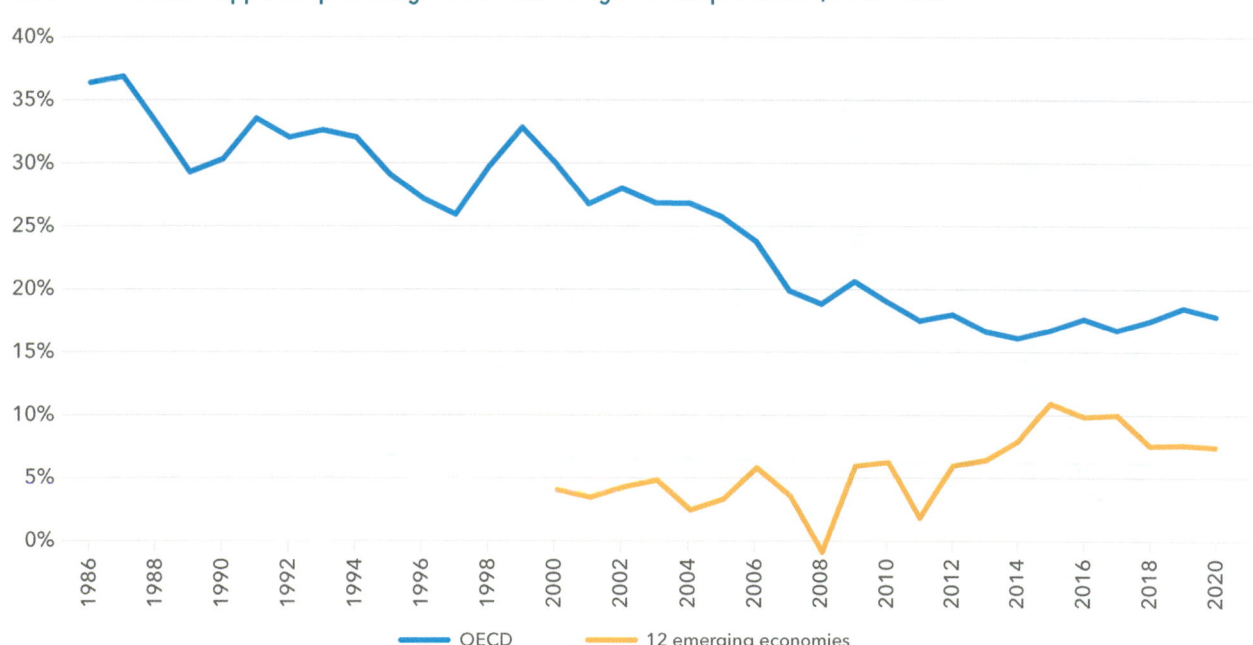

Source: Data from UNCTAD Stat 2021.

bowl" of competing trade rules that increase transaction costs for businesses through variable tariffs, complicated rules of origin, and various business requirements.[24] For example, recent regulations put forward by the European Commission would prohibit imports of some agricultural products, such as soy, beef, palm oil, cocoa, and coffee, if they are shown to contribute to deforestation.[25] Smaller developing countries may be forced to choose among major trading partners or risk being left further behind, as they do not have the capacity to meet multiple sets of standards.

DOMESTIC SUPPORT

Trade-distorting domestic support is another impediment to eliminating hunger and malnutrition by 2030, and the funds could be better used for both nutrition and climate goals. Under the AoA, trade distorting domestic support for WTO members was capped and reduced, and members were encouraged to shift domestic support measures to minimally production- and trade-distorting programs (so-called green box measures). By 2008, producer support among OECD members (as measured as a percentage of the value of production) had fallen by almost 50 percent from 1986–1988 levels, and many countries had shifted producer support away from highly distorting forms, such as market price support, to less distorting programs that do not link support to outputs or input usage (see Chapter 2) (Figure 4).

Unfortunately, policy reforms in OECD countries have largely stagnated since 2008, and farm support has risen in recent years as countries have recoupled policies to production and prices or have provided supplemental support when farm incomes have fallen (for example, due to the adverse impacts from the COVID-19 pandemic).[26] Moreover, domestic support levels have been rising among the large emerging economies, including India and China.

Not only does domestic support potentially distort production and trade, it also can have deleterious impacts on the environment, climate, and nutrition. Recent studies have examined the impacts of removing agricultural support (including border measures) and found that elimination of agricultural support does reduce GHG emissions, but the reductions are small (less than 2 percent).[27] Moreover, the past 25 years of WTO negotiating history suggest that elimination of agricultural support is highly unlikely and, if not offset by less distortionary forms of support, could have a deleterious impact on nutrition.[28]

While eliminating agricultural support would likely do little to protect the climate, many believe that repurposing agricultural support toward development and adoption of climate-smart agricultural practices would contribute to global climate goals of sustainability, resilience, and reducing GHG emissions (see Chapter 2).

EXPORT COMPETITION

Lastly, as we saw during the price spikes of 2007/08 and 2010/11 and more recently during the COVID-19 pandemic, countries sometimes try to buffer the impact of global price shocks on domestic markets by restricting exports.[29] Such export restrictions exacerbate world price volatility by shorting world supplies, which then encourages other exporting countries to follow suit with their own restrictions.[30] This in turn can lead to market-disruptive behavior (preemptive purchases, hoarding, or tactical tariff and consumption tax reductions) on the part of net food importers, further increasing world prices. With production volatility apt to increase because of climate change, countries will be more likely to impose export restrictions if WTO disciplines are not established to prevent such responses. Despite agreement by G20 members in 2011 not to impose export restrictions on humanitarian food aid being procured by the World Food Programme,[31] efforts to expand this agreement to the entire WTO membership have failed.[32]

PROMOTING POLICIES THAT FACILITATE, NOT IMPEDE, TRADE

Since the groundbreaking achievements of the AoA, subsequent reform efforts have largely stalled in the WTO. Little progress has been made in increasing market access for agricultural products, reining in domestic support, or ending export restrictions. As countries commit to reducing GHG emissions, adopting sustainable agricultural practices, and promoting positive nutrition outcomes, the multilateral system may be further challenged to square domestic and international policy goals with the WTO's overarching goal of trade liberalization.

The following actions by WTO members could facilitate trade and help meet the goals of eliminating hunger and malnutrition while addressing the challenges of climate change.

REDUCE TARIFF AND NONTARIFF BARRIERS. While substantial gains have been made in expanding market access through preferential trade agreements, many poorer developing countries remain outside of those agreements and have thus suffered from trade diversion and loss of competitiveness. To the extent that bilateral and regional trade agreements can remain open to new entrants, the door is also open for further plurilateralization and, ultimately, integration into the multilateral system. Increased market access facilitates both the exports and imports of food-insecure countries, which can mitigate the impacts on food security of increased production variability caused by climate change.

ENSURE THAT STANDARDS ARE APPLIED EQUITABLY AND ARE SCIENCE-BASED. With the potential proliferation of border measures aimed at achieving climate or nutrition outcomes, these measures must be implemented in a way that does not discriminate arbitrarily against or among foreign suppliers. For example, the Codex Alimentarius Commission, established by the Food and Agriculture Organization (FAO) and World Health Organization (WHO) in 1963, develops harmonized international food standards, guidelines, and codes of practice to protect the health of consumers and ensure fair practices in food trade. The Codex Alimentarius is well recognized and integrated into the WTO Sanitary and Phytosanitary Agreement. Establishing a similar set of standards for border measures (such as CBAMs or product labeling) — based on scientific research on GHG emissions and agricultural production practices — could ensure that border measures contribute to GHG reductions without reducing food security.

REDUCE AND REPURPOSE HARMFUL DOMESTIC SUPPORT. Domestic support levels have declined significantly since 1995 but they remain high. Capping all trade-distorting domestic support at current applied levels would make it more difficult for countries to backtrack on reforms. To the extent possible, efforts should be made to encourage the repurposing of subsidies to provide more public goods that would support climate resilience and mitigation, such as investments in R&D to increase GHG-efficiency and other minimally trade-distorting climate-smart policies.

ENCOURAGE CLIMATE-SMART POLICIES THAT REDUCE GHG EMISSIONS BUT DO NOT DISTORT TRADE. Climate-smart agriculture is an important tool to mitigate agriculture's contribution to GHG emissions and help producers and communities better adapt to the new conditions brought on by climate change. Further, free and open trade should be seen as integral to any climate-smart strategy because, globally, it can lead to a more efficient use of resources that helps reduce GHG emissions from global agricultural production. Thus, on a climate-smart basis, efforts at further liberalization of agrifood trade through reductions in tariff and nontariff barriers, trade-distorting domestic support, and export subsidies and restrictions should be vigorously pursued.

PROMOTE CLIMATE MITIGATION STRATEGIES THAT DO NOT IMPEDE TRADE. Export restrictions are a poor strategy for global climate adaptation or mitigation. In times of high prices and crop shortfalls, which may be caused by climate conditions, export restrictions can significantly exacerbate food price volatility and undermine confidence in the world trading system. Despite recent experiences, little consensus has emerged on how to discipline export restrictions. A modest step would be to adopt language from the G20 *Action Plan on Food Price Volatility and Agriculture* that would prohibit export restrictions on humanitarian food aid procured by the World Food Programme.[33] Stronger measures would prohibit restrictions on exports to poor, net-food-importing countries, as those countries are most vulnerable to the impacts of climate change and related crises.[34]

With 2030 less than eight years away, meeting the goal of eliminating hunger and malnutrition presents a formidable challenge, one that demands renewed efforts to increase productivity and incomes to bring billions out of poverty. Trade has played a central and growing role in supplying consumers with needed calories and nutrients as regions have benefited from increased imports. As the impacts of climate change on productivity will be highly variable and uneven across regions,[35] trade is expected to continue expanding and become more important for ensuring a supply of nutritious food, particularly for low-income countries. Without this expansion, low-income countries will face higher food costs and far more households are likely to be food insecure as climate-related disruptions become more common. Promoting climate-smart policies that are minimally trade-distorting can help reduce GHG emissions without penalizing foreign suppliers of important food products, particularly those in developing countries.

CHAPTER 4

Research for the Future
Investments for Efficiency, Sustainability, and Equity

GERT-JAN STADS, KEITH WIEBE, ALEJANDRO NIN-PRATT, TIMOTHY B. SULSER, RUI BENFICA, FASIL REDA, AND RAVI KHETARPAL

Gert-Jan Stads is a senior program manager, **Keith Wiebe** is a senior research fellow, **Alejandro Nin-Pratt** is a senior research fellow, **Timothy B. Sulser** is a senior scientist, and **Rui Benfica** is a senior research fellow, all in the Environment and Production Technology Division, International Food Policy Research Institute, Washington, DC. **Fasil Reda** is coordinator of the UNIDO Program for Country Partnerships, Ethiopia. **Ravi Khetarpal** is executive secretary of the Asia-Pacific Association of Agricultural Research Institutions (APAARI), Bangkok.

KEY MESSAGES

- Research and innovation are critical not only to increase agricultural productivity in the face of climate change, but also to transform global agrifood systems through improved efficiency and resilience in achieving social, economic, nutritional, and environmental goals.

- Investment in agricultural research and innovation by larger middle-income counties has expanded substantially in recent decades, but investments, especially in smaller low- and middle-income countries, are too small to address future impacts of climate change across food systems.

Steps can be taken to ensure that R&D contributes to greater productivity, sustainability, and equity:

- Increase research investments for food systems in the context of climate change, both on and beyond the farm in low- and middle-income countries. Sustained growth in public and private investment in food innovation will depend on large middle-income countries, such as China, Brazil, and India, and accelerated growth in other countries with large research systems.

- Build cooperation for sharing innovations. Greater integration of food research at the regional and global levels will allow countries with limited domestic research capacity to benefit from the gains achieved by countries with more developed systems.

- Promote both public and private sector investment. Governments should provide an enabling environment for private sector investment in agrifood innovations, but a clear and critical role also remains for increased public investment to achieve broad food system goals.

Food systems everywhere are facing major new challenges. Shocks caused by COVID-19 have currently seized our attention, but the pandemic has also accentuated persistent problems of poverty, hunger and malnutrition, population growth, and pressure on natural resources, notably land, water, and biodiversity. Adding to these challenges, climate change poses a serious threat to food security and livelihoods as greenhouse gas (GHG) emissions continue to rise. Changing temperatures, highly variable precipitation, shifting growing seasons, and extreme weather events are already making agricultural yields and prices more volatile, with rural areas across the world feeling the effects most profoundly. Yet, as the world's population moves toward 9 or 10 billion by 2050, unprecedented increases in global food production — of at least 60 percent over 2005-2007 levels — will be needed to meet growing demand.[1]

Innovation is essential to address these challenges and to ensure more inclusive access to food and decent livelihoods for future generations. Innovation will be needed in agricultural technologies to increase and diversify production in ways that make more efficient use of resources. It will also be needed in the infrastructure, institutions, and services that support food systems in order to make them more inclusive, resilient, and sustainable.[2] Some innovation will happen autonomously (for example, as producers, consumers, and other private sector actors respond to changing market conditions), but much more research and investment is needed if long-lasting, broad-based sustainability is to be achieved.

Investing in the agriculture sector — and particularly in agricultural research — can be a highly effective pathway for both reducing poverty and hunger and addressing climate change impacts on food systems.[3] A recent modeling exercise found that raising agricultural productivity enough to reduce global hunger to 5 percent would require additional investments of $52 billion annually until 2030 in agricultural research, resource management, and infrastructure.[4] Investment in agricultural research is particularly effective: past R&D investments by CGIAR and by national agricultural research systems in low- and middle-income countries (LMICs) have shown a 10 to 1 benefit-cost ratio.[5] This suggests that $1 invested in research today will yield, on average, a stream of benefits equivalent to $10 over future decades (in present value terms). Moreover, studies consistently show that spending on agricultural research has a greater impact on agricultural productivity than other types of public expenditures, regardless of the mode of investments, timeframe, and specific targets chosen. Agricultural research spending has also performed best or second-best in reducing poverty, whether the comparison is with other agriculture spending, such as irrigation, soil conservation, and farm subsidies, or with investments in other rural areas, such as health, education, and roads.[6]

Nevertheless, many LMICs consistently underinvest in agrifood systems research. This neglect can be attributed in part to long lags between investments and reaping benefits at the farm level,[7] widely diffused benefits, and the "abstract" nature of research and innovation compared with more tangible investments in physical infrastructure. Concerted action to increase LMIC investment will be crucial to accelerating innovation and addressing challenges that are beyond the ability of individuals (and even individual countries) to manage on their own. Moreover, addressing today's nexus of threats will require a holistic research agenda for food systems (beyond agricultural production) to better understand the biological, economic, social, environmental, and health aspects of interlinked areas — from crop and animal production and their inputs, yields, and emissions to storage, transport, food processing, packaging, and marketing, as well as food consumption and waste.

In this chapter, we review how patterns of research investment for food systems have evolved over the past half century and how research and innovation will need to evolve to address climate change and the host of challenges facing food systems in the decades ahead.

THE CHANGING AGRICULTURAL RESEARCH ENVIRONMENT

Over the past five decades, LMICs have benefited from considerable improvements in agricultural productivity, with positive impacts on poverty reduction and nutrition.[8] During the Green Revolution of the 1960s and 1970s, large public investments in crop genetic improvements and yield-enhancing inputs — built on scientific advances made in high-income countries and adaptation to LMIC conditions — prompted significant productivity increases, especially for rice, wheat, and maize.[9] Since then, global public agricultural research investment has continued to grow, doubling between 1981 and 2016 (Figure 1).

Just 20 years ago, high-income countries still accounted for the bulk of this research spending, but rapid increases in spending by large middle-income countries, coupled with stagnating spending growth in high-income countries, has shifted the global balance. By 2016, LMICs accounted for nearly 60 percent of global agricultural research spending. However, China, India, and Brazil alone accounted for more than half of LMIC spending, while sub-Saharan Africa's share in global public agricultural R&D spending has stagnated at about 5 percent. Private sector involvement in agricultural research also shifted the balance in investment. Private spending tripled from $5.1 billion to $15.6 billion globally between 1990 and 2014, outpacing growth in public spending. Though most private R&D expenditures originate in high-income countries, more than a quarter of these expenditures (including those by seed and fertilizer multinationals and tropical fruit companies) directly target commodities or research areas relevant to LMIC farmers.[10] However, overall, private research has focused on just a few commodities, notably cereals, soybeans, horticulture, meat, cotton, aquaculture, and oil and sugar crops,[11] and has neglected many commodities that are economically and nutritionally important in LMICs, including roots, tubers, legumes, and indigenous crops (such as teff in Ethiopia). Given that 800 million people around the world still faced hunger in 2020[12] and many more consume low-quality diets that cause micronutrient deficiencies and diet-related obesity and noncommunicable diseases[13] (see Chapter 8), a critical role for public agricultural research remains.

PRODUCTIVITY GROWTH REMAINS A PRIORITY

Agricultural productivity growth will remain a priority not only to meet food needs, reduce poverty, and improve nutrition[14] but also to address climate change. To meet projected food demand by 2050, global

FIGURE 1 Long-term trends in agricultural research spending

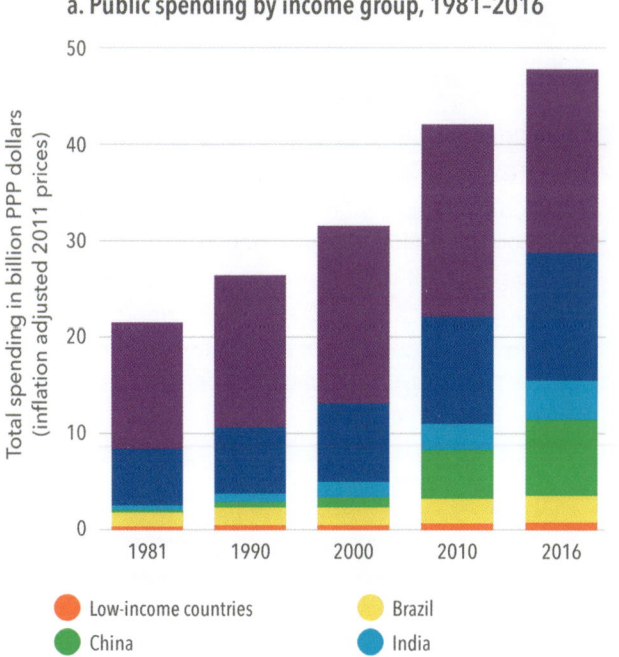

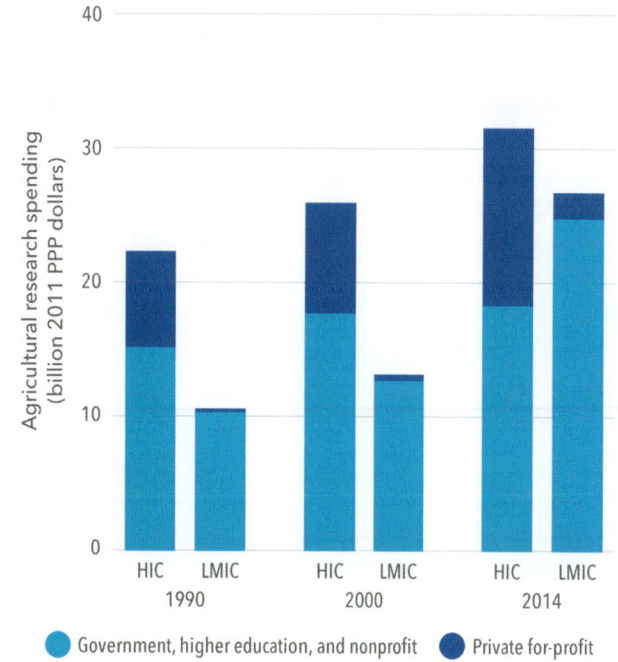

Source: Public sector data compiled from ASTI (https://www.asti.cgiar.org/data); private sector data from K. Fuglie, The Growing Role of the Private Sector in Agricultural Research and Development World-Wide," *Global Food Security* 10 (2016): 29–38.

Note: Income group classifications are based on the situation in 2019. HIC = high-income countries; LMIC = low- and middle-income countries.

agricultural productivity must grow at an average rate of at least 1.28 percent annually.[15] Since 2010, global agricultural productivity has grown at 1.51 percent per year, but growth in low-income countries has averaged only 0.96 percent. With such sluggish growth rates, LMICs will only meet a fraction of their increased food demand through productivity improvements.[16] Moreover, climate change is already eroding earlier productivity gains, necessitating urgent shifts in research goals and priorities toward adaptation, especially in the most vulnerable countries.[17] The world can no longer rely on the main drivers of past agricultural growth — namely, expansion of cultivated land area and exploitation of natural resources (Figure 2) — which have contributed to GHG emissions and resource depletion. Agricultural productivity must be boosted through yield increases, more efficient use of scarce resources, and a reduction in crop losses, rather than greater use of natural resources. Increasing investment in R&D to support innovation in agricultural technology and other segments of food value chains, as well as a strong enabling environment to achieve rapid and wide-scale adoption of sustainable technologies, is therefore a top priority.

GREATER AGRICULTURAL R&D INVESTMENT IS NEEDED IN LMICS

Agricultural productivity growth will continue to be crucial to LMICs and is inextricably linked to investments in R&D that generate improved technology for more precise breeding and input use efficiency,[18] as well as other investments *in* agriculture, such as extension, irrigation, and input distribution policies, and *for* agriculture, such as rural roads and electricity.[19] Yet, a recent global estimate of underinvestment in agricultural R&D suggests that LMICs achieved just 50 percent of attainable investment levels in 2016,[20] with underinvestment most prevalent among countries with small and medium-size research systems.[21] Given that private R&D investment — while significant — cannot fully close the investment gap, agrifood

FIGURE 2 Drivers of past agricultural growth

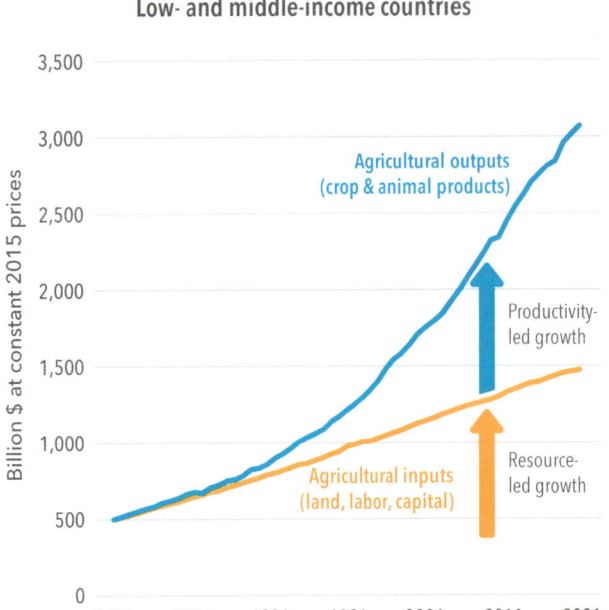
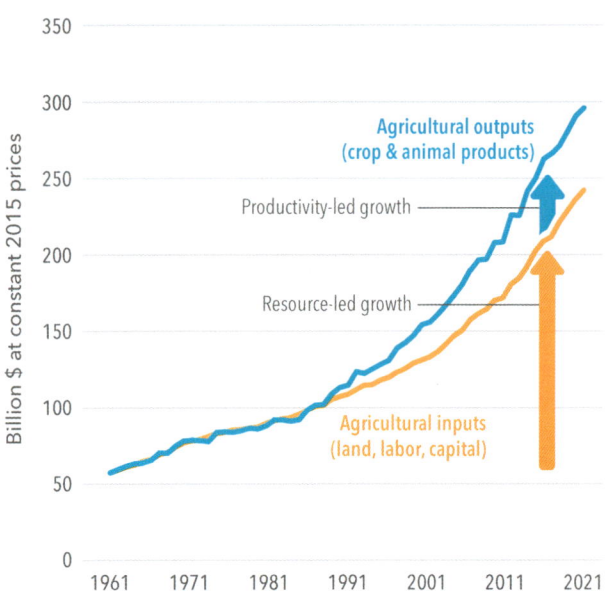

Source: USDA-ERS, accessed 2021.

innovations will continue to rely heavily on public agricultural research. The findings suggest that closing the LMIC investment gap will depend on sustained investment growth in large countries, such as Brazil, China, and India, and accelerated growth in other countries with large research systems. This will allow countries with lagging research systems to benefit from the gains made by countries with more advanced systems and similar agroecologies. National and international public research institutions must play a large role, particularly in areas where economic incentives for private research are weak.

R&D MUST ALSO TARGET SUSTAINABILITY AND RESILIENCE

Higher productivity alone will not be sufficient to achieve sustainable and inclusive agrifood systems. Food systems are major drivers of changes in land use, depletion of freshwater resources, and pollution of aquatic and terrestrial ecosystems,[22] and the production of food (especially animal-source foods) generates more than a third of the anthropogenic GHG emissions that cause climate change.[23] In turn, climate change will have major impacts on the future quality, quantity, and distribution of food. At present, only about 7 percent of LMIC spending on agricultural innovation targets sustainable agricultural intensification.[24] Going forward, increased investment will need to be directed to research and innovation focused on healthier and more sustainable diets, improvements in technologies and management, reductions in food waste and loss, mitigation of GHG emissions, and increased smallholder resilience and adaptation to climate change, to name but a few areas.[25]

INNOVATION IN AGRICULTURAL TECHNOLOGIES IS CRUCIAL

ADAPTATION. Innovations in breeding and in production and management systems will play a crucial role at the farm level in adaptation to climate change impacts. Promising agricultural technologies such as precision agriculture, biofertilizers, and genome editing are already accelerating productivity growth without adding to pressures on natural resources (see Chapter 12). These technologies offer novel uses and applications that address environmental conditions and climate change effects.[26] In addition, new breeding techniques can help crops and animals become more tolerant of heat stress and pests.[27] Innovative farming systems can counter soil erosion and improve moisture and nutrient retention. These include crop and animal diversification; integration of livestock systems, forestry, and crop production; changes in feeding practices; and shifting livestock and crop production locations.[28]

The impact of technological advances will vary with the diverse contexts in which the technologies are applied.[29] For example, drought-tolerant, early maturing varieties will provide greater benefits to farmers and consumers if climate change brings sharp decreases in growing season rainfall. Likewise, nitrogen-use efficiency is only useful in cropping systems where nitrogen availability is a constraining factor. Investments and innovations therefore need to be tailored to the specific context in which they are being applied.[30]

MITIGATION. To achieve mitigation goals, R&D must be stepped up to develop new agricultural technologies that can reduce GHG emissions. The technologies and practices currently available are insufficient to mitigate global warming. For example, practices like alternate wetting and drying in paddy rice and expansion of agroforestry systems can only provide a portion of the mitigation required in agriculture to reduce the pace of global warming.[31] Land-based mitigation technologies play an important role, but are in turn dependent on negative emissions through afforestation/reforestation, intercropping, agricultural productivity improvements, and shifts in food demand. To complement land-based mitigation, potent non-CO_2 gas emissions like methane must be reduced, given the potential for rapid reduction of these shorter-lived GHGs in the atmosphere. For example, feed additives and supplements can reduce methane emissions from livestock production, while alternative approaches to water, soil carbon, nitrogen, and land management provide options to reduce emissions from rice production. Furthermore, food waste and loss, which generate 8 to 10 percent of global GHG emissions,[32] must be addressed. On a per capita basis, postharvest losses are highest in low-income regions due to poor infrastructure, storage, and handling. Innovations that enhance operational management of harvest, transport, and storage of agricultural commodities could therefore have a significant impact on food security and GHG emissions (see Chapter 11).

MORE FOCUS IS NEEDED ON DOWNSTREAM VALUE CHAINS

Most agrifood system research on climate change adaptation and mitigation has focused on agricultural production, leaving the implications of climate change for downstream components of food systems largely unexplored. These include the effects of extreme events and sea level rise on agriculture-related services, transportation, infrastructure, and storage facilities, as well as the effects of regulatory policies and energy and GHG mitigation policies on the adaptive capacity of domestic food systems.[33] Innovations in processing, packaging, logistics, and commercializing technologies are also important.[34] These supply chain innovations facilitate profitable marketing of output by farmers, creating returns to new technologies, and thus widespread adoption of the new technologies. R&D investment for downstream technologies in the food system will need a much higher profile in the context of climate change and development of food systems.

COMPLEX INTERPLAY OF CLIMATE CHANGE AND INNOVATIONS

There are complex interconnections between climate change impacts and the effectiveness of increased investments in R&D and specific innovations. Climate change blunts the impact of agricultural R&D and complementary investments, making it that much more important to increase these investments, as Figure 3 illustrates. With climate change, progress in reducing hunger slows compared to a world without climate change. By 2030, the number of hungry in the developing world drops by 186 million (36 percent) with investments under no climate change while making the same investments under climate change reduces hunger by only 165 million (28 percent).[35]

Challenges that come with gradual changes in temperature and precipitation patterns can mostly be met, on average globally, through normal adaptation

FIGURE 3 Impact of investments in agricultural R&D, water management, and market access infrastructure on hunger reduction (% reduction in 2030 compared to no climate change scenario in 2030)

Without climate change

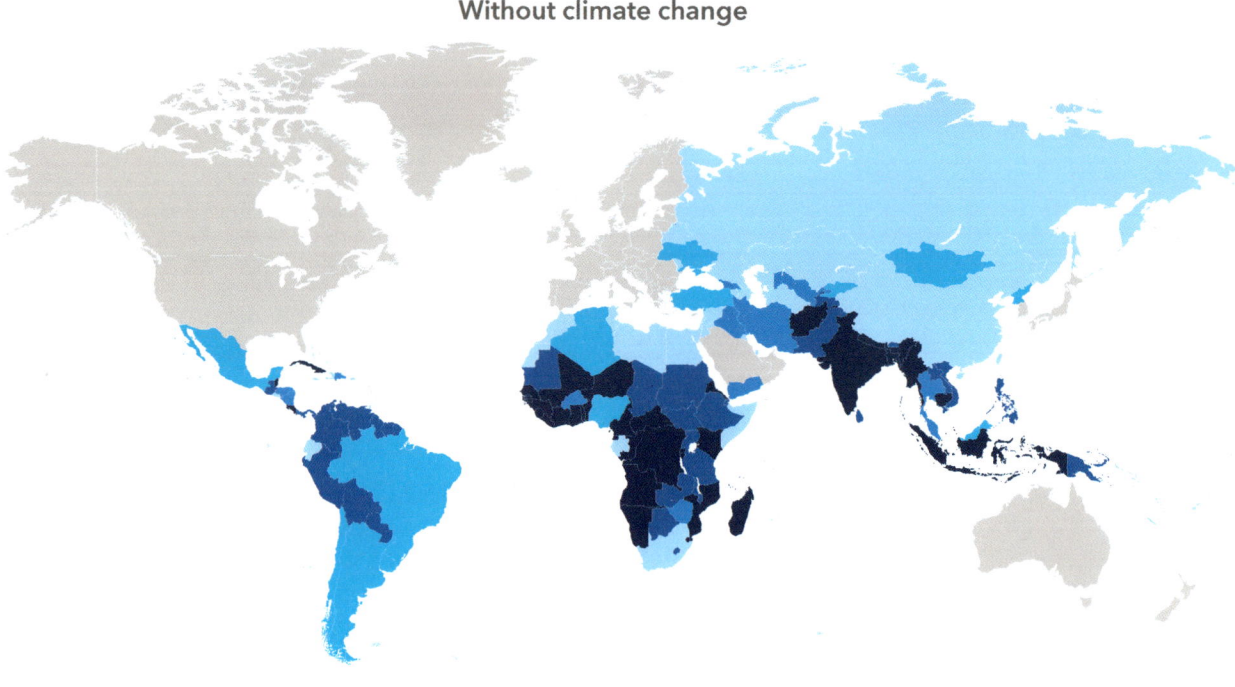

With climate change

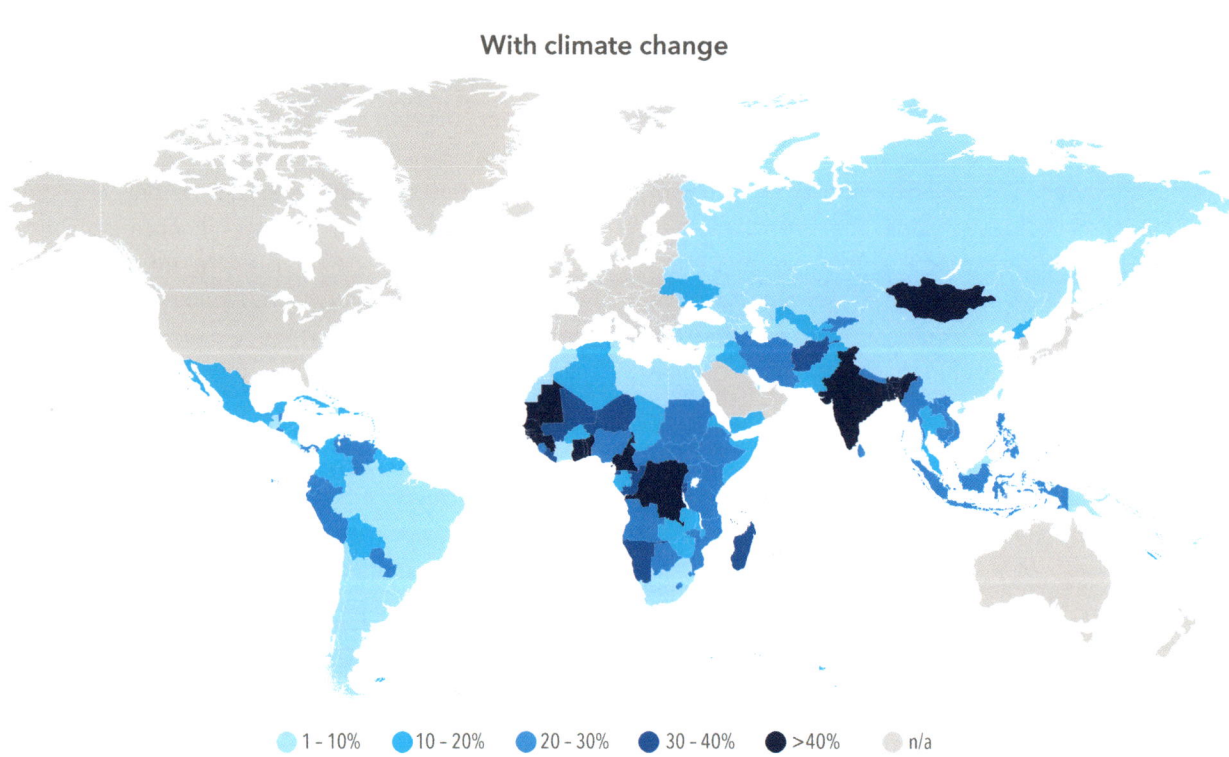

● 1 – 10%   ● 10 – 20%   ● 20 – 30%   ● 30 – 40%   ● >40%   ● n/a

Source: Data mapped from T. Sulser, et al., *Climate Change and Hunger: Estimating Costs of Adaptation in the Agrifood System*, Food Policy Report (Washington, DC: IFPRI, 2021).

Note: Assumes middle-of-the-road changes in population and income (IPCC socioeconomic pathway SSP 2). Climate change is modeled based on IPCC emissions pathway RCP 8.5 with the HGEM general circulation model. See Sulser et al. 2021 for details.

processes in the food system with funding levels following historical trends.[36] In fact, the impacts on production and food security of expected changes in population and per capita incomes are projected to outweigh the impacts of average changes in climate out to mid-century.[37] However, extreme and sudden shocks[38] and the combination and interaction of these driving forces — that is, changes in longer-term climate trends and socioeconomic development, plus increasing frequency of shocks and extreme events — present major challenges. The impact of climate change depends on the resilience of communities, with wealthier societies and communities better able to withstand shocks and recover afterward. Likewise, the effectiveness of different interventions will depend on socioeconomic status, the severity of climate change, and other factors.[39] All these complex, simultaneous challenges call for more balanced research and innovation agendas concerned with environmental sustainability, climate change adaptation, and mitigation and with nutrition and inclusion, as well as the more traditional focus on productivity growth.

INSIGHTS AND PRIORITIES FOR ACTION

Agriculture and food systems need to provide sufficient and nutritious food for a growing global population, while at the same time minimizing environmental impacts, enabling producers to earn a decent living, and adapting and responding to climate change. The traditional focus of agricultural research and innovation systems has been on enhancing productivity. Looking ahead, this focus must broaden to include the larger set of social, economic, nutritional, and environmental goals that are becoming increasingly important to ensure the sustainability of global food systems.

RAISE LMIC INVESTMENT ACROSS FOOD SYSTEMS AND BUILD COOPERATION. Collectively, LMICs spend less than 0.5 percent of their agricultural GDP on agricultural R&D.[40] A recent recommendation suggests that countries allocate at least 1 percent of their food-system-related GDP to food systems research.[41] Although detailed information on global food research investment is unavailable, the data on LMIC agricultural R&D spending imply that this ambitious investment target will require considerable effort. Certainly, not every country will be in a position to invest 1 percent, while others can easily invest more. This highlights the need for closer integration of research activities at the regional (and global) level — in an increasingly globalized world, technologies and innovations can more easily spill across borders.

LMICs are a very diverse group, and not every LMIC is underinvesting. Many large LMICs (notably China, India, and Brazil) have well-developed systems producing world class innovations. In contrast, most small low-income countries, challenged by low capacity and limited ability to take advantage of economies of scale and scope, have overall been less effective in driving food systems transformation. The scarce resources of smaller LMICs are spread thinly over a wide range of commodities and agroecological zones. As a result, they generally record much lower returns to R&D investment compared with their larger counterparts.

To allow countries with lagging innovation systems to benefit from the gains made in countries with more advanced systems and similar agroecological conditions, a closer integration of agricultural R&D and innovation at the (sub)regional level is required. Smaller countries will need to collaborate with countries with large research systems that share mutual research needs and goals, as well as with regional and global R&D institutions, in order to acquire the knowledge and technologies that will support agricultural development and growth in the coming decades. In Africa, for example, the larger systems of South Africa, Kenya, Nigeria, and a few other countries can become the regional drivers of innovation, while smaller African systems should focus on coordinating their innovation activities with these leading players and investing to maximize "spill-ins." In addition, better coordination and a clear articulation of mandates and responsibilities among national, (sub)regional, and global R&D and innovation players (including the establishment of regional centers of excellence and/or specialization) are essential to ensuring that scarce financial, human, and infrastructure resources are optimized, duplications minimized, and synergies and complementarities enhanced. Continued support to and growth of regional bodies, networks, and mechanisms will further aid in supporting agendas that target issues of

> **BOX 1 SETTING R&D PRIORITIES IN EGYPT**
>
> A recent analysis in Egypt shows how investment in a range of innovations can be the most effective way to tackle the complex problem of climate change. Egypt is expected to see particularly strong climate change impacts, with a 10 percent decline in food crop yields by mid-century. Our modeling work shows that adverse climate effects on some crops (such as fruits and vegetables, potatoes, rice, and wheat) can be fully offset with increased investments in drought- and heat-tolerant crop varieties plus combinations of investments in soil fertility improvement, water management, and crop protection. Other crops prominent in Egypt's agricultural portfolio (maize, oil crops, pulses, and sugar) are projected to experience more severe impacts from climate change. For these, combinations of technologies will be crucial for minimizing the negative climate impacts.
>
> **Source:** N. Perez, Y. Kassim, C. Ringler, T. Thomas, H. ElDidi, and C. Breisinger, *Climate-Resilience Policies and Investments for Egypt's Agriculture Sector: Sustaining Productivity and Food Security*, IFPRI Food Policy Report (Washington, DC: IFPRI, 2021).

(sub)regional interest. The restructured One CGIAR can play a constructive role in this regard.

SET CLEAR R&D INVESTMENT PRIORITIES. Countries and regions will need to identify priority innovation areas that are sustainable, equitable, inclusive, and scalable, and consider where additional spending has the largest impact in terms of both productivity and climate change adaptation and mitigation. Investments must target innovations not only in primary production, but also in the postharvest handling, storage, processing, distribution, and consumption of food and agricultural commodities.[42] Even at the farm level, a combination of technologies may be the most effective way to boost productivity and meet climate change goals (Box 1). Moreover, national and regional investments in food systems innovations need to be aligned with broader public and private investments — such as those in infrastructure, financial services, information technology, and digital services — for synergistic adoption and inclusion in agrifood systems.

PROMOTE BOTH PUBLIC AND PRIVATE INVESTMENT IN RESEARCH AND INNOVATION. Although private sector investment in agrifood innovations is increasing, greater private involvement is needed to tackle key emerging challenges, particularly in the postharvest stages of value chains (see Chapter 11). Cultivating private R&D funding requires that national governments provide a more enabling policy environment through tax incentives, protection of intellectual property rights, and regulatory reforms to encourage the international spill-in of technology (see Chapter 10). In addition to stimulating agribusiness research funding (see Chapter 5), countries also need to forge and leverage new mechanisms and partnerships that bring together different investors, including small-scale farmers.

Even with an improved enabling environment and increased private investment, a clear role remains for increased public investment to achieve broader food system goals for which private incentives are insufficient. The current low levels of investment in many LMICs are striking, considering the high potential returns on these investments, particularly compared with payoffs to other types of investment (see Chapter 2). Looking ahead, the challenge of mobilizing public support for shared investment to achieve shared food system goals may well be greater than the challenge of innovation itself. But it is not a challenge we can afford to ignore.

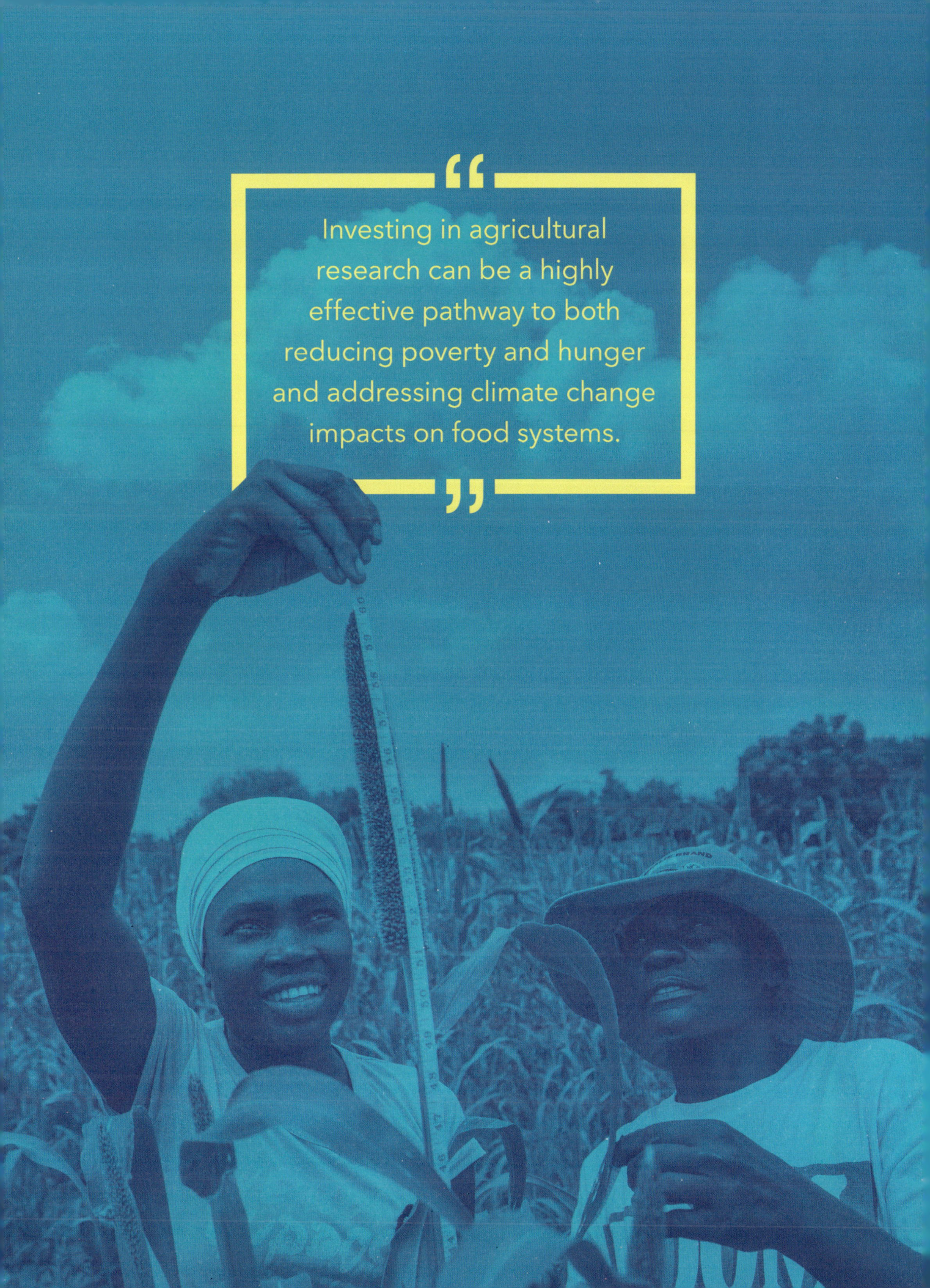

CHAPTER 5

Climate Finance
Funding Sustainable Food Systems Transformation

EUGENIO DÍAZ-BONILLA AND RUBEN ECHEVERRÍA

Eugenio Díaz-Bonilla is a special advisor, Inter-American Institute for Cooperation on Agriculture, and a senior visiting research fellow, International Food Policy Research Institute, Washington, DC. **Ruben Echeverría** is a senior advisor, Agricultural Development, Bill & Melinda Gates Foundation.

KEY MESSAGES

- Reorienting financial flows to address climate change adaptation and mitigation is one of the key objectives of the Paris Agreement on climate change, and global interest in its important role for transformation of the world's food systems is growing. Six main flows of funds are relevant: expenditures by consumers, sales by food systems operators, international development flows, public budgets, banking systems, and capital markets.

- Current financial flows for climate change in the agriculture, forestry, and land use (AFOLU) sector, one of the components of food systems, amount to US$20 billion annually — less than 4 percent of total climate finance. Estimates of additional funds needed for a climate-positive transformation of food systems plus meeting other Sustainable Development Goals range up to $350 billion per year to 2030.

- Reorienting funds requires further analysis of this financial gap, and identification of existing counterproductive investments as well as appropriate climate-positive activities and potential finance sources.

Important steps can be taken now to increase funding for climate change mitigation and adaptation in food systems:

- Establish effective incentive frameworks, as well as enabling macroeconomic and trade environments, for climate-positive food systems transformation. Governments can consider legislating net-zero carbon targets, pricing of climate externalities, development of carbon markets, and disclosures of climate risks as ways to create an effective incentive framework.

- Guide consumption- and production-related financial flows in food value chains. To reorient consumer spending, governments can influence the food environment using fiscal tools, regulations, and information; investments by food systems operators can be influenced by consumer demand and by taxes, subsidies, and regulations.

- Use international development funds strategically. Multilateral and bilateral development agencies should be held to their climate commitments, and their investments can be used to leverage private funds from global capital markets.

- Improve the allocation of national public budgets. To achieve the greatest impact, public funds should be targeted to research and innovation for sustainable intensification of agriculture, as well as investments in science across entire food value chains and the consumer environment.

- Steer banking systems and capital markets toward climate-positive operations. Climate mitigation and adaptation projects constitute a miniscule share of bank lending and private sector investments, while investments in counterproductive activities remain high.

- Ensure banking systems and capital markets support inclusive transformation by identifying investable opportunities and targeting credit lines to disadvantaged groups. Small farmers and businesses, women, and youth are most affected by climate change but often lack access to investment funds.

The transformation of food systems is crucial for achieving multiple global objectives, including the climate change mitigation, adaptation, and resilience goals established in the 2015 Paris Agreement.[1] This international treaty committed signatories to "making finance flows consistent with a pathway towards low greenhouse gas emissions and climate-resilient development."[2] Major global initiatives in 2021, including COP26 and the United Nations Food Systems Summit (UNFSS), have reiterated the Paris Agreement objective of reorienting finance flows and have focused international attention on the critical role of food systems transformation in meeting global climate goals (Box 1).

Food systems – from agricultural production to food value chains, including inputs, processing, transportation, and retail, to consumers – must retool their operations to cut greenhouse gas (GHG) emissions, adapt to climate change, and enhance climate resilience. This transformation will require new technologies, institutions, infrastructure, policies, and other interventions, many discussed in this report, to achieve climate-related

> **BOX 1 FINANCIAL ISSUES AT UNFSS AND COP26**
>
> Finance for the transformation of food systems to address climate change appeared prominently in discussions at two major events of 2021: the United Nations Food Systems Summit (UNFSS) and the climate change Conference of Parties in Glasgow (COP26). This suggests there is growing global commitment to addressing mitigation and adaptation in food systems and to finding innovative ways to fill the finance gap. The UNFSS was a broad consultation, involving governments and many other stakeholders, with no legally binding commitments. The Summit was framed around five Action Tracks, which presented several financial proposals. A Finance Lever Group created to inform the UNFSS (with the World Bank, IFPRI, the Food and Land Use Coalition, and other participants) issued a document on Food Finance Architecture: Financing a Healthy, Equitable and Sustainable Food System,[a] which identifies core "imperatives" needed to optimize public spending and mobilize private capital for a global food systems transformation. In addition, several coalitions and initiatives were formed around finance, including the Coalition of Action for Inclusive and Sustainable Food System Finance: The Public Development Banks Initiative and the Good Food Finance Initiative. Further work on finance will be taken up by those or other coalitions, and by specific national plans.
>
> The COP26 meeting was a negotiation among governments and the resulting agreements have legal implications. (There were also voluntary commitments, with different stakeholders.) The main resolution reached by the participating countries – the Glasgow Climate Pact – includes several sections on finance. In addition to reiterating the Paris Agreement commitment to reorienting finance flows to address climate concerns, the Glasgow Pact highlights the need for additional adaptation finance for developing countries. It urges developed countries to double the amounts available by 2025, and multilateral institutions and the private sector to provide additional support. In view of the increasing financial needs of developing countries in the context of COVID-19, the agreement calls for more funds from multilateral development banks and other financial institutions and increased transparency in climate finance. In addition, a "Glasgow Dialogue" was initiated with countries, relevant organizations, and stakeholders to discuss funding for activities to avert and address loss and damage associated with climate change in developing countries.

goals. Many of these interventions need climate finance.[3] In particular, investment in sustainable intensification of agriculture, mainly in developing countries, will play a key role in the solution to climate change.[4] Sustainable intensification can potentially play a triple role in addressing climate change: reducing emissions from crop and livestock production via climate-smart practices; helping capture more GHGs through efficient agriculture and landscape management; and adapting and building resilience to more challenging climate and weather conditions.[5]

FLOWS OF FUNDS

To meet national and global climate goals, existing flows of funds must be reoriented and financial support must be mobilized for the broad range of mitigation and adaptation activities needed across food systems. Taking a broad view of "financing," six main flows of funds need to be oriented and scaled up toward meeting global climate change objectives (Figure 1).[6] Two are "internal" to food systems: food and food-related expenditures by consumers, which constitute the sales/revenues of operators in food value chains (including flows between different subcomponents of the value chains). Four are "external" to food systems: international development flows (concessional and nonconcessional loans, grants, and donations); public budgets (expenditures and revenues); banking systems; and capital markets. Different options to reorient and scale up each of those flows are discussed in the section below on what can be done.

To improve financing flows for climate change mitigation and adaptation in food systems, we must consider: 1) how large is the gap between current financial flows for mitigation and adaptation activities and the expenditures needed for climate-positive food systems transformation?; 2) are existing financial flows currently supporting activities detrimental to the

FIGURE 1 Flow of funds for food systems

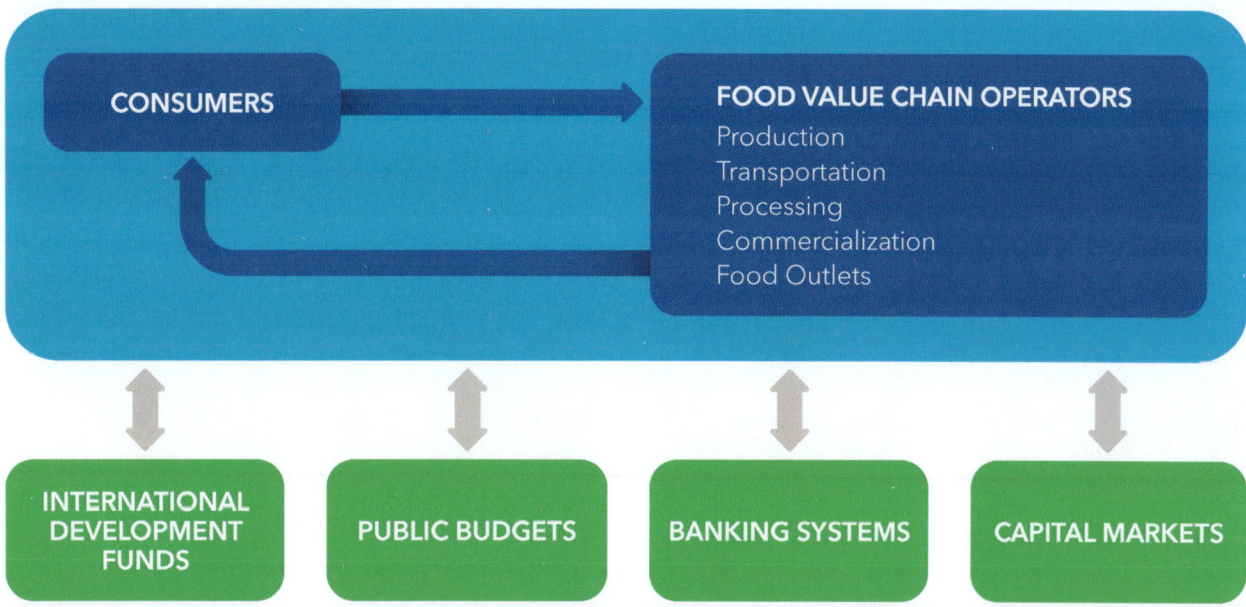

Source: Based on E. Diaz-Bonilla, J. Swinnen, and R. Vos, "Financing the Transformation to Healthy, Sustainable, and Equitable Food Systems," in *Global Food Policy Report 2021: Transforming Food Systems after COVID-19*, 20–23 (Washington, DC: IFPRI, 2021).

desired objectives?; and 3) what financial flows can be used to support different types of climate-positive activities for food systems? We discuss these three issues in more detail below. Then, in the following section, we focus on what can be done in the short and medium term to steer financial flows toward mitigation and adaptation at the scale needed, and away from counterproductive investments that risk locking food systems into climate-negative trajectories.

FINANCE GAP

The Standing Committee on Finance (SCF), established at COP16, has taken some initial steps toward measuring the finance gap. The SCF's most recent report on existing financial flows for all mitigation and adaptation activities presents a high-end estimate of US$775 billion and a low-end estimate of about $574 billion.[7] The lower estimate derives from a Climate Policy Initiative report and database[8] that provide a useful disaggregation of these financial flows, allowing us to analyze both all climate-related finance and the share specifically for mitigation and adaption in agriculture, forestry, land use, and natural resource management (AFOLU) in terms of the six flows of funds (Table 1).[9]

The bulk of these existing climate funds go to renewable energy (43 percent), energy efficiency (30 percent), and sustainable transportation

TABLE 1 Climate finance, low-end estimates of annual average 2017–2018 (US$ millions)

	Households and individuals	Corporations	International development flows	Public budgets	Banking system	Capital markets	Total
Total climate finance	52,651.3	155,988.6	84,195.9	58,835.5	205,801.1	16,783.6	574,255.9
Only AFOLU	3.5	34.0	9,128.0	4,130.5	7,117.5	107.0	20,520.5

Source: Data from Climate Policy Initiative, *Updated View of the Global Landscape of Climate Finance 2019* (London: 2020); and personal communication from Baysa Naran (regarding data for AFOLU).

Note: AFOLU = agriculture, forestry, and other land use.

(18 percent), while the share for AFOLU is only 3.6 percent of the total.[10] However, some expenditures, particularly for adaptation and mitigation in the energy and transport sectors, may also contribute to climate objectives in food systems beyond the AFOLU component.

Total gross disbursements from all developmental sources (bilateral, multilateral, and private) to all recipients and for all purposes, average about $256 billion a year (2015–2018).[11] Of these, funds with identifiable mitigation and adaptation uses averaged about $84 billion for 2017–2018, of which about two-thirds were from multilateral development banks (MDBs). In the case of AFOLU, international development funds are $9.1 billion, and MDBs account for more than two-thirds of that total.[12] Multilateral climate funds, such as the Global Environment Facility, with their complex web of differing governance structures, modalities, and objectives, provide just $3 billion yearly (average 2017–2018), of which only $793 *million* is for AFOLU.[13]

Estimates of the financial resources needed in coming years to meet climate-related goals in food systems range from an additional $15 billion to $350 billion per year to 2030, depending, among other things, on the components of food systems targeted and on how the mitigation and adaptation objectives in food systems are defined.[14] Addressing climate change over larger sectors of the economy would have higher price tags — for example, the additional transition costs for the entire energy sector are estimated at about $5 trillion annually until 2030.[15]

In designing food system transformation plans, constraints on financing will need to be considered. Some of those constraints are global, such as the total amount of world savings, which are distributed very unevenly across regions.[16] Other resource constraints will be evident at the level of specific flows, such as international development flows or public budgets in individual countries. Any proposal to increase investments in climate-positive activities will require changes either in investment for other activities or in savings and consumption, with systemic implications and potentially economywide repercussions that must be considered.[17]

COUNTERPRODUCTIVE INVESTMENTS

Some existing financing operations are contributing to GHG emissions and hindering attainment of climate-related objectives and the 2030 Sustainable Development Goals (SDGs). For instance, some governments subsidize fossil fuels, food producers clear forests with support from food processors or agricultural subsidies, and so on. Development funds continue to finance some investments with high GHG emissions, such as coal-based energy. An estimated $3.9 billion (annual average, 2016–2017) of international development flows finance fossil-fuel activities, exceeding climate-positive financing from multilateral climate funds. However, this may change with the commitment at COP26 of 25 countries and public finance institutions to end financing of projects abroad with unabated fossil-fuel energy (that is, without carbon-capture) by the end of 2022.[18] Stopping those counterproductive uses of funds would both reduce the negative climate effects and, potentially, free up resources for more climate-positive investments. Reallocating these funds would help to cover some of the costs of adaptation and mitigation activities needed in food systems, but may not be sufficient. Additional resources may have to be reallocated from other, climate-neutral uses. But trying to reallocate funds from uses that are counterproductive from a climate-change perspective will not be easy, given the political economy of vested interests of long-term support programs (see Chapter 2).

IDENTIFYING FINANCE SOURCES IN LINE WITH THE ACTIVITIES ENVISAGED

Determining the appropriate climate-positive activities and associated costs will help identify the actors involved and the relevant sources of financing and decision-makers. For example, if adaptation and mitigation objectives imply costs for agricultural extension services or social safety net programs, the financial resources will most likely come from public budgets, which depend on direct government decisions. However, private investments in sustainable agricultural intensification by farmers or in energy efficiency and food waste management by processing firms may require loans from banks, which have their own priorities and decision-making processes.

WHAT CAN BE DONE?

Several important steps can be taken now to identify financing needs and to increase funding for climate change mitigation and adaptation in food systems from the six sources discussed above.

FILL DATA GAPS ON FINANCING AND EMISSIONS. The data-gathering work of the SCF must be strengthened to help fill the remaining data gaps on climate finance.[19] In particular, data-gathering must expand beyond AFOLU to include review of financial flows to all segments of food systems. More detailed data are also needed on international financial commitments to address deforestation linked to food systems, food waste and loss, and the energy matrix and GHG emissions from primary production, processing, transportation, commercialization, and consumption in food systems (see Chapter 11 on value chains). A formal process of voluntary reporting on climate-related financial risks for different types of firms has been promoted by the Task Force on Climate-Related Financial Disclosures (TCFD). These disclosure recommendations could have broader applications for generating data and changing incentives for private and public sector investments (Box 2).

ESTABLISH EFFECTIVE INCENTIVE FRAMEWORKS FOR FINANCING OF CLIMATE-POSITIVE FOOD SYSTEMS TRANSFORMATION. Governments have a variety of policy and intervention options to provide effective frameworks to promote investment in mitigation and adaptation, and to reorient financial flows away from unhealthy foods and environmentally damaging production and toward health- and climate-positive options. National macroeconomic and trade policies define the general business environment for agriculture and food production and marketing, including aspects such as price stability that can facilitate investment (see Chapter 3 on trade). Governance of carbon emissions, including legislated sectoral and net-zero targets, and adequate pricing of carbon and other externalities will be crucial to guiding decisions of consumers, producers, and other agents in food systems.[20] Governments should work to improve policies, regulations, and infrastructure to ensure that carbon markets function properly, following the recent COP26 decisions on operationalizing carbon credit trading, a carbon market, and a framework for non-market-based approaches (see Chapter 3, Box 1, on carbon border adjustment mechanisms).[21] Sound carbon markets that allow for carbon credits, with the possibility of using the related income stream to support dedicated bonds, could help reduce GHG emissions from food systems and related energy systems.[22] Also, other

---

BOX 2 DISCLOSING CLIMATE-RELATED INVESTMENT RISKS

Requirements for public disclosure of climate-related financial risks by firms, banks, and other investors could help to reduce detrimental investments and provide useful data on risky investments. The Task Force on Climate-Related Financial Disclosures (TCFD), created by the G20's Financial Stability Board, has developed a set of voluntary disclosure recommendations for publicly listed companies regarding the climate-related financial risks they face. These recommendations have recently been updated and are now being considered by several standard-setting organizations, including the International Financial Reporting Standards (IFRS) Foundation and the International Organization of Securities Commissions (IOSCO). They are also being taken up by some individual countries (the United Kingdom, Japan, and others), which are starting to require domestic organizations to report according to the TCFD recommendations, and other countries are considering legislation to require those disclosures as part of the pledges made during COP26. As yet, disclosure requirements refer mainly to publicly traded companies. Central banks and regulators are also reviewing their own monetary and financial supervision functions in the context of climate change, which may have larger impacts on the reorientation of financial flows.[a] By increasing transparency about climate change impacts, wider coverage of, and regulatory requirements for the disclosure of climate risk can help generate needed information and change investment incentives.

instruments such as payments for environmental services (PES) could help stop deforestation and promote ecosystem restoration and ecosystem services.

GUIDE CONSUMPTION- AND PRODUCTION-RELATED FLOWS OF FUNDS IN FOOD VALUE CHAINS. The internal flows of funds in food systems, namely food expenditures by consumers — which are also the sales/revenues of the agents in the food system — amount to about $8 to $10 trillion, or 9.2 to 11.4 percent of world GDP.[23] Consumers want affordable, convenient, good-tasting, safe food. Also, they often express a preference for healthy food and environmental sustainability, although their choices may not reflect these preferences, in some cases because healthy and sustainably produced foods may be too costly. To reorient consumer spending, governments can influence the food environment — including prices, incomes, preferences, and the market structure that frame consumers' decisions — using taxes and subsidies, income support, nutritional information, and regulations (for example, labeling requirements) (see Chapter 8). For producers and food value chain operators, governments already influence investment decisions through regulations and controls related to health, nutrition, and food safety (see Chapter 11). Other such interventions on both the producer and consumer sides will be needed to address climate objectives, such as stopping deforestation and reducing food loss and waste. For example, food companies could be charged for the environmental costs of their waste. In addition, private sector initiatives, such as pledges by food companies to reduce their carbon footprint, may require public mechanisms for monitoring and enforcement. Wider adoption of the TCFD climate-risk disclosures would help in this regard (Box 2).

USE INTERNATIONAL DEVELOPMENT FUNDS STRATEGICALLY. At COP15 in 2000, developed countries pledged to provide $100 billion annually for mitigation and adaptation, a pledge that was renewed at COP26. Yet international development funds are clearly falling short, and some support counter-productive investments. Overall, international development funds going to agriculture and food systems, climate change, and related uses are small compared with the needs. The $100 billion commitment made to developing countries must be fulfilled, with a larger grant component and sufficient resources allocated to food systems. To support a positive shift in financing, bilateral and multilateral development agencies should improve disclosure of their operations with common reporting methodologies, comply with commitments linked to zero deforestation and no financing of coal and other high-emissions projects, and clearly align their operations with climate goals and the SDGs.

These international public resources also should be used more strategically to leverage and mobilize the vast liquidity in global private capital markets. Examples include blended and parallel financing, guarantees to de-risk specific projects, and socially or environmentally themed bonds that can mobilize private investments addressing larger humanitarian and development objectives. This potential for leverage is also relevant to the current debate about possible uses of the International Monetary Fund's latest issue of Special Drawing Rights (SDRs) (about $650 billion, of which about 60 percent goes to developed countries). Reallocating a larger share of these SDRs to developing countries for climate finance could help mobilize private capital, especially if used to guarantee the issuance of perpetual bonds.[24]

IMPROVE ALLOCATION OF PUBLIC BUDGETS FOR THE TRANSFORMATION OF FOOD SYSTEMS. Countries' public budgets provide another important source of funds for climate-change related investments (Table 2).

For the largest impact, public funds should be directed to investments in research and innovation for sustainable intensification of agriculture to ensure that interventions are appropriate for local contexts and science-based (see Chapter 4). Unfortunately, expenditures in agriculture R&D in developing countries are low (excluding China),[25] and focus mainly on productivity. Only 7 percent of total agricultural R&D spending has explicit environmental objectives, and even less includes social aspects of sustainable intensification. Scaled-up investments in science and technology are also needed across the whole food value chain and the consumer environment. A recent proposal recommends that countries' investments in these areas should reach at least 1 percent of food-system-related GDP.[26]

TABLE 2 Public budgets: Total, agriculture sector, climate change, and fossil fuel subsidies (US$ billions)

PUBLIC BUDGETS	Total	Developed countries	Developing countries	Developing countries w/o China
Total a/	26,913.9	19,768.4	7,145.5	4,625.6
Of which:				
Agriculture, forestry, and fisheries a/	515.9	159.2	356.6	117.2
Climate change b/	58.8	na	na	na
(of which AFOLU) b/	4.1	na	na	na
Fossil fuel subsidies c/	826.0	na	na	na

Source: a. Average 2010s, updated from E. Diaz-Bonilla, *Financing SGD2 and Ending Hunger*, Food Systems Summit Brief (Bonn: UNFSS, 2021) with data from FAOSTAT; b. Average 2017-2028, from Climate Policy Initiative, *Updated View of the Global Landscape of Climate Finance 2019* (London: 2020); c. Data for 2020, from I. Parry, S. Black, and N. Vernon, "Still Not Getting Energy Prices Right: A Global and Country Update of Fossil Fuel Subsidies," IMF Working Paper WP/21/236 (IMF, Washington, DC, 2021).

Note: These figures should be seen as approximations of the orders of magnitude and should not be added across categories, given that sources of the data and time periods vary, and categorization of "developed" and "developing" countries varies across datasets. Agriculture, forestry, and fisheries covers the primary sector, but not expenditures that can support production (such as rural infrastructure) or other parts of the whole food system.

One possibility under discussion is to repurpose the agricultural support measures (about $620 billion), which currently include a variety of expenditures and transfers to producers and consumers that often promote unsustainable production practices and unhealthy diets (see Chapter 2). Of this, some 35-40 percent are subsidies (concentrated in Europe and China) that could be repurposed toward the provision of environmental public goods and R&D to support more sustainable diets.[27] Quantitative estimates of those potential reallocations show the possible complementarities across SDGs, but also highlight trade-offs that must be considered in repurposing those expenditures. Public budgets also include large subsidies to fossil fuels (expenditures and tax exemptions), amounting to about $826 billion.[28] The Glasgow Climate Pact agreed at COP26 commits countries to "accelerating efforts towards the phase-down of unabated coal power and inefficient fossil fuel subsidies, recognizing the need for support towards a just transition." Eliminating those subsidies would reduce incentives for fossil-fuel use, and the funds could be reallocated to adaptation and mitigation in food systems and other sectors.

Public expenditure reviews can help determine whether the level and composition of government spending is adequate — as well as efficient, effective, and equitable — to meet climate change objectives and other SDGs. Expenditure reviews can help define budget reallocation, better targeting, and programmatic improvements for public expenditures within the existing budget limits.

However, these and other reallocations may not be enough to reach the levels of additional funding needed; expenditures may have to be increased and financed by more public revenues. The latter would require better tax administration and tax reforms to enlarge countries' tax bases. For example, proper pricing of externalities of fossil-fuel use would have the double effect of changing incentives and generating fiscal resources that could help to reach all SDGs and the climate objectives.[29] However, it is important to note that many developing countries are already facing fiscal stress as a result of the COVID-19 pandemic, particularly because they have limited access to unconventional monetary instruments and to low-cost borrowing in global markets. These countries will require strong support from international financing agencies to design and fund their pandemic recovery programs in the short term and a just climate transition in the medium term.

STEER BANKING SYSTEMS AND CAPITAL MARKETS TOWARD CLIMATE-POSITIVE INVESTMENTS. The banking system and capital markets are significant sources of funds, but support for climate mitigation and adaptation projects make up a miniscule share of their investments. While overall flows of new bank loans (of more than one year) averaged about $1.6 trillion annually (2015-2019),[30] the flow of loans for climate-related activities averaged only about $206 billion (2017-2018), of which $158 billion was from national public banks and $48 billion from commercial banks.[31] Banking flows for AFOLU climate activities were significantly lower, at $7.1 billion, and were provided largely by national development banks. In capital markets, long-term bond issues were valued at $27.3 trillion and global equity at $826.8 billion in 2020,[32] but capital markets provided less than $17 billion for climate financing generally and only $107 *million* for AFOLU, which was largely invested in developed countries.[33] The amounts invested in agriculture and the transformation of food systems, both specifically to address climate change and more broadly for other SDGs, appear small.

At the same time, banks and investors continue to finance fossil-fuel operations and activities linked to deforestation, though the scale of such climate-negative investments is unclear.[34] Policies requiring disclosure of climate-related risks could make banks and investors – like companies – provide a full accounting of GHG emissions from both their own operations and supply chains and from the companies that they finance. Such disclosure may prompt these financial institutions to raise interest rates and charges for companies contributing to climate change or not do business with them. Some relevant accounting standards have been developed, including the Greenhouse Gas Protocol and the Partnership for Carbon Accounting Financials, as well as the TCFD recommendations (Box 2).

While disclosures can reduce funding for activities with high GHG emissions, further work may be needed to mobilize additional private investments to support a climate-positive transformation of food systems and to achieve other SDGs. In capital markets, a growing number of actors now look for positive social and environmental outcomes as well as profits. To this end, some banks and other investors have made pledges and formed coalitions such as the Glasgow Financial Alliance for Net Zero (GFANZ) and the Climate Finance Leadership Initiative. GFANZ has argued it can mobilize $100 trillion through 2050 for climate-positive investments, with an annual flow of about $3–4 trillion.[35] However, standards for such nonfinancial objectives vary, as do monitoring and reporting activities. For these pledges to be effective in reorienting investment, the macroeconomic, regulatory, and incentives framework discussed above will be essential, including the legislation of net-zero emissions targets, pricing of externalities, development of carbon markets, and risk disclosures. Together these will generate demand and create markets for the necessary financial flows.

ENSURE BANKING SYSTEMS AND CAPITAL MARKETS SUPPORT INCLUSIVE TRANSFORMATION. Food system actors most affected by climate change include small farmers, small and medium enterprises (SMEs), native communities, women, and youth. Action will be needed to create incentives for funding the *inclusive* transformation of food systems, particularly investments led by these groups. First, a robust pipeline of investable opportunities (including individual projects, impact investment funds, green bonds, and other instruments) must be developed. These must have an adequate risk and reward profile to attract investors, and clear, measurable, and monitorable impact objectives aligned with climate goals and other SDGs. To develop this pipeline, a proposal has been put forward to establish a CGIAR unit to link private capital and banks with investable opportunities for small farmers and rural populations in socially and environmentally relevant activities.[36]

Second, central banks, within adequate monetary programs that considers inflation objectives, can offer specific lines of credit to financial entities, which in turn can provide loans for climate-positive activities[37] targeted to small farmers and SMEs, including women and youth, in food value chains. Well-managed public development banks,[38] which already play a key role in climate finance, can be powerful instruments for addressing market failures that affect agricultural and rural financial markets and climate finance. They can also crowd-in private sector funds from commercial banks and private investors by using blended finance and de-risking arrangements with their own public capital.[39]

Third, continued innovation in financial instruments is needed, both on the lending side and for savings and other financial services used by food value chain actors.[40] "Fintech" options based on digital services can help to better reach small farmers, SMEs, and rural populations (see Chapter 12 on digital innovations). New instruments such as sustainability-linked loans and bonds can also support climate finance. These are being used to finance decarbonization transition plans, with interest rates that fluctuate depending on the attainment of emission-reduction goals or supply chain sustainability metrics. However, further innovations are needed to mobilize funds from banks and capital markets on the scale needed, and in ways that consider the special needs of small farmers and disadvantaged groups in food systems.

WAYS FORWARD

Although, as shown, current climate finance does not reach the scale needed, the quantitative estimates suggest that, in the aggregate, there are sufficient potential financial resources available to fund a climate-positive transformation of food systems as well as to achieve other SDGs. However, much more work is needed to create the necessary incentives to mobilize these funds. Change must start with the macroeconomic and overall incentive framework, including the legislation of net-zero carbon targets, pricing of climate externalities, development of carbon markets, and disclosures of climate risks.

Beyond these frameworks, there are several options for guiding financial flows away from climate-negative activities and toward climate-positive ones. Implementing these options will require a sustained global effort. The UNFSS and COP26 have advanced two promising institutional approaches.

First, these two global events sparked the formation of several *coalitions* of interested stakeholders, public and private, including those focusing on net-zero deforestation, net-zero carbon finance, good food finance, reduction of food waste and loss, and sustainable livestock production. To be effective, these coalitions must clarify their governance, funding, and operational structures and approaches, and those with overlapping topics may need to consolidate around common structures.

Second, both global meetings considered mechanisms to design and implement *national plans* for climate-positive food systems transformation. The UN Secretary-General announced at the UNFSS that there will be UN-appointed resident country coordinators to help manage the work of the UN organizations around national food systems transformation programs (the "national pathways").[41] Furthermore, as part of the climate change negotiations, countries must present their objectives for mitigation (Nationally Determined Contributions, NDCs) and may develop plans for adaptation (National Adaptation Plans, NAPs). In addition, at COP26, suggestions were made for setting up national platforms to coordinate private investors around country-focused plans for the energy transition.[42]

These global approaches can only be implemented if countries structure their own national committees to design and coordinate their national plans for sustainable food systems transformation, integrate them with their NDCs and NAPs, and ensure that all bilateral and international organizations operate in accordance with those plans. The establishment of international mechanisms to help countries design, finance, and implement such national programs could strengthen institutionally weak initiatives. The fiscal constraints created by the current pandemic increase the need for such support and more creative use of international development funds. Most important, investments in science, technology, and innovation (including best practices for system transformation and institutional change) must be increased to achieve the desired objectives for food systems in developing countries, where there are large opportunities to reduce emissions, capture more carbon, and improve adaption and resilience in order to support sustainability and long-term food security and nutrition for all.

CHAPTER 6

Social Protection
Designing Adaptive Systems to Build Resilience to Climate Change

DANIEL O. GILLIGAN, STEPHEN DEVEREUX, AND JANNA TENZING

Daniel O. Gilligan is deputy division director, Poverty, Health, and Nutrition Division, International Food Policy Research Institute, Washington, DC. **Stephen Devereux** is a research fellow, Institute of Development Studies, Falmer, UK, and the Centre of Excellence in Food Security, University of the Western Cape, South Africa. **Janna Tenzing** is a PhD candidate, Department of Geography and Environment, and the Grantham Research Institute, London School of Economics and Political Science, London.

KEY MESSAGES

- Social protection programs are a central component of national strategies in low- and middle-income countries (LMICs) to increase incomes for poor households and protect them from livelihood shocks.

- Social protection is also vital to effective climate change responses; it supports adaption to more frequent extreme weather events and can support mitigation.

- These measures are especially important in LMICs, where most program beneficiaries are poor rural households engaged in agriculture and where climate change may drive major economic disruptions.

- Adaptive social protection (ASP) is an integrated approach that addresses the challenges of climate change by combining social assistance programs with humanitarian assistance and disaster risk reduction strategies.

The following steps can strengthen the role of social protection systems in climate adaptation:

- Expand coverage of existing social assistance to immediately improve resilience of the most vulnerable, reduce hardship, and promote economic inclusion.

- Reform social protection systems by strengthening coordination between conventional social assistance, humanitarian response, and disaster risk reduction.

- Undertake risk and challenge assessments, including of contextual factors and climate forecasting, to inform social protection adaptation.

- Reform program modalities to support household coping strategies, such as using digital transfers that are accessible during local seasonal migration.

- Make social protection "climate smart," such as through innovative insurance and productive inclusion initiatives.

Social protection programs are a central component of national strategies in low- and middle-income countries (LMICs) to increase incomes for poor households and protect them from shocks to their livelihoods. Social protection programs currently reach more than 2 billion people worldwide and are found in every country in sub-Saharan Africa.[1] Social protection systems comprise a wide variety of programs that include targeted cash and food transfers, food vouchers, school meals, public works, old age pensions, and public sector insurance, as well as the policy, administrative, and funding mechanisms to deliver these programs. Numerous studies that draw extensively on rigorous impact evaluations have documented substantial short-term impacts of social protection programs, especially cash and in-kind social assistance, on food security and asset formation,[2] as well as on education, health, and dietary diversity.[3] However, evidence on the impact of social protection systems designed to sustainably reduce poverty by responding to large-scale shocks is more limited. For example, many national social protection systems in the poorest and most vulnerable countries of sub-Saharan Africa and South Asia dating to the 2000s started with targeted standing safety net programs and then later integrated measures to provide humanitarian assistance, scale up temporary transfers to better respond to shocks, and promote household and community *resilience* — the ability to avoid or escape from chronic poverty in the face of myriad stressors and shocks.[4] The potential for social protection to help address the challenges of climate change has been recognized for more than a decade,[5] but the expansion of social protection programs designed to address climate change is relatively recent.[6]

This chapter outlines the policy justification for including social protection in climate change responses in LMICs. We then briefly introduce potential approaches for designing social protection systems to address some of the most serious risks and consequences of climate change. Finally, we propose five steps that governments and their partners can take now to strengthen the role of social protection in climate change adaptation.

WHY IS SOCIAL PROTECTION NEEDED FOR CLIMATE CHANGE RESPONSES?

Social protection measures have been recognized as a vital component of effective climate change responses, as they provide protection from the more frequent extreme weather events caused by climate change (and other shocks such as COVID-19). Contemporary social protection approaches can also play a proactive role in climate change mitigation when they are purposefully combined and integrated with wider climate-related policies.

Adaptive social protection (ASP) is an integrated model that best encapsulates the current approach to addressing the challenges of climate change through social protection.[7] ASP can be used to design social protection systems that build the resilience of poor and vulnerable households to a range of covariate shocks, including natural disasters and slow-onset hazards that are accelerating and intensifying due to climate change, as well as forced displacement and pandemics like COVID-19. This approach integrates (1) social assistance programs like cash transfers, which raise the incomes of poor households and help them respond to shocks, with (2) humanitarian assistance, which provides targeted temporary aid to households that experience a large covariate shock, and (3) disaster risk reduction strategies, which make investments to reduce the effect of future shocks.

There is a strong policy justification for social protection to feature prominently in responses to climate change in LMICs. Most social protection programs target poor households in rural areas of LMICs, where many livelihoods are directly or indirectly linked to agriculture. Major weather events, including droughts and floods, are among the most prevalent and destructive shocks to these economies. In many places, climate change makes such events more frequent and severe. A substantial body of rigorous research documents the impact of social protection programs, showing that they increase household food security and assets,[8] reduce poverty,[9] increase savings,[10] increase education,[11] and promote resilience[12] to economic shocks, including severe weather events.[13]

In addition to this rationale, the projected impacts of climate change underscore the importance of social protection in national response strategies. Climate change has the potential to drive sweeping economic changes, including declining returns to labor in agriculture, a shift toward rural non-farm employment, and rural out-migration to urban centers or other countries (Box 1). Recent evidence provides examples of effective "cash plus" approaches, where, for example, cash transfers combined with agricultural advisory services increased crop production and livestock ownership.[14] Studies that test the impact of cash transfers and job training show some positive effects on off-farm employment under the right conditions. These studies suggest potential for programs providing social assistance and complementary trainings, investment, or services to strengthen impacts on crop diversification, investment in soil and water management, or off-farm employment generation that could help address immediate risks to livelihoods from climate change.

As targeted interventions reaching large populations of poor households, social protection programs will make climate change responses more inclusive. ASP programs can serve as a platform for other components of the climate change response, such as improving access to service delivery in agriculture and health and ensuring that poor and socially marginalized groups are included in climate change initiatives in these sectors.[15] The potential of these programs to include women is especially important, as women are disproportionately affected by climate change.[16] In many societies, women have limited control over household resources, hold less savings and credit, have access to smaller or more marginal lands, and face restrictive norms that limit their ability to rely on others in times of need. As a result, women are more vulnerable to the increasing shocks and stresses brought on by climate change, and they can face increased marginalization and inequality. Transfers from social assistance programs are often targeted to women, which help them build savings and assets, and can strengthen their autonomy over livelihood choices. This could bolster their resilience to climate change-related shocks. However, many programs are not otherwise designed to address the challenges faced by women or their climate risks.

By reforming program modalities to improve resilience, ASP programs have the potential to make climate change response strategies more socially

BOX 1 A ROLE FOR URBAN SOCIAL PROTECTION IN THE CLIMATE CHANGE RESPONSE

Adaptive social protection (ASP) can play a central role in the climate change response in urban areas, but an effective urban approach will differ from the better-known modalities used in rural areas. One reason is that poor households face unique challenges in an urban setting, including the need for employment outside the farm or home, deplorable housing conditions, and high costs for food and basic necessities. The urban context also benefits the poor in some ways, such as affording greater mobility and increased opportunities for employment and human capital investment.

Climate change will contribute to worsening conditions for the urban poor in several ways. First, the number of urban poor is likely to grow, perhaps substantially, as accelerating weather shocks reduce the returns to agriculture, driving migration to cities from rural areas. Also, climate change is likely to increase episodes of extreme heat in urban areas, to which the urban poor are highly vulnerable: they often work outside with little ability to avoid work on hot days, and their housing can be dangerously hot.[a] These heat shocks increase the risk of health problems, but healthcare is often unaffordable for urban poor households.

Urban ASP programs can help address these challenges by conducting vulnerability assessments, improving housing conditions for the poor, improving access to public transit, and providing cash transfers to allow poor households to make decisions that overcome their specific constraints.[b] The centrality of the employment problem explains why urban social protection programs often include job skills training or employment matching and support. However, many government-supported skills training programs fail in LMICs, so innovative approaches are needed. In addition, effective urban social protection programs sometimes include mobile phone-based transfers, vouchers, or smart cards for free public transportation and subsidized childcare to assist women in accessing more employment opportunities.

and gender inclusive. To achieve this, policymakers must recognize that, despite their poverty, beneficiaries of social assistance programs have agency and often work in sectors like agriculture that are most vulnerable to climate risks. Evidence shows that these households will respond to incentives and invest in strategies to improve their well-being when resources and opportunities are available. Thus, the challenge is to identify effective designs for social protection programs that strengthen climate resilience. India's Mahatma Gandhi National Rural Employment Guarantee Scheme, for instance, has taken steps to support climate change adaptation outcomes through its public works.[17] Furthermore, it has the potential to be socially transformative through its inclusion of historically excluded groups such as rural laborers, Scheduled Castes, and women, though the extent of its inclusiveness varies substantially across states, depending on how committed local governments are to its implementation.

POTENTIAL APPROACHES FOR ADAPTIVE SOCIAL PROTECTION TO ADDRESS CLIMATE CHANGE

Recent experience suggests several approaches for ASP programs to address specific aspects of the challenges posed by climate change:[18]

IMPROVING CLIMATE CHANGE ADAPTATION AND MITIGATION. Cash transfers may facilitate the adoption of more climate-resilient crops or related farming practices. A recent study in Malawi used a randomized controlled trial design to test the effect of cash and input transfers, with and without a program of intensive agricultural extension. Both cash and input transfers increased the value of production, and production gains were largest for the group receiving transfers and extension services, though these effects were not sustained after one year.[19] Although this model tested agricultural intensification, it also suggests a promising approach to testing strategies that use cash transfers to promote climate mitigation strategies. In another example, Ethiopia's Productive

Safety Net Program (PSNP) provides transfers to more than 8 million people, primarily through a system of labor-intensive public works (often including women and youth) to rehabilitate land and natural resources on community-held property. A recent study showed that the PSNP increased tree cover by 3.8 percent from 2005 to 2019, plausibly contributing to reduced global warming.[20]

APPLYING INNOVATIVE APPROACHES FOR RISK MANAGEMENT. Insurance-based solutions are a natural response to the challenge of climate risk management, but the private sector has been largely unable to develop crop insurance or related instruments (such as risk-contingent credit) to share risk with rural households. Digital innovations are helping to reduce those barriers, but climate change is raising new challenges by shifting the distribution of outcomes in unpredictable ways. ASP strategies offer possible solutions through planned temporary assistance that is designed like insurance, but with premiums paid by the public sector. One example is state-contingent cash transfers that are pre-committed to be paid to targeted households and delivered in response to climate triggers.[21] In Kenya and Ethiopia, index-based livestock insurance (IBLI) schemes have enabled pastoralists to manage climate risks.[22] This allows households to undertake riskier, but more profitable, livelihood activities, knowing that assistance will arrive in the event of a significant weather shock.

INCENTIVIZING EMPLOYMENT TRANSITIONS AND SUPPORTING GEOGRAPHIC RELOCATION. Social assistance programs can also support economic transitions caused by climate change, such as changes in employment. These changes often represent a moment of economic vulnerability, particularly for the extremely poor. In Egypt, the government introduced FORSA, an innovative program of asset transfers or job training and employment services combined with temporary maintenance transfers, to support beneficiaries of the country's Takaful safety net program as they transition off social assistance. Similar approaches could be applied to support job transitions from sectors declining due to climate change. Evidence of the effect of job training programs in LMICs is mixed, but programs can be successful if they provide skills that are in demand in the labor market. Migration from rural to urban areas is expected to accelerate with climate change, but extremely poor households often lack the resources or financing to undertake the cost of migration or the risk that urban employment will not materialize. Research from Bangladesh showed that a modest transfer to support seasonal migration had very large returns for households facing extreme seasonal food insecurity.[23]

These approaches are promising examples of social protection strategies to address the challenges and constraints imposed by climate change, but evidence of their effectiveness is limited. As governments adapt their social protection systems in response to climate change, research will help to identify which approaches have the most potential. Meanwhile, the effectiveness of many social protection designs is supported by substantial rigorous evidence; these point to steps that governments can take now to strengthen their response to climate change.

FIVE STEPS GOVERNMENTS CAN TAKE NOW

The following steps could be taken now to strengthen the climate change response of their social protection systems.

EXPAND COVERAGE OF EXISTING SOCIAL ASSISTANCE TO IMMEDIATELY IMPROVE RESILIENCE. Substantial, rigorous evidence confirms the significant impact and cost-effectiveness of many common social protection modalities, including the popular program modality of targeted monthly unconditional cash transfers. As an immediately available policy option, governments around the world can increase program enrollment among those most vulnerable to the effects of climate change due to increasingly frequent weather shocks or the changing economic trends described above. This recommendation to expand coverage of social protection programs raises a concern about their fiscal sustainability. Government responses to climate change will face trade-offs between increasing access to transfer programs and investing in sector-specific mitigation and adaptation strategies. Social protection programs cost 1.5 percent of GDP on average, a substantial investment.[24] However, where these investments enable poor households to recover from shocks more quickly or avoid drawing down productive assets, the cost-effectiveness of transfers from well-implemented programs may be substantial.

REFORM SOCIAL PROTECTION SYSTEMS TO BE MORE ADAPTIVE. Strengthening coordination and integration between conventional social assistance, humanitarian response, and disaster risk reduction approaches will reduce risk and improve resilience throughout the social protection system. In a crisis like a severe drought or flood, social protection programs should be able to undertake the following "shock-responsive" adjustments, according to the type of shock and who is affected: *vertical expansion* to increase the size or frequency of payments to existing program beneficiaries if they are most at risk, *horizontal expansion* to provide payments to newly vulnerable households, and *piggybacking* to draw on existing administrative structures to provide new forms of assistance.[25] Many countries applied at least one of these adjustments during the COVID-19 pandemic.[26] Even before the pandemic, Ethiopia's PSNP was a good example of an ASP system with well-documented evidence of success. The PSNP includes a contingency financing facility in which district officials develop detailed plans to respond to shocks. This enables them to conduct a rapid assessment and respond quickly with appropriately scaled and targeted assistance, whether to existing PSNP beneficiaries or other vulnerable households. A recent study found that during a severe drought in 2015, the PSNP helped avoid a significant food security crisis. The program reduced the length of the food insecure period by 57 percent and facilitated the full recovery of household food security within two years.[27]

UNDERTAKE RISK AND CHALLENGE ASSESSMENTS TO INFORM SOCIAL PROTECTION ADAPTATION. Effective approaches to reducing the risk from climate hazards are likely to depend on context, including vital economic sectors, past exposure to shocks, and dimensions of social and gender inequality. Assessments should also include careful climate change forecasting to anticipate the scope and timing of worsening climate risks. Government officials can then evaluate which risks can be addressed through existing social protection mechanisms and which require new approaches. Governments should guard against the tendency to link program designs too closely to specific climate hazards. Instead, they should design climate-smart social protection systems that consider multiple drivers of vulnerability and strengthen household and community resilience to a variety of shocks.

REFORM PROGRAM MODALITIES TO SUPPORT HOUSEHOLD COPING STRATEGIES. For example, a transition from manual collection of food aid or cash to digital vouchers or cash transfers using "mobile money" platforms would allow recipients to receive payments promptly, anywhere in the country.[28] This would overcome the problem in which geographically-based safety net transfers to rural households prevent them from migrating to cities for more profitable employment. This problem could also be addressed by strengthening urban safety nets, which are likely to expand as the effects of climate change worsen.

MAKE SOCIAL PROTECTION "CLIMATE SMART." Enhance the absorptive, adaptive, and transformative capacities of people facing climate change by scaling up weather-indexed insurance schemes to reach both crop farmers and livestock producers; environmentally friendly public works projects such as community-based watershed management; and productive inclusion initiatives such as the Sahel Adaptive Social Protection Program.[29]

Many governments are working to adapt the design of social protection programs to improve their effects on the resilience of vulnerable households to economic conditions. Climate change increases the urgency of these adaptations. Use of ASP approaches is likely to make national climate change responses more effective and inclusive.

CHAPTER 7

Landscape Governance
Engaging Stakeholders to Confront Climate Change

RUTH MEINZEN-DICK, WEI ZHANG, HAGAR ELDIDI, AND PRATITI PRIYADARSHINI

Ruth Meinzen-Dick and **Wei Zhang** are senior research fellows and **Hagar ElDidi** is a research analyst, all with the Environment and Production Technology Division, International Food Policy Research Institute, Washington, DC. **Pratiti Priyadarshini** is senior program manager, Foundation for Ecological Security, Anand, India.

KEY MESSAGES

- Confronting climate change requires governance at the landscape level where actions are taken that support natural resource management for adaptation and mitigation. Landscape-level governance can foster a shared vision and coordinated actions among people with diverse livelihoods, resource uses, and interests.

- Integrated, landscape-level coordination and incentive structures must actively involve multiple communities, government, and the private sector. Different geographic and time scales require different formal and informal institutions for governance — from local user groups to state institutions to nationally determined property rights.

- Integrated landscape approaches are still being developed, with much to be learned. Treating these pilots as "learning labs" can help to build knowledge quickly.

- Human and social capital, such as leadership and trust, are important assets for landscape management that should be fostered to increase stakeholder engagement. Addressing power inequities among landscape-level stakeholders is essential for knowledge-sharing, cooperation, and participation.

- Multistakeholder platforms (MSPs) can facilitate coordination and engagement in knowledge exchange and decision-making for landscape governance.

Policy priorities should include:

- Develop MSPs to build collective action on climate change. Effective MSPs require investment of time and resources to build a shared vision among actors, as well as to strengthen community and government capacity for cooperative landscape management.

- Strengthen resource rights to support long-term investments. Tenure security, including both individual rights and collective rights over shared resources such as forests and irrigation systems, is essential to create incentives for long-term investments.
- Devolve resource rights and management responsibilities. Shifting rights and responsibilities to communities can promote locally appropriate actions on climate change and encourage governments to collaborate across sectoral divisions and with communities and the private sector to help address climate change and other goals.

Confronting climate change requires action at all levels, from the individual to the global. While there are campaigns to change individuals' behavior and calls for global and national government action, more attention is needed to governance at the landscape level. The natural resources and ecosystem services that meet the material and nonmaterial needs of communities and form the basis of our agrifood systems, as well as the various types of land users and stakeholders with different land ownership and use rights, are intertwined in landscapes (Box 1). Much of the debate and policymaking around the interconnected challenges to agrifood systems – climate change, biodiversity loss, environmental degradation, and food insecurity – happen at the global and national scales. However, integrated landscape approaches offer great promise for helping countries to meet their Nationally Determined Contributions to greenhouse gas (GHG) emissions reductions by managing resources to reap multiple benefits and balance economic, social, and environmental goals.[1] In the case of climate change, decisions on how and where to reduce GHG emissions and how to adapt in ways that can address other critical goals, including food and livelihood security, must take place at the landscape level. The real outcomes will be determined by the cumulative actions of many local stakeholders within particular landscapes, with differing, but potentially complementary interests that can support sustainable use of resources within their specific environmental, cultural, and socioeconomic contexts.[2] In this chapter we present a framework that highlights the importance of coordinated action and then look in more detail at approaches to strengthening this coordination for integrated landscape approaches, particularly polycentric governance systems and multistakeholder platforms.

Numerous initiatives are recognizing that the landscape scale is appropriate for coordinating the actions of people with diverse livelihoods and interests.[3] For example, water storage is key to reducing both floods and droughts, but increasing storage requires identifying sites in a watershed that are technically suitable and will meet the needs of different groups, and ensuring that the storage is not only constructed but also maintained over time (see Chapter 9 regarding location of clean energy projects). Integrated systems thinking can bring stakeholders together for more effective and equitable climate change responses that could achieve major increases in carbon storage in forests and soils and support scaling-up of other nature-positive solutions for food systems. While identifying effective actions and technologies to address climate change is essential, identifying appropriate governance structures and policies to promote inclusive, effective landscape management for mitigation and adaptation is likewise critical.

Reaching climate change adaptation and mitigation goals will require governance that fosters cooperation and capacity among the range of actors driving resource use and farming practices at the landscape level. Such local institutional arrangements, supported by an enabling state policy framework, must account for both the spatial and time scales of each response. Figure 1 provides examples of several common strategies involving natural resource management practices, with those primarily related to adaptation in orange and those more relevant to mitigation in green; the relative roles of state and collective action are illustrated by the triangles on the right-hand side.

The spatial scale ranges from the individual farm to the global scale. Actions at the farm level, such as planting a drought-resistant crop or adopting agroforestry and similar plot-level climate-smart practices, generally require little institutional coordination. However, coordination may be needed at higher levels to support climate-smart practices, for example, to produce the new crop varieties and develop seed systems to distribute them and extension programs to promote them.

At the group or community level, climate change response options – such as constructing a small reservoir for irrigation to promote resilience to droughts and floods or landscape restoration and diversification to increase biodiversity, carbon sequestration, and other ecosystem services – all require some form of coordination. At this local level, collective action

BOX 1 WHAT ARE LANDSCAPE APPROACHES?

The interconnected nature of natural resources has drawn attention to the importance of coordination of actors across physical landscapes. Recognition of the complexity of sociotechnical interactions within landscapes has given rise to *integrated landscape approaches* (ILAs). Yet neither landscapes as a scale for action nor ILAs have a precise definition,[a] and the concept remains open to interpretation despite, or perhaps in part due to, the wealth of literature on the topic.[b] It is generally agreed that ILA is an umbrella term for integrated strategies that seek to address "wicked" problems at the landscape level, where complex challenges such as food security, climate change mitigation and adaptation, sustainable resource management, and poverty alleviation often intersect, and that ILAs attempt to bridge disciplinary, sectoral, and governance divides to better balance and ideally improve landscape-level social and environmental decision-making and outcomes.[c]

FIGURE 1 Role of coordination institutions and property rights in climate change mitigation and adaptation responses

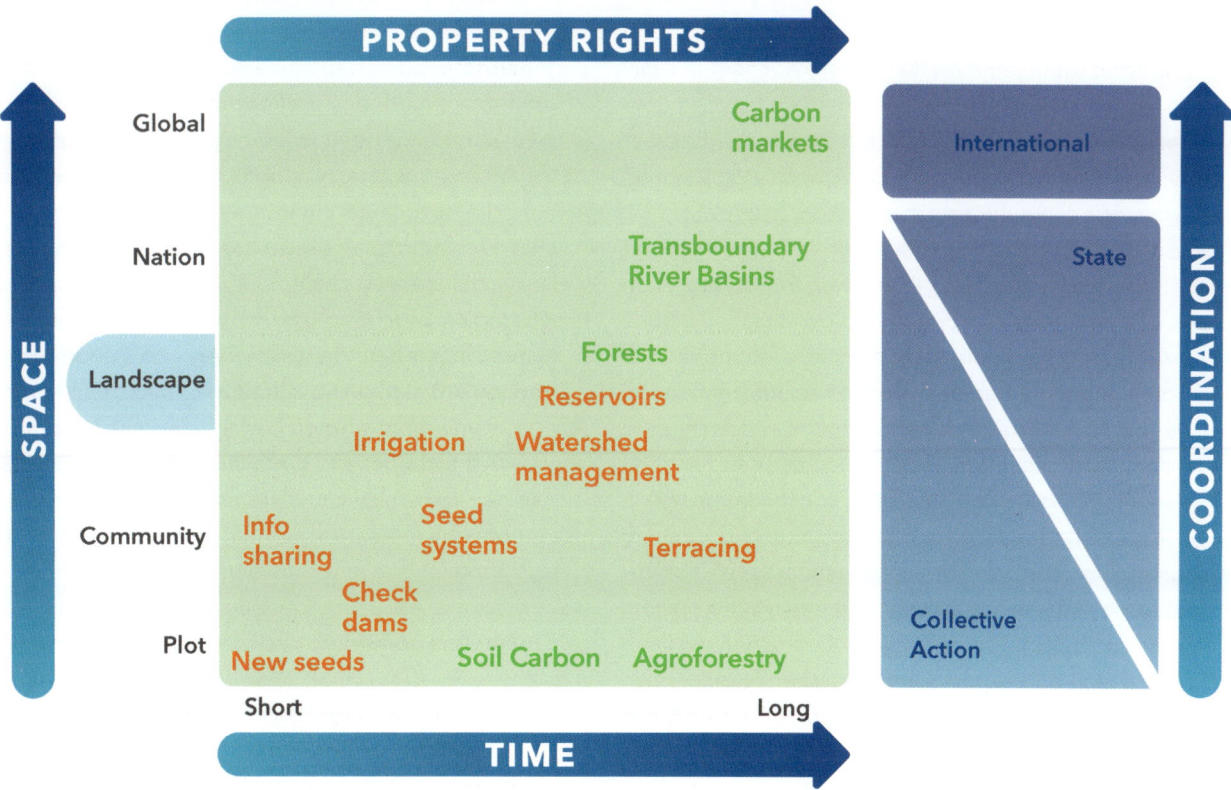

Source: Adapted from R. Meinzen-Dick, Q. Bernier, and E. Haglund, "The Six "Ins" of Climate-Smart Agriculture: Inclusive Institutions for Information, Innovation, Investment, and Insurance," CAPRi Working Paper No. 114 (IFPRI, Washington, DC, 2013).

institutions such as water user associations or forest user groups are often the most appropriate governance mechanisms. Some state institutions may also be relevant. Examples of roles for state institutions include providing technical advice to farmers' groups on constructing or operating an irrigation system and incentivizing other actors in the value chain, such as operators of food storage and processing enterprises to deliver co-benefits to mitigation, for example by reducing postharvest food loss.

Moving up the spatial scale, government agencies become increasingly important for coordination, although collective action institutions may still be relevant. In Nepal, for example, the National Federation of Forest User Groups coordinates between local forest user groups and the government. When the relevant scale for policies or action is the global level, then international institutions are usually required for coordination, either through existing international bodies such as UN agencies or new institutions created for the purpose, such as carbon credit exchanges. However, even for climate change programs with a national or international scope, much of the actual action takes place at the landscape level, between the community and the (sub)national levels, where coordination of investments with local land use and resource use planning calls for interaction between the state and a variety of formal and informal groups. Experience has shown that sustainable action at these levels cannot be achieved by fiat; rather, it requires engaging a wide range of stakeholders.[4] Multistakeholder platforms, discussed below, offer a promising structure for inclusive landscape management.

The timeframe for actions also affects the nature of institutional arrangements needed, particularly the strengthening of rights to land and resources required to foster action. While climate change response schemes are urgent, some will show results in the short term (a year or two), others in two to ten years, and some have a much longer time horizon. Compared to planting an annual crop, for example, planting trees is a long-term investment. The longer the time span, the more difficult it will be to gain and maintain support for action and to monitor progress, unless benefits are clearly apparent. Among the investments shown in the figure, many with longer time horizons require attention to property rights. To ensure that the individuals or communities making the investments (such as planting trees or protecting forests) have the incentives and authority to make and protect these investments, we must consider whether individual farmers have tenure security or if forest and water user groups are recognized as custodians of these resources. In addition, some actions will be intermittent, such as responses to crises like drought or flooding. Crises call for institutional structures designed for preparedness and quick responses, but these do not need to operate all the time.

KEY ELEMENTS OF INTEGRATED LANDSCAPE APPROACHES

With the growing urgency of climate change, integrated landscape approaches (ILAs) have gained increasing traction among funders of aid programs, conservation organizations, and researchers,[5] as well as credibility and legitimacy as an emerging discourse at the UNFCCC.[6] Broadly speaking, ILAs are landscape approaches that consider how interconnected components of the landscape can be managed to reap multiple benefits and balance economic, social, and environmental goals. Recognition of the potential of ILAs has sparked a proliferation of initiatives to address sustainable development issues at a landscape scale,[7] but many of these initiatives have not achieved their goals.[8] Researchers have developed principles and drawn lessons to guide the design and implementation of ILAs,[9] but much uncertainty still remains about how to successfully operate an ILA and how to measure its outcomes and impacts.[10] However, the expanding work on polycentric governance, learning labs, and new modes of knowledge co-production provide useful guidance on approaches.

POLYCENTRIC GOVERNANCE SYSTEMS

The governance systems of successful ILAs provide the means of integrating decision-making and management across sectors.[11] Since administrative boundaries are seldom aligned with landscape boundaries and landscapes most often display a mosaic of natural resources (such as forests, streams, and grazing lands), *polycentric governance* – governance with multiple actors and centers of authority working at different scales, both hierarchical and overlapping[12] – offers a potentially powerful solution to scale mismatches and effective landscape governance.[13] For instance, in India, natural resources are under the jurisdiction of different state agencies (including the forest and revenue departments), as well as local government.[14] They are also collectively managed by such state agencies on different administrative levels, various local user groups (like water committees and forest communities at the village level), and joint programs such as forest management and watershed management programs. In practice, the different organizations and groups have substantial autonomy, each working to mobilize resources for governance of natural resources, though coordination is not always sufficient. Polycentric governance, complemented by practices of joint deliberation among different actors, can build a knowledge base that draws on different perspectives and expertise, while enhancing the levels of trust and credibility among the actors involved.[15]

LEARNING LABS

For decision-makers and stakeholders, the potential advantages of ILAs make them an appealing way to introduce landscape thinking into planning and management approaches.[16] However, there will be no single recipe for effective implementation of ILAs. Experiences from one place or context may not be directly transferable to another, and management responses will need to continually evolve as climate change and other challenges continue to unfold.[17] To address this need for context-specific approaches, landscapes should be considered as "learning laboratories," where the impacts of climate and other environmental as well as socioeconomic changes will be felt and the responses tested.[18] ILA initiatives can learn from recent approaches such as "living labs."[19] These are a form of collective governance and experimentation with growing applications to address sustainability challenges and opportunities created by urbanization in European cities.[20] In ILA initiatives, living labs can serve as stepping-stones to catalyze system transformation by involving various stakeholders in co-designing and testing integrated social and technical innovations.[21]

KNOWLEDGE AND LEARNING

These design and implementation considerations require new modes of knowledge production and collaborative learning[22] to ensure local ownership of initiatives and integration of indigenous knowledge, improve governance systems, and bridge the science-practice-policy gaps.[23] There is a need to invest in human and social capital (such as leadership, trust, and collective vision) as assets of the landscapes, including strengthening the capacity of local stakeholders for effective and equitable participation in the design and implementation of ILAs. Participatory engagement approaches, such as multistakeholder platforms, that enable engagement of representatives from the relevant stakeholder groups can provide the space for collaboration, learning, and joint decision-making.[24] Building such institutional capacity does not happen overnight nor through blueprint approaches, but requires time and adaptive approaches.

MULTISTAKEHOLDER PLATFORMS

Coordination among actors is essential and may require interventions for facilitation. Multistakeholder platforms (MSPs), dialogues, and processes are gaining prominence as a means of achieving ILAs.[25] MSPs are sustained, intentionally created spaces to promote dialogue, deliberation, and collaborative action among stakeholders – including civil society (especially communities), government, private sector, and other relevant groups and organizations – who stand to be meaningfully affected by landscape policies, or changes to resources or resource access.[26] Through MSPs, stakeholders can take part in collaborations, negotiations, knowledge exchange, and decision-making regarding the uses and governance of landscapes. This space is especially important in contexts of high diversity of user groups and common-pool resources within a landscape[27] – for example, where the more privileged groups can dominate, or in landscapes where forest users, watershed users, pastoralists, and farmers may compete over resources and land use. Effective MSPs recognize the interests and contributions of diverse actors, including local communities, and provide coordination and space for social learning. By creating this structure for cooperation, MSPs foster information-sharing and collaboration among diverse stakeholders to influence climate change policies and actions across multiple governance levels.[28] A recent study reports on six MSP cases where landscape initiatives to protect ecosystem functions resulted in restoration improvements.[29] More broadly, MSPs can help build an enabling policy environment for climate change adaptation and mitigation policy.

DESIGN FOR INCLUSION

MSPs are not panaceas, however, and need to be appropriately and effectively designed and implemented to support the coordination needed for integrated landscape management. For local communities, the importance and likelihood of participation in MSPs for resource management increases with their dependence on the resources for their livelihoods.[30] Too often, key groups are either excluded or not interested in engaging, which can stymie an MSP.[31] Stakeholders may not be willing to participate when trust is lacking among the actors, or when they do not perceive benefits to participation.[32] Power inequalities can affect voice and participation. For example, a water catchment MSP in Zimbabwe initially included only large commercial farmers who were involved in water policy reform processes; when smallholders from communal areas were finally invited, they were surprised to find that they were now expected to pay for water, and were not equipped with the skills to negotiate. Recognizing and addressing power inequalities and barriers to inclusion is key for effective knowledge-sharing and dialogue.[33]

Procedural rules and facilitation strategies can be set up to increase engagement and help mediate power relations, such as consensus or rotational leadership of MSPs to empower all stakeholders through the process.[34] The "action resources" that actors can utilize for participation and negotiation matter in enabling effective participation. Action resources include information, technical knowledge, negotiation skills, social networks and capital, money, authority, and the ability to form alliances, all of which give actors agency and help strengthen their claims in stakeholder engagement processes.[35] Strengthening communities' capacity to participate by building on their action resources is key for effective engagement. For example, using communities' customary ecological knowledge for planning, instead of relying only on so-called "scientific" knowledge and models for decision-making, can lead to greater equity in participation. It is also important to temper the role of government actors, so that they provide legitimacy and enable follow-up action without controlling the process.

POLICY PRIORITIES

Climate change cannot be addressed by government alone, or by individuals. Achieving the coordination and incentive structure needed for widespread adoption of many mitigation and adaptation strategies such as reducing deforestation, increasing vegetative cover, and improving water availability requires integrated,

---

BOX 2 MSP BENEFITS TO COMMUNITY AND SOCIETY: BUILDING COLLABORATION IN INDIA

The Foundation for Ecological Security (FES), an Indian NGO, works with rural communities to restore common lands and improve livelihoods through establishing and strengthening local institutions. FES also helps establish subdistrict-level multistakeholder platforms (MSPs) to support effective landscape governance. These MSPs provide a space for intercommunity collaboration, strengthening local voices, and building trust among stakeholders.

In Angul district in Odisha state, for example, forests provide village communities with ecosystem services that mitigate climate change impacts and help communities adapt. These ecosystem services include food, fodder, and firewood, provision of inputs for making agricultural tools and farm fences, soil and moisture retention, and regulation of hydrological and nutrient flows. Over the past few decades, however, rapid industrialization and land fragmentation have caused biodiversity losses and acute water scarcity, with severe impacts on the agriculture-based livelihoods of communities. Forest fires exacerbated by rising temperatures and irregular rainfall have further increased the vulnerability of the region's small and marginal farmers.

FES worked with village institutions in Angul to initiate the Krushak Mela (Farmers' Fair) in 2005 as a multistakeholder platform to promote dialogue among diverse stakeholders around the issues of conservation and livelihood improvement. Around 3,000 people participate, including village communities, local leaders, government officials from different departments, civil society organizations, political representatives from the region, ecologists, agriculture specialists, and microfinance institutions. Community members showcase traditional seeds and crop varieties, new and climate-smart practices in agriculture, and other innovations. Government representatives share information on programs available to farmers; and NGOs share information on best practices in land use management and soil and moisture conservation.

Multiple benefits have emerged from this MSP. Collective initiatives were undertaken to strengthen local markets, thus improving economic opportunities for the village communities. The MSP helped revive traditional adaptation practices such as seed exchanges. Community members evolved rules to regulate open grazing, with a positive impact on biomass. The MSP also provided a platform for cross-learning among farming communities, as well as opportunities to engage with external actors including government agencies and the private sector for diversifying their livelihoods, and improved their access to entitlements, services, and government programs.

Over the years, the MSP has helped to bridge the communications and trust gap between community and government actors. It also sparked conservation action plans across villages at a landscape level. This has addressed landscape-level challenges larger than a single community, but still allows for sufficient interaction and connectedness to realities on the ground, and avoids high-level "detached" planning.[a] FES is now taking into account the lessons from Angul and other MSPs throughout India in places where they are working to scale up such initiatives, both working directly and through other NGOs or government programs.

landscape-level approaches that actively involve multiple communities along with the government and private sector. Investments need to be made not only in physical infrastructure, but also in governance to coordinate and incentivize people's actions over the long term.

DEVELOP MSPS TO BUILD COLLECTIVE ACTION ON CLIMATE CHANGE. Generally, collective action is easier and more effective within the social boundaries of small villages. But the natural boundaries of resources like forests and watersheds go beyond these social and administrative boundaries, and management on the larger landscape level is needed for natural resources that have interconnected biophysical aspects and face interlinked socioeconomic pressures.[36] While there is no single prescription to achieve such collaboration, growing experience with MSPs indicates that they can instigate conservation and governance action plans at a landscape level — clustered around watersheds, forest patches, or rangelands.[37] Developing effective MSPs often requires an investment of time and resources in facilitation to build trust and a shared vision among actors, as well as in strengthening community and government capacity to engage meaningfully in landscape management. Government policies can help provide the enabling conditions for stakeholder engagement in landscape management and promote landscape-level adoption of climate-smart change.

STRENGTHEN RESOURCE RIGHTS TO SUPPORT LONG-TERM INVESTMENTS. Secure tenure is vital to provide incentives for individuals, households, and communities to invest in sustaining or improving natural resources and ecosystem services. Policies that strengthen recognition of these rights can thus improve incentives for investment in sustainability. For example, tenure reforms in Ethiopia and Rwanda have been shown to increase local investment in tree planting and soil conservation,[38] and a systematic review of studies finds strong evidence of positive effects from land tenure security on such environmentally beneficial agricultural investments.[39] But tenure policies need to be implemented in a socially inclusive and equitable manner so that all social groups, including women as well as men, have incentives to invest.[40] Moreover, increasing tenure security must include not only individual rights but also collective rights over the commons — shared natural resources such as forests, rangelands, wetlands, and irrigation systems.

DEVOLVE RESOURCE RIGHTS AND MANAGEMENT RESPONSIBILITIES. Policies that devolve natural resource rights and management responsibilities from states to communities are particularly important in this regard — by explicitly recognizing community rights over such common resources, devolution can increase tenure security. By shifting rights and responsibilities, devolution policies can also encourage government agencies to engage with communities in better ways; instead of seeing the state as the "owner" or "custodian" of resources, devolution should encourage government agencies to collaborate across sectoral divisions and with communities as well as private sector actors. However, this will require longer-term processes to change mindsets and ways of working.

Faster action and more transformative changes are needed for the achievement of SDGs and to confront climate change. Engaging stakeholders through integrated landscape management has been shown to be a key ingredient to the solutions and warrants greater attention and investment by the development and conservation communities. It is time to shift the mindset from technology-oriented solutions to people-centric transformative changes.

CHAPTER 8

Nutrition and Climate Change
Shifting to Sustainable Healthy Diets

MARIE T. RUEL AND JESSICA FANZO

Marie T. Ruel is division director, Poverty, Health, and Nutrition Division, International Food Policy Research Institute, Washington, DC. **Jessica Fanzo** is Bloomberg Distinguished Professor of Global Food and Agricultural Policy and Ethics, Johns Hopkins University, and director of the Johns Hopkins Global Food Ethics and Policy Program, Baltimore, Maryland.

KEY MESSAGES

- Ensuring that everyone has access to – and consumes – sustainable healthy diets is one of the most significant challenges for today's food systems.

- Climate change and environmental and natural resource constraints are putting enormous stress on food systems. In turn, the way we produce, process, and move food, and the global changes in diets toward greater (often excessive) consumption of animal-source foods and ultra-processed foods (UPFs) are contributing to climate change.

- The effects of climate change on food systems, diets, nutrition, and health disproportionately impact marginalized populations in low- and middle-income countries (LMICs).

- Shifting to sustainable healthy diets that protect both human and planetary health will present several challenges:

- *Adoption* of sustainable healthy diets will require major changes in consumption patterns globally, with changes varying by region and country.

- *Adaptation* of a reference diet, if developed, will need to accommodate different countries, contexts, and population groups.

- *Affordability* of these diets will need to be ensured; current examples of potentially sustainable healthy diets are unaffordable for a large proportion of the population in LMICs, as are many nutritious foods.

- Achieving these goals will require the development and implementation of policy packages in LMICs that include multipronged, coherent, and mutually reinforcing actions. Priority policy packages should include consumer education approaches along with fiscal measures and food environment policies. Examples of priority actions could include:

- Education approaches to inform, educate, nudge, and influence dietary choices (public awareness campaigns, nutrition counseling, mass media, social media, and so on); promotion of breastfeeding; and development and updating of food-based dietary guidelines.
- Fiscal measures to discourage consumption of unhealthy UPFs, and/or incentives to retailers to subsidize and boost consumption of nutritious foods.
- Food environment policies, including food labeling and certification, and regulation of marketing and promotion of unhealthy foods to children to enhance demand for healthy diets.

One of the most significant challenges for food systems today is ensuring that every individual has access to – and consumes – *sustainable healthy diets*. These are defined as nutritious, healthy (meaning they help prevent disease), safe, affordable, and culturally acceptable diets that support optimal nutrition and health and cause low environmental pressure and impact.[1] The food systems tasked with producing these healthy diets are under significant stress due to environmental and natural resource constraints, as well as climate change.[2] Moreover, the types and amounts of foods we consume – and the way we produce, process, and move these foods – compromise the stability and resilience of natural resources and biodiverse ecosystems and contribute to climate change.[3]

Both now and in the future, climate change is expected to adversely affect diets, nutrition, and health through impacts on the quantity, quality (nutrient content), diversity, safety, and affordability of produced food.[4] Production constraints, in turn, will continue to cause the loss of livelihoods, income, and food security for food producers, processors, and their families and will jeopardize their diets, nutrition, and health.[5] Combined with the impacts of climate change on disease patterns, these effects will continue to disproportionately impact marginalized populations in low- and middle-income countries (LMICs), including those with the least access to resources and tools for adaptation.[6] Shifts toward healthy diets must therefore focus on the dual goals of protecting and improving the nutrition and health of populations while also meeting environmental goals in an equitable way.[7]

This chapter reviews current challenges in ensuring sustainable healthy diets for all and identifies priority policy measures to achieve this goal, given current trends in food demand and supply, the rapidly modernizing and resource-intensive state of food systems, and the climate change crisis.

SHIFTING DIETARY PATTERNS, CLIMATE CHANGE, AND ENVIRONMENT

CHANGES IN FOOD SUPPLY AND DEMAND

Over the past few decades, diets in LMICs have changed as a result of rapid urbanization, rising incomes, the increased participation of women in the labor force and the related need for convenience foods, and the modernization of food retail, including the aggressive marketing of ultra-processed foods,[8] snacks, and beverages (referred to collectively as "UPFs" in this chapter).[9] Some of these changes have been positive, such as the increase in food supply quantity and diversity and the rises in consumption of fruits and vegetables (F&V) and animal-source foods (ASF) (the latter of which is positive for populations in LMICs as they increase their consumption from very low levels to moderate amounts).[10] Other changes, however, are of great concern for health, nutrition, and the environment, such as the shift from whole grains to refined cereals, overconsumption of red meat as incomes rise, and increased consumption of UPFs that lead to excess intake of calories, saturated fats, salt, and added sugars.[11]

Between 1990 and 2015, the global production of ASF rose by more than 60 percent and global demand increased by more than 40 kg per person per year.[12] ASF demand rose in all regions (with the exception of red meat consumption, which declined by 10 kg per person per year in industrialized countries); but the types of ASF and amounts consumed varied widely (Table 1).[13] These national consumption figures, however, obscure large inequalities in access within countries, especially in LMICs: although consumption increased among wealthier populations, regular access to ASF remained out of reach for marginalized groups, even the relatively small amounts of ASF needed to meet their nutrient requirements.[14]

For F&V, the global supply does not align with current or future needs, which are based on the WHO

TABLE 1 Change in demand for animal-source foods, 1990–2020 (kg/person/year)

Region	Fish, Seafood	Milk	Eggs	Poultry	Red Meat
Eastern Asia	21.0	21.3	12.4	11.4	28.4
Southern Asia	3.1	50.1	2.2	2.8	-1.5
Southeast Asia and Pacific	18.1	10.0	3.9	10.5	7.7
West Asia and North Africa	3.3	29.7	1.1	9.5	5.5
Sub-Saharan Africa	-0.2	4.9	0.0	3.1	-0.4
Latin America	1.0	26.7	5.8	26.0	9.5
Industrialized Countries	0.8	17.7	0.6	15.4	-10.1
World	6.1	19.7	3.6	8.5	2.9

Change in kg/capita/yr (1990 – 2020)
-50.0 to 50.0

Source: M.T. Herrero, D. Mason-D'Croz, P.K. Thornton, et al., *Livestock and Sustainable Food Systems – Status, Trends, and Priority Actions*, UN Food Systems Summit 2021 (Geneva: United Nations, 2021). Reproduced with permission of the publisher.

Note: All regional definitions use UN definitions.

FIGURE 1 Sales of ultra-processed food (kg/capita), 2006–2024

a. By region

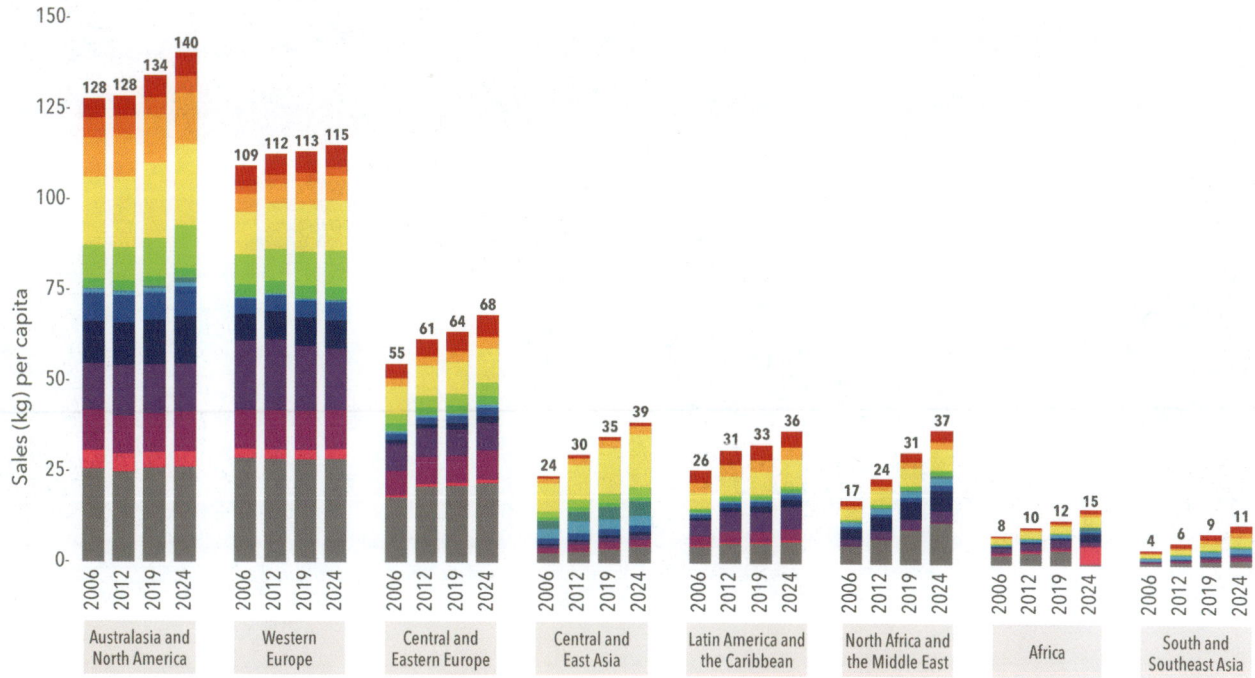

b. By socioeconomic group

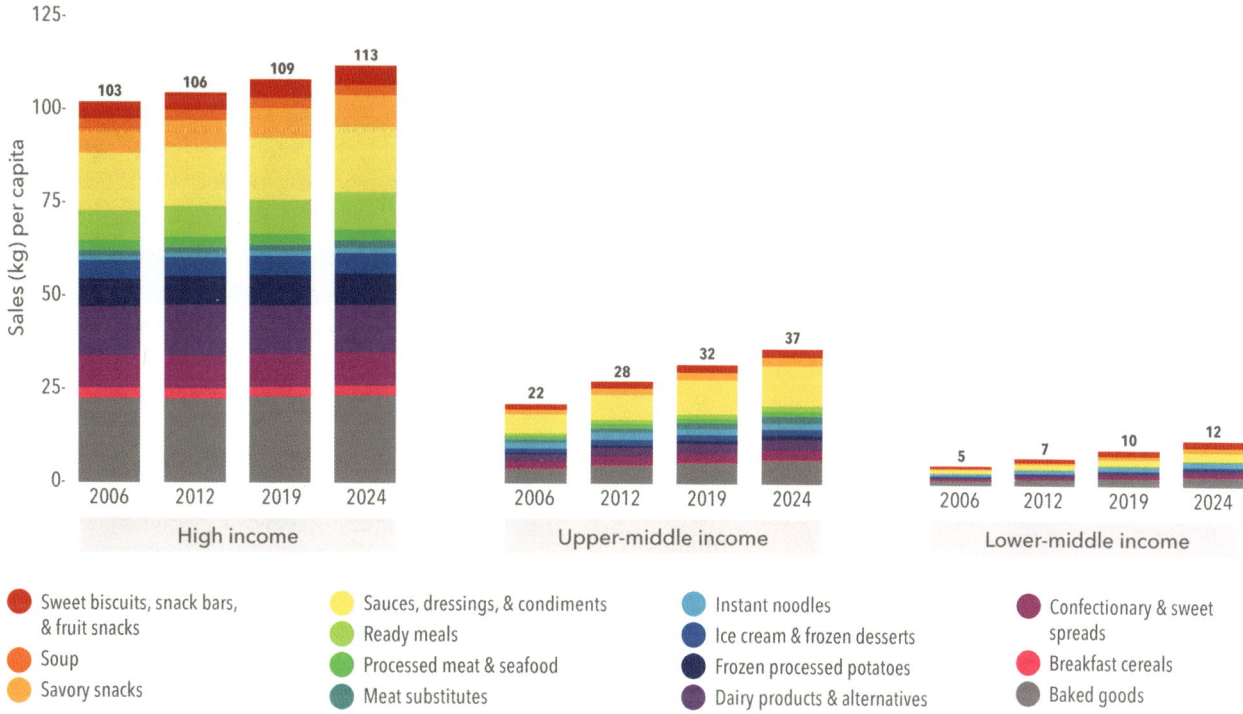

Source: P.I. Baker, P. Machado, T. Santos, et al., "Ultra-Processed Foods and the Nutrition Transition: Global, Regional and National Trends, Food Systems Transformations and Political Economy Drivers," *Obesity Reviews* 21, 12 (2020): e13126. Reproduced with permission of the publisher.

minimum recommended consumption target of 400 grams per day.[15] Consumption of F&V has risen globally over the past 40 years, but this increase primarily occurred in high-income countries (HICs). In some countries, particularly in sub-Saharan Africa, F&V consumption levels remain at the same low or declining levels as 40 years ago.[16]

Euromonitor data on per capita sales of UPFs show steady increases in all regions of the world since 2005 (Figure 1a). Sales have more than doubled in LMICs, while still remaining at much lower levels than in higher-income regions (Figure 1b). In LMICs, markets for these foods are growing rapidly because of their long shelf life and hyper-palatability and the large investments in commercial marketing.[17]

CHANGING DIETS AND ENVIRONMENTAL IMPACT

The way that foods are produced, including the agroecological and farming system characteristics, contribute to their environmental impacts.[18] Figure 2, which presents data on the environmental pressures of different food groups for six environmental outcomes,

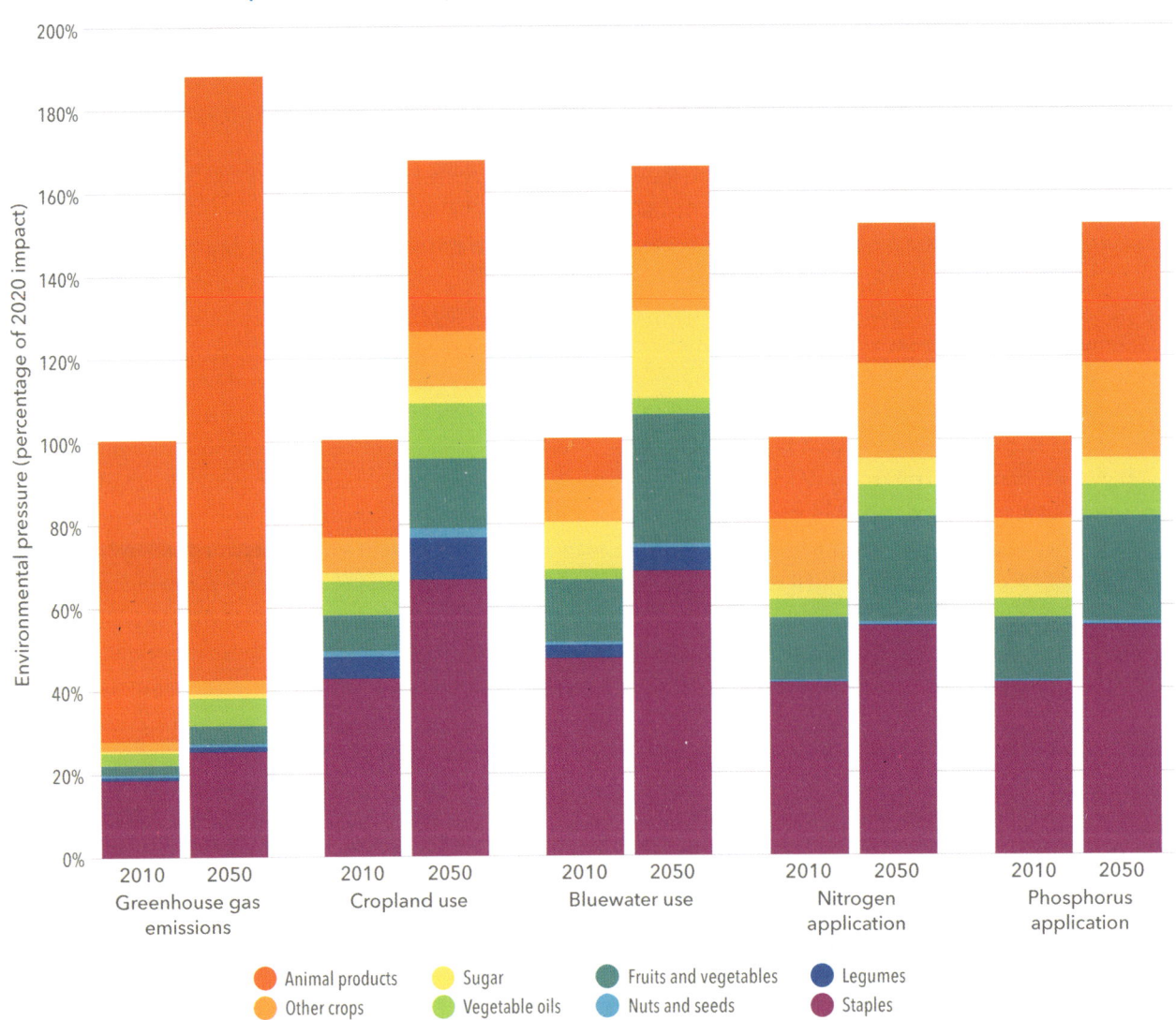

FIGURE 2 Environmental pressures from food groups: 2010 and a "business as usual" 2050 projection

Source: M. Springmann, M. Clark, D. Mason-D'Croz, et al., "Options for Keeping the Food System within Environmental Limits," *Nature* 562 (2018): 519–525.

Note: "Bluewater use" includes fresh water in streams, rivers, lakes, and aquifers.

shows that each food group contributes to different environmental stresses. The figure demonstrates that ASF production emits a significant amount of greenhouse gas (GHG) emissions compared to other food groups.[19] However, GHG emissions vary widely among ASF types, with meat, fish, and eggs contributing more than milk and cheese, and beef accounting for more than four times the amount of emissions from pork, chicken, or fish.[20] Overall, plant-based foods cause fewer adverse environmental impacts, but staple food production has significant impacts on environmental outcomes, largely because of the sheer volume produced and the land needed to feed both humans and animals. It should be further noted that the environmental footprint of different foods, especially perishables such as ASF and plant foods, also depends on farm management practices and off-farm processes, including manufacturing, transport, processing, and waste disposal.[21]

The environmental impact of UPFs remains understudied,[22] but a recent review shows that they are responsible for a significant amount of food-production-related energy use, GHG emissions, land and water use, biodiversity loss, fertilizer use, and food loss and waste.[23] These impacts are compounded by the extensive use of plastic in UPF packaging and the production systems used for certain ingredients in these foods, such as sugar or oil palm that are grown in monoculture or plantation-type systems.[24] An analysis of food purchases in Brazil shows that overall diet-related GHG emissions and land and water use increased by approximately 20 percent over the past 30 years. The increase in environmental pressure was much larger for UPF purchases than for less processed foods, which showed some reduction in environmental impact over time.[25]

THE CHALLENGES OF SHIFTING TO SUSTAINABLE HEALTHY DIETS

DESIGNING A SUSTAINABLE HEALTHY DIET AND IMPLICATIONS FOR GLOBAL DIETARY SHIFTS. In addition to proposing global targets for diets and the environment, the EAT-Lancet Commission recommended a "Healthy Reference Diet" that includes an abundance of F&V, diverse plant sources of protein and other essential nutrients, low consumption of most ASF, and limited amounts of refined cereals and UPFs.[26] This theoretical diet, if produced sustainably, is an example of a potential *sustainable healthy diet* that could protect both human and planetary health. Using this diet as a benchmark, the Commission shows that major changes in consumption patterns would be required to shift to sustainable healthy diets on a global level and that the specific nature of the changes would vary by region. The proposed global dietary shifts would help prevent approximately 11 million premature adult deaths per year and, if combined with sustainable production practices and reductions in food loss and waste, would allow food production to remain within safe planetary boundaries.

ADAPTING THE SUSTAINABLE HEALTHY REFERENCE DIET TO DIFFERENT CONTEXTS AND POPULATION GROUPS. Recommended intakes for the EAT-Lancet diet align with energy and nutrient requirements for an average adult man. Such a diet would be inadequate for population groups with increased nutrient requirements, including pregnant and lactating women, infants, young children, and adolescents as well as for populations in many LMICs who lack access to large-scale food fortification or targeted supplementation.[27] Without these complementary measures, plant-based diets would likely be too low in micronutrients that are either absent (such as B12), low, or less bioavailable (such as calcium, iron, and zinc) in plant foods. For groups with increased needs, these diets would have to be complemented with additional ASF or supplemented with lacking nutrients.[28]

MAKING SUSTAINABLE HEALTHY DIETS AFFORDABLE FOR ALL. The unaffordability of sustainable healthy diets, especially for marginal and nutritionally vulnerable populations, represents another major constraint. A global analysis of the cost of the EAT-Lancet reference diet, for example, shows that the cost exceeded the total per capita household income for 1.6 billion people, mostly in sub-Saharan Africa and South Asia. On average, the cost of the diet represented 89 percent of mean income in low-income countries (LICs), as compared to 27.5 percent in HICs.[29] Additional insights come from an analysis of the relative caloric price (RCP) of 21 food groups across 176 countries (compared to starchy staples).[30] The findings show that

F&Vs and ASF have the highest RCPs and the largest variations across countries and regions, with higher RCPs in LMICs than HICs. Results for milk and eggs were particularly striking, with RCPs of 2.0 and 2.6 respectively in HICs and approximately 11 in LICs. F&V and ASF tend to be costlier (per kcal) than staple cereals or legumes, largely because of their perishability, seasonality (for most F&V), and food supply constraints.[31] Moreover, food prices will likely continue increasing as a result of climate change, especially the price of staple grains. Given that these foods are the primary components of diets for humans and animals, rising prices will have secondary effects on the price of ASF.[32] By contrast, fats and oils, sugar, and UPFs are universally low-cost options because of their lower production costs and losses, ease of storage and transport, and long shelf life.[33]

POLICY OPTIONS TO SHIFT CONSUMPTION TOWARD SUSTAINABLE HEALTHY DIETS

Policies to shift consumption toward sustainable healthy diets must focus on 1) increasing the availability, access, and affordability of nutritious foods; 2) discouraging excessive consumption of ASF (especially red and processed meat) and UPFs; and 3) encouraging demand for nutritious foods and sustainable healthy diets. We identified five broad categories of demand-side and food environment policy actions that could be included in a comprehensive and coordinated policy package to promote sustainable healthy diets,[34] along with supply-side policies to stimulate efficiency, sustainability, and equity in the production of nutritious, safe, and affordable foods. These include social protection policies to support marginalized populations (see Chapter 6) and policies intended to reduce food waste (see Chapter 11). This chapter concentrates on consumer-focused policies to stimulate demand for sustainable healthy diets, fiscal policies to address affordability constraints, and food environment policies to support consumers in making healthier and more sustainable food choices.

INFORM, EDUCATE, AND GUIDE CONSUMERS TO DEMAND SUSTAINABLE HEALTHY DIETS

Consumer information, education, awareness creation, and guidance should be included in all comprehensive strategies that aim to shift consumer demand toward healthier and more sustainable dietary choices. These approaches are particularly important for nutritious foods that have relatively low income or price elasticities,[35] including F&V, legumes, nuts, and whole grains.[36] Approaches used to inform, educate, nudge, and influence dietary choices include public awareness campaigns, behavior change communication, mass media, "edutainment,"[37] social media, mobile nutrition and health services, cooking demonstrations and classes, and school-based nutrition education. Evidence from LMICs and rigorous evaluations of these approaches is limited, however.[38] An exception is breastfeeding promotion, an intervention known to improve breastfeeding practices in LMICs.[39] This "win-win," double-duty action addresses multiple forms of malnutrition and conveys numerous benefits for mother and child,[40] while also being more environmentally friendly than breastmilk substitutes.[41] Some evidence shows that nutrition education approaches may be more effective when combined with other complementary strategies, including policies to relieve income and affordability constraints (such as social protection programs) and food environment policies to stimulate demand for sustainable healthy diets.[42]

National food-based dietary guidelines (FBDGs) are a key policy instrument used by countries to translate global evidence on healthy (and sustainable) diets into practical, culturally appropriate, and context/population-specific recommendations.[43] A recent review of the 90 FBDGs available worldwide (with only 7 from African countries) showed that while most emphasized the consumption of nutritious food groups and the need to limit salt, sugar, and fat intake, very few included considerations for environmental sustainability. More work is needed to support LMICs in developing context/population-specific FBDGs that are quantified (for example, including upper limits for the consumption of unhealthy foods/ingredients) and optimize achievement of sustainable healthy diets. These guidelines must consider the country's state of food fortification and the unique needs of different population subgroups (depending on age, gender, physiological status, and activity levels).[44] School-age children and adolescents should be prioritized, given the importance of establishing habits and preferences for healthy eating and lifestyles early in life.

STIMULATE DEMAND FOR NUTRITIOUS FOODS OR DISCOURAGE CONSUMPTION OF UPFs OR HIGH-GHG-EMISSION FOODS

Fiscal policies targeted to the retail sector can increase affordability and stimulate demand for nutritious foods. These policies include subsidies and the removal of taxes, which facilitate consistent access to affordable nutritious foods by encouraging the retail sector to supply, promote, and reduce the price of these foods in food environments. Experience suggests that fiscal measures can effectively change behavior related to both the purchase and consumption of targeted nutritious foods, although much of the evidence comes from simulation studies and upper middle- and high-income countries.[45] In Chile, a modeling study concluded that combined fiscal policies, which included taxing less nutritious foods and removing taxes on F&V, would have the most impact on promoting healthier diets.[46]

Fiscal policies have been extensively used to reduce the consumption of unhealthy foods, especially UPFs,[47] with up to 40 countries having implemented a tax on sugar-sweetened beverages by the end of 2020.[48] Overall, the evidence on unhealthy food and drink taxes has shown decreases in purchases of these foods.[49] There is less experience with implementing taxes on high-GHG-emission foods, but modeling evidence suggests that both health and the environment could significantly benefit if the price of red and processed meat included the health-related costs to society of consuming these foods.[50] Findings from a simulation exercise that added a tax to high-emission food commodities (such as red and processed meat) in HICs and applied those tax revenues to health promotion showed that doing so could avert the negative health and climate impacts disproportionately affecting marginalized populations.[51] A modeling study of the European Union diet also demonstrated that food group-specific taxes can be effective in reaching nutrition and environmental goals, but high (possibly unrealistic) tax increases would be needed to meaningfully shift consumption toward less ASF and UPFs and more F&V.[52]

MAKE FOOD ENVIRONMENTS MORE SUPPORTIVE OF HEALTHY AND SUSTAINABLE FOOD CHOICES

Food labeling and certification can also support consumers in making healthy, sustainable food choices. In contrast to most HICs, the labeling of packaged foods is not mandatory in LMICs, except for some Latin American countries.[53] As a prime example of such a policy, Chile established a black front-of-package warning to identify foods high in energy and added salt, sugar, or saturated fats.[54] This policy was part of a comprehensive set of regulations that included school bans and marketing controls, including prohibition of cartoons or similar kid-friendly logos on breakfast cereals, over several years. These regulations helped to significantly reduce the exposure of children and adolescents to such advertisements (44 percent and 58 percent, respectively),[55] and several Latin American countries followed Chile's example, possibly leading to a broader positive impact on the whole region. In some meta-analyses and systematic reviews, food labeling was shown to reduce energy and total fat intake and increase vegetable consumption among consumers in HICs.[56] Labeling was shown to convey other benefits as well, such as a positive industry response to product reformulation that resulted in reductions in sodium (8.9 percent) and trans fatty acid (64.3 percent) content.[57] While labeling policies appear promising, designing and implementing them in LMICs will require robust commitments, time, efforts, and funding, as well as major adaptations to ensure that labels are tailored to literacy levels and cultural factors in these countries.

Nascent efforts to establish and legislate certifications and carbon footprint labels focus on consumers in HICs, with the goal of increasing literacy and awareness of the environmental sustainability of different food commodities. Reviews suggest that more educated, wealthier consumers are more likely to have a positive attitude toward carbon footprint labels and be willing to pay more for foods identified as being sustainably produced.[58] Recommendations include designing friendlier label systems for carbon footprints and avoiding over-labeling (for example, on different food characteristics or targets).

Governments can *regulate the marketing of UPFs to children* by banning or limiting advertising of these foods on media platforms or by limiting sales in and around schools.[59] Evidence suggests that mandatory bans are more effective than self-regulation at reducing children's exposure to the advertisement of foods high in fat, salt, and sugar.[60] As with food labeling, marketing restrictions may spur efforts by private companies to reformulate UPFs and reduce unhealthy, unsustainable ingredients and foods.[61] Currently, however, only 16 countries have regulations in place to ban the marketing of UPFs to children through media or schools.

SETTING PRIORITIES

Evidence is accumulating on the range of policy options to stimulate demand for sustainable healthy diets. To achieve both the sustainability and health outcomes of the world's diets, there is an urgent need to develop and implement policy packages in LMICs that include multipronged, coherent, and mutually reinforcing actions. But there is also a need to prioritize. Priority policy packages to generate demand for sustainable healthy diets should include, at a minimum, consumer education approaches along with fiscal measures and food environment policies that support positive changes.

Education measures should start with aggressive breastfeeding promotion as a "win-win" solution for both maternal and child health and for the environment. FBDGs in LMICs should be prioritized and updated regularly to incorporate environmental sustainability considerations and to align with the 21st century challenges of rapidly changing diets and nutrition and disease risk profiles.

Fiscal measures should be adapted and rigorously tested in a greater number of LMICs to document their impacts on discouraging consumption of unhealthy UPFs or on boosting consumption of nutritious foods by incentivizing retailers to subsidize these foods. For the latter, fiscal measures should be paired with behavior change communication, especially for foods with low income or price elasticity (such as F&V, legumes, and nuts).

Similarly, there is a need to continue testing and adapting innovations in *food environment policies* (such as labeling, promotion, and marketing) in LMIC contexts and to document their usefulness in helping consumers make healthy, sustainable dietary choices and maintain them over their lifetime.

Renewed efforts are needed to reverse the rapid deterioration of diets in LMICs and the related risks for nutrition, health, and the environment. More attention and effort are required to implement innovative and evidence-based demand-side policies, but these also must be accompanied by supply-side policies to reduce the loss and waste of nutritious and perishable foods across supply chains and to ensure sustainable agricultural production.

> Shifts toward healthy diets must focus on the dual goals of protecting and improving the nutrition and health of populations while also meeting environmental goals in an equitable way.

CHAPTER 9

Rural Clean Energy Access
Accelerating Climate Resilience

CLAUDIA RINGLER, ALEBACHEW AZEZEW BELETE, STEVEN MATOME MATHETSA, AND STEFAN UHLENBROOK

Claudia Ringler is deputy director of the Environment and Production Technology Division, International Food Policy Research Institute, Washington, DC. **Alebachew Azezew Belete** is a senior energy researcher, Ministry of Water and Energy, Ethiopia. **Steven Matome Mathetsa** is a senior scientist, Eskom Research, Testing and Development, Johannesburg. **Stefan Uhlenbrook** is strategic program director, Water, Food and Ecosystems, International Water Management Institute, Colombo.

KEY MESSAGES

- Rural livelihoods in low- and middle-income countries are doubly jeopardized by energy poverty and climate change. Lack of access to affordable clean energy reduces agricultural productivity, affects health and nutrition outcomes, and adds to environmental degradation, which in turn, further contributes to climate change.

- Reliable access to clean energy can protect rural households against adverse climatic events and support new off-farm economic opportunities. Accelerating a rural clean energy transition will thus be key not only to reducing climate change but also to improving rural lives and livelihoods.

- Existing rural lending mechanisms are often unsuitable for the purchase of clean energy technologies, such as photovoltaic solar panels.

Several actions can accelerate rural access to clean, sustainable energy for all:

- Identify locations where promising energy and water sources and productive uses are in close proximity; this can jointly support energy, water, and food security without compromising ecosystem health.

- Create an enabling environment for accelerated clean energy development. This requires integrated governance across the water-energy-food-environment sectors, including institutions that can help identify synergies or trade-offs with natural resources and livelihoods and thus grow positive impacts. It also requires equitable access to energy through investments, incentives, and direct support for poor farmers and entrepreneurs.

- Develop appropriate financial incentives to expand dissemination of clean energy technologies to underserved rural populations, for example, credit at lower interest rates linked to climate mitigation and productive uses of clean energy. Implementation of these financial incentives will require capacity building for both credit suppliers and smallholder farmers.

- Strengthen women's agency in rural clean energy systems. Women and men experience energy and water poverty differently, and energy technologies are often aimed at men. Promoting a women-centered clean energy program can trigger multiple social, economic, and environmental benefits in rural communities.

Globally, the energy sector accounts for almost three-quarters of total greenhouse gas (GHG) emissions[1] and is thus responsible for the majority of adverse climate change impacts on rural livelihoods, including growing water, energy, and food insecurity and environmental degradation. According to a recent report from the Intergovernmental Panel on Climate Change, annual investments of US$2.4 trillion (2010 dollars) in energy systems are needed to limit global warming to 1.5°C.[2] Such investments would support decarbonizing the largest polluters and improving energy efficiency. More and cheaper clean energy technologies and greater energy efficiency are equally critical for accelerating access to energy in underserved rural areas in ways that promote ecosystem health and inclusivity.

COSTS OF ENERGY POVERTY

Rural livelihoods in low- and middle-income countries are doubly jeopardized by climate change and energy poverty. Climate change diminishes water resources and thus reduces opportunities for food production and access to water for other uses. At the same time, households that lack access to energy cannot pump or otherwise access sufficient water for crops, animals, and domestic use. Moreover, without cleaner and more efficient energy sources, these households must rely on fuelwood, crop residues, or cow dung for cooking, which degrades remaining forests and agricultural landscapes and has negative health impacts. This can result in a downward spiral of growing ecosystem degradation and vulnerability. Thus, water, energy, food security, and environmental sustainability are closely interlinked, and their interactions are experienced most acutely in rural areas (Figure 1).

Energy poverty is widespread in rural areas. As of 2020, close to 600 million people in rural sub-Saharan Africa and 170 million people in South Asia lacked access to electricity; and many millions more suffered from unreliable access. Even more people – 2.5 billion, most of whom are in rural areas – do not have options for clean cooking.[3] In addition, the COVID-19 crisis has further worsened energy poverty. Accelerating a rural energy transition will thus be key not only to reducing climate change but also to improving rural livelihoods. Reliable clean energy, such as solar power used for irrigation, can protect rural households against droughts and other adverse climatic events. For example, the Dhundi solar cooperative in Gujarat, India, allows farmers to sell back unused solar power to the grid at a guaranteed price. This constant source of risk-free income has incentivized

FIGURE 1 The linkages between rural energy access, water and food security, and ecosystem health

Source: IFPRI.

farmers to conserve groundwater and allows them to better weather drought and other adverse agricultural events.[4] A study in Bihar in eastern India assessed the impact of various drought-proofing programs on agricultural productivity and farmer welfare. It found that drought-relief programs and safety net programs were ineffective, as were subsidies on diesel for irrigation, due to delays, uncertainties, and high transaction costs. Solar-powered pump sets, however, were an effective drought-proofing strategy that allowed farmers to maintain yields during drought conditions.[5] Access to clean energy can also support new off-farm economic opportunities in agro-processing and other economic sectors that contribute to more diversified, resilient rural livelihoods.[6]

The cost of energy poverty is considerable, both for human and planetary health. Although its full impacts in rural areas remain unknown,[7] studies have shown that lack of clean energy access hampers agricultural productivity and stunts agriculture sector growth. It also stymies agribusiness growth; limits the production, storage, and consumption of nutritious, high-value foods; and places immense time burdens on women, contributing to their disempowerment.[8] Lack of access and use of clean energy also jeopardizes overall ecosystem health and biodiversity by directly contributing to deforestation, land degradation, and GHG emissions. One of many such pathways is through agriculture: lack of access to energy, including clean energy, reduces agricultural productivity,[9] while intensified agricultural systems lower emissions.[10]

Several studies show direct links between lack of electricity and poor health and nutrition outcomes. The use of traditional fuels for cooking causes more than 1.5 million deaths every year, mostly of women and children. Moreover, women and children are also often responsible for collecting fuel for cooking,[11] a time obligation that can deprive them of education or income-generation opportunities. While research on the linkages between energy access and nutrition remains limited, a few salient findings are emerging. A study in Nigeria looked at the relation between children's nutrition outcomes and energy access. Using nighttime light intensity as an indicator of electricity access and urbanization, the study showed that nighttime light is a significant predictor of stunting and chronic malnutrition (height-for-age z-scores [HAZ]) for children under the age of five through welfare impacts,[12] even after controlling for other factors known to influence stunting. Likewise, in rural Bangladesh, research found that electricity access can improve children's nutrition (using HAZ as the indicator).[13]

At the same time, use of fossil and biomass fuels rather than clean energy sources contributes directly to climate change. Reliance on fossil-fuel-driven pumps is a common response to clean energy poverty that contributes to climate change. In India, for example, fossil-fuel driven groundwater irrigation is estimated to account for 8 to 11 percent of total national GHG emissions.[14] On the other hand, increasing access to clean energy can increase agricultural productivity — and also reduce deforestation and forest degradation, thus helping to preserve biodiversity and even lessen zoonotic disease risk by reducing interactions between wildlife and humans — without adding to GHG emissions.[15]

CLEAN ENERGY INNOVATIONS

Clean energy innovations suitable for rural areas, including farming and small enterprises, are becoming more readily available. One important innovation is small-scale solar power technologies that can be used to pump water for domestic uses and for irrigation, which can support the production of high-value vegetables and fruits.[16] New solar technologies can also be used to cool poultry houses as well as milk containers for storage and transportation. These cold-chain advances can improve productivity and food safety and extend the shelf life of nutritionally dense and high-value foods such as milk, eggs, and green leafy vegetables, especially as temperatures are rising.[17] For rural entrepreneurs, solar driers can improve the quality and safety of harvested products, including fish and fruits. However, the upfront costs of solar technologies remain too high for most smallholder farmers, and rural lending is dominated by business and finance models that are designed for the purchase of seeds and fertilizers (which have lower costs and quicker returns on investment), but are not suitable for higher cost, longer-term investments in solar or other clean energy equipment by small farmers or business owners.

Decentralized or distributed renewable energy (DRE) systems are another clean energy solution that is increasingly available and suitable for rural areas. These systems can provide locally generated electricity for farm production and drinking water systems. They can also provide electricity for public buildings such as medical facilities and schools. Microgrids can integrate electricity inputs from multiple sources, including solar photovoltaic panels, micro-hydropower systems, and diesel generators. With proper integration of storage and load management, systems operating at a local scale can power economic activities beyond the farm, substantially benefiting local economies. Such systems, however, generally require outside finance for initial establishment and creative finance models to ensure sustainability.[18]

OPPORTUNITIES TO ACCELERATE CLEAN ENERGY ACCESS IN RURAL AREAS

The costs of clean energy technologies have declined substantially over the last two decades,[19] but technology uptake has remained low in many rural areas that would particularly benefit from better energy access. In addition to substantial upfront investment costs, weak supply chains and services, inadequate financing mechanisms and financing ecosystems, complex technologies, and the high cost of borehole drilling services have been identified as limiting uptake.[20]

Food production remains the primary rural livelihood activity in low- and middle-income countries, and thus the most obvious entry point for the development of a thriving energy technology market. However, for widespread adoption to occur, critical investments are needed to establish enabling frameworks that can match potential users with appropriate water and energy systems; to develop institutions across the water-energy-food-environment sectors to ensure that adverse environmental impacts are reduced or avoided; to increase access and equity by developing appropriate finance mechanisms that are accessible to smallholder farmers and small entrepreneurs; and by strengthening women's agency in clean rural energy systems.

To accelerate rural access to clean, sustainable energy, we propose the following five steps:

IDENTIFY LOCATIONS FOR PRODUCTIVE USES THAT CAN JOINTLY SUPPORT ENERGY, WATER, AND FOOD SECURITY WITHOUT COMPROMISING ECOSYSTEM HEALTH.
The identification of appropriate productive energy uses in the agrifood sector, particularly in places where promising energy and water sources and producers and next users are in close proximity, can accelerate clean rural energy access. Various recent efforts have focused on co-locating energy investments and productive users, primarily irrigation operations or agro-processing centers. To identify potential locations, the International Water Management Institute (IWMI) has developed an online solar suitability tool[21] using a GIS-based multi-criteria evaluation technique that accounts for solar irradiation, slope, groundwater levels, aquifer productivity, groundwater storage, groundwater sustainability, population, roads, and travel time to markets. A study conducted by IFPRI researchers analyzed the economic feasibility of solar irrigation across Africa, considering cropping patterns, costs of solar-powered pumps and alternative pumps, and a set of biophysical factors. Groundwater-fed solar irrigation was found to be cost-effective in southern and central Africa, but less so in countries that subsidize diesel fuel (Angola, Nigeria, and Sudan). Solar panels were also more economical than diesel pumps for more water-intensive crops.[22] Another study identified priority areas for on- and near-farm electricity using geospatial analysis.[23] And a study in Ethiopia focused on the economic benefits from alternative productive use investments, such as horticulture irrigation, grain milling, *injera* baking, milk cooling, bread baking, and coffee washing — estimating a joint potential to generate $4 billion annually following electricity rollout by 2025.[24] Additional income would be generated from the purchase of mechanized equipment. However, studies such as these are seldom incorporated into planning for energy systems, which tends to focus narrowly on optimizing energy systems rather than considering their impacts on rural livelihoods and well-being.

DEVELOP INSTITUTIONS ACROSS THE WATER-ENERGY-FOOD-ENVIRONMENT SECTORS TO

STRENGTHEN SYNERGIES AND REDUCE TRADE-OFFS. Many countries continue to develop energy strategies without considering potential synergies or trade-offs with either natural resources, including water and land, or other sectors and actors, such as food production. Single-sector strategies are likely to miss important synergies and ignore trade-offs associated even with cleaner energy technologies. This can lead to unnecessary costs and environmental damage. For example, the installation of large fields of solar panels can compete with agricultural production areas or natural habitats, thus affecting food security, livelihoods, and biodiversity. Similarly, unfettered development of solar-powered irrigation can lead to overexploitation and degradation of groundwater resources. With fuel-based technologies, fuel costs rise with the amount pumped and thus create an incentive for water conservation.[25] For solar technologies, however, there are no additional financial costs to pumping more water – so well-designed strategies and strong institutions are needed to protect against unsustainable water withdrawals and related environmental degradation.[26] To address this, institutions that jointly consider food, energy, water and environmental systems are needed. The Niger Basin Authority in West Africa is one example. The agency works directly with ministries of water, energy, agriculture, and environment in its nine basin countries and is currently developing a legal document to actively consider trade-offs across these sectors for more effective implementation of the Niger Basin Shared Vision for sustainable development.[27]

ENSURE EQUITABLE ACCESS TO ENERGY THROUGH INVESTMENTS, INCENTIVES, AND DIRECT SUPPORT. In addition to minimizing environmental damages from energy development, an enabling framework for broadening clean energy access to poorer farmers and entrepreneurs is needed. This is a tall order. Both Ethiopia (Box 1) and South Africa (Box 2) have been working toward accelerating clean energy access, with mixed results.

Ethiopia aims to ensure energy access for all by 2025. Despite this ambitious goal and considerable investment, more than half of Ethiopia's population still lacks access to reliable electricity, especially in rural areas, which remain dependent on fuelwood and kerosene. In South Africa, the Renewable Energy Independent Power Producer Programme (REIPPP)[28] allows private industries to produce clean electricity for both non-grid and grid systems (including feed-in of surplus energy generated). This has increased the development of photovoltaic solar energy systems for both individual and industrial use. Rural areas, however, lag behind in adopting these systems due to lack of funding, inappropriate business models, rural poverty, and a relative neglect of these areas. This is a missed opportunity, given that South Africa's rural areas offer both abundant space and sunlight to generate and use solar power.

BOX 1 THE CHALLENGES OF ACCELERATING ENERGY ACCESS IN ETHIOPIA

Most of Ethiopia's 115 million people live in rural areas, where only 29 percent have access to electricity.[a] The country has one of the lowest electricity consumption levels in the world, at just 80 kWh hours per capita, compared with 12,154 kWh in the United States.[b] Although the government is implementing various programs and strategies to achieve universal electricity access and has plans to become an important energy exporter, there are major challenges in extending access to rural areas. These include capacity, technological, and economic limitations, underdeveloped rural infrastructure, poor information sharing, and political barriers. The energy sector in the country lacks indigenous organizations that adapt international technologies or develop their own options for use in more remote areas.[c] Moreover, heavy regulation of the sector has limited private investments in this capital-intensive sector. Finally, transboundary political challenges linked to Ethiopia's flagship energy project, the Grand Ethiopian Renaissance Dam (GERD), have slowed other rural energy development projects.

> **BOX 2** SOUTH AFRICA'S ENERGY PROGRAMS
>
> South Africa's Integrated National Electrification Programme (INEP), launched in the post-apartheid era, increased electricity access from 35 percent of the population in 1994 to approximately 88 percent by 2018.[a] The program, which includes free basic electricity for poor households, has reduced rural households' reliance on firewood and cow dung as energy sources. However, energy supply has not kept pace with growing demand, leading to frequent supply interruptions or "loadshedding," as parts of the electric grid are temporarily shut off. The country is also grappling with the need to reduce dependence on fossil fuel technology as well as with rising costs of electricity production.[b] One response has been the Renewable Energy Independent Power Producer Programme (REIPPP), which supports the use of clean technology for both small- and large-scale rural-based activities, including food production, postharvest processing, domestic water use, and water pumping for irrigation. However, poor farmers are unable to meet the program's requirements, slowing the expansion of rural clean energy. As a result, some rural communities have returned to using wood and dung as domestic fuels.

DEVELOP APPROPRIATE FINANCIAL INCENTIVES FOR UNDERSERVED RURAL POPULATIONS. In most rural areas of low- and middle-income countries, finance systems have not developed appropriate mechanisms to support investment in clean energy technologies by smallholder farmers and small entrepreneurs. Most credit facilities are designed for crop production — loans are small and require repayment within or following a single growing season. Energy technologies such as solar-powered pumps require larger loans and longer payback periods to be affordable by poor farmers. Also, given the potential of these technologies to lower the risk of default due to climate shocks, credit should be offered at lower rates. Few credit providers, however, have developed financial products targeted at clean energy technologies for rural clients. The development of appropriate financial incentives will require capacity building for both suppliers and smallholder farmers. Pay-as-you-go systems, where users pay for use time rather than the solar system, have become common for home solar systems, and are now being piloted for smallholder irrigation systems as well. However, the high seasonal variability of irrigation requires greater flexibility than is needed for home systems. A set of business models has been proposed for solar pumps in Ethiopia, including an outgrower or insurance scheme where agribusinesses working with contract farmers develop flexible payment mechanisms or even provide the pump for free, if the cost could be recouped through increased production.[29]

STRENGTHEN WOMEN'S AGENCY IN CLEAN RURAL ENERGY SYSTEMS. Women and men experience energy and water poverty differently because of their different assets and culturally and socially determined divisions of labor. Women are most often responsible for securing both water and energy sources for domestic use, affecting their availability for care work, income generation, and leisure.[30] However, technologies to secure access to water and energy for productive uses, such as agriculture and livestock rearing, continue to be aimed at male farmers. For example, most solar and other mechanized irrigation pumps are managed by men; this can contribute to women's disempowerment and lower incomes from the crops and livestock that women manage.[31] Depending on how energy, water, and irrigation systems are designed, implemented, and managed, women and men will benefit differently, with the burden on women potentially increasing rather than decreasing. Thus, for rural energy technologies to achieve their full potential, women-centered clean energy programs are needed. As an example, the Self-Employed Women's Association, a trade union of women working in India's informal sector, is running a solar irrigation pump program geared to its farmer members. To overcome women's key challenge of upfront down payments for energy equipment, the program facilitates separate loans for down payments sourced from a second bank; this not only increases access by women farmers to this technology but also reduces banks' overall lending risks.[32] Similarly, GROOTS Kenya, a grassroots women's organization, developed a biogas program to support women members who could not afford the cost of electricity for cooking in selected rural districts.

UNLOCKING THE BENEFITS OF CLEAN ENERGY

Accelerating rural clean energy investments can unlock access to water resources, increase food security, expand rural employment, increase incomes, and build climate resilience by contributing to both adaptation and mitigation. Reducing rural energy poverty is also critical for social justice, human development, and planetary health. Despite growing evidence of the synergistic role of energy strategies in the water-energy-food-environment nexus, most energy plans continue to be developed in siloes, and therefore lead to sub-optimal outcomes. A stronger focus on processes and institutions, as well as incentives for social inclusion could help accelerate access and increase benefit streams and, importantly, reduce environmental damage by supporting clean energy access for all.

CHAPTER 10

Bio-innovations
Genome-Edited Crops for Climate-Smart Food Systems

JOSÉ FALCK-ZEPEDA, PATRICIA BIERMAYR-JENZANO, MARIA MERCEDES ROCA, EDINER FUENTES-CAMPOS, AND ENOCH MUTEBI KIKULWE

José Falck-Zepeda is a senior research fellow, Environment and Production Technology Division, International Food Policy Research Institute, Washington, DC. **Patricia Biermayr-Jenzano** is an adjunct professor, Georgetown University, Washington, DC. **Maria Mercedes Roca** is executive director, Bioscience Thinktank, Monterrey, Mexico. **Ediner Fuentes-Campos** is co-founder and executive director, The Bridge Biofoundry, and member of the Sistema Nacional de Investigación, Panama City. **Enoch Mutebi Kikulwe** is a scientist, Alliance of Bioversity International and CIAT, Kampala.

KEY MESSAGES

- New genome-editing (GEd) technologies, including CRISPR-based tools designed to edit genes, will play a critical role in addressing climate change adaptation and mitigation in agriculture. GEd allows researchers to rapidly develop climate-resilient and climate-adaptable crop varieties tailored to low- and middle-income countries (LMICs).

- Pursuing pragmatic approaches that enable convergence of GEd applications with ecologically and environmentally sustainable production systems is a prudent and valuable approach.

- Public and private sectors in LMICs both have a role to play in developing GEd products to address climate change but will need a robust enabling environment to support this development.

- Functional and streamlined regulatory frameworks are an important component of any robust enabling environment to create and support incentives for product development and deployment. Lessons learned from earlier technologies will be critical to successfully advancing GEd products through the process of approval, transfer, and adoption.

- Transparency across GEd research and development (R&D), regulation, and deployment will be essential to ensure social "buy-in" from a broad range of stakeholders. Achieving buy-in will require more comprehensive assessment methods to build evidence on GEd tools, as well as the prioritization of strategic communication and outreach.

- Start-ups and small and medium enterprises (SMEs) can help drive the democratization of GEd in LMICs, as they have been more agile in implementing GEd R&D processes. Partnerships that enable technology transfers or even generate spin-offs or SMEs are a promising strategy to rapidly deliver new climate-resilient applications for LMICs.

- GEd crops can help small farmers — including women and youth farmers, indigenous people, and other vulnerable groups — to increase farm productivity and adapt to climate change. Understanding the needs and preferences of farmers, including gendered needs, for different crop traits, delivery methods, and extension services is essential to the successful development and deployment of GEd products.

As growing populations, changing diets, and climate change affect growing conditions for crops, our agriculture and food systems must increase production and productivity to ensure access to healthy and diverse diets for all. The expanding demands on agriculture and food systems must be met without increasing pressure on the environment, and while accomplishing other development goals and objectives. Addressing this complex problem requires the identification of game-changing interventions that can drive sustainable, equitable agriculture for food system transformation. "Bio-innovations" will be one key set of interventions. Bio-innovations encompass biotechnology-based tools and product innovations, as well as innovations in their governance, regulation, and social and business contexts. They hold potential to contribute to food system transformation by accelerating productivity growth and reducing agriculture's environmental footprint, as well as contributing to climate change adaptation and mitigation.[1]

Within the portfolio of potential bio-innovations, this chapter focuses on second-generation biotechnologies, specifically genome-editing (GEd) tools and the products developed through their use. We group GEd tools and products together under the term "GEd applications," which includes newer genome-editing tools among other applications (Box 1). To date, however, the global experience with bio-innovation tools has almost exclusively been with first-generation applications, including genetically modified organisms (GMOs). This experience will likely offer lessons for newer GEd applications that are soon to enter the market and can inform the present discussion around their innovations and their commercialization, governance, and social acceptance. Our discussion focuses exclusively on crops, although it is also relevant for genome-edited animals, microorganisms, and industrial and pharmaceutical applications.

GEd is a revolutionary and disruptive technology for crop improvement.[2] Its applications can effectively unlock existing genetic value by introducing new traits into crops, while reducing the time necessary to develop new varieties. In the case of maize, time needed to deliver hybrids can be cut by as much as half, assuming that regulatory scrutiny is streamlined (for example, CIMMYT's maize lethal necrosis–resistant project using GEd[3]). Furthermore, this technology enables researchers to improve crops that have been difficult to enhance with conventional tools, including crops of interest to low- and middle-income countries (LMICs) like cassava, bananas, and sweet potatoes.

GEd applications can be particularly important for climate change mitigation and adaptation. These crops can be engineered to increase productivity, thereby reducing the amount of land needed for agriculture and thus reducing GHG emissions. Introduced traits can also improve crop resilience and nutrition, reduce pesticide and fertilizer runoff and leaching, and enhance soil health, all contributing to climate change mitigation.[4] GEd crop applications can also support adaptation to unexpected changes in environmental factors, including precipitation, temperature, extreme climatic events, and increased pest and disease

---

BOX 1 DEFINITIONS

The two distinct terms "gene editing" and "genome editing" are currently used almost interchangeably in the regulatory and popular science literature, with the supposition that they mean the same thing. In molecular biology, however, these two concepts have very different definitions, denoting distinct and increasingly complex levels of genetic structure in an organism. There are three genetic structure levels: gene, genome, and epigenome.

The **gene** is a basic unit of genetic information. The **genome** is the complete set of genetic information, including all genes present in an organism, which provides all the information that the organism requires to function. This includes regulatory sequences that operate like "on" or "off" switches. The **epigenome** is an even higher level of genetic structure or expression, which involves chemical modifications to the DNA and proteins that regulate the expression of genes within the genome. The three genetic structure levels give rise to gene, genome, and epigenomic editing processes. Gene and genome editing are now entering the product pipeline, and epigenomic editing will enter this pipeline soon.

The first-generation tools of genetic engineering randomly insert genetic material or genes from different or the same species into a host genome, resulting in **transgenic** or **cisgenic** organisms, respectively. These are also known as **genetically modified organisms** (GMOs). In contrast, second-generation tools such as CRISPR allow a more precise type of genetic engineering in which a gene or regulatory sequence can be identified and located at a specific site of an organism's genome. This gene or regulatory sequence can be silenced, deleted, modified, or replaced. These techniques are also known in some places as **new plant breeding techniques** (NPBTs) or **precision genetic technologies**.

Note: For formal definitions of gene, genome, and epigenome, see P. Portin and A. Wilkins, "The Evolving Definition of the Term 'Gene,'" *Genetics* 205, 4 (2017): 1353-1364; D. Goldman and L.F. Landweber, "What Is a Genome?" *PLoS Genetics* 12, 7 (2016): e1006181; and M. Ridley, *Genome* (New York: Harper Perennial, 2006).

incidence. Examples of GEd plants include drought- or salt-resistant crops, which result in fewer crop losses and less yield deterioration.[5]

As a result of their greater precision, ease of use, efficiency, and productivity,[6] these applications are the most promising bio-innovation available to foster food system transformation that addresses climate change, food security, nutrition, and livelihoods in LMICs. From the standpoint of achieving several goals and objectives, pragmatic approaches that use multiple approaches and applications – combining sustainable ecological approaches with genetically improved seeds – can be effective.[7] Applying such approaches to GEd in a broader context appears to be prudent.

In addition to the R&D investments necessary to use GEd tools and create new products (see Chapter 4), context-appropriate policies, regulatory frameworks, and programs will also be needed to create an enabling environment for the creation, assessment, and adoption of GEd applications.[8] This enabling environment will be particularly critical for the inclusion of vulnerable groups, particularly smallholders, women farmers, and other marginalized populations in LMICs, who often have limited access to new agricultural technologies and related knowledge.[9] Policy, regulatory, and other governance actions must facilitate widespread adoption and ensure equitable access to appropriate, beneficial, and safe GEd applications.

Agricultural innovations and technology adoption are not without cost. There can be "winners" and "losers" among both technology "adopters" and "non-adopters." However, technology use or consumption outcomes may be hard to predict.[10] Adoption processes can have unintended outcomes and their temporal and dynamic nature may induce both positive and negative impacts, creating unforeseen trade-offs.

To address this uncertainty, researchers have proposed an approach for proactively identifying potential exclusionary and other negative effects emerging from the adoption of transformational technology applications.[11] This approach includes promoting alternatives to mitigate negative effects, while ensuring compliance with responsible innovation pathways as much as possible. All potential outcomes are considered, such as the impact of farmers not having access to a technology that could address food security and environmental impacts from climate change. Furthermore, as GEd technologies become easier and cheaper to use, and democratized through widespread use for everyone, the dependence on multinational companies for seeds and pesticides will decline significantly, and risks related to climate change and production can be addressed with safer GEd products.

Within the scientific and regulatory community, consensus is growing that, in principle, newer GEd applications are safer for use than first-generation bio-innovations and even conventional plant breeding.[12] This consensus is based on evidence from GEd products that contain only small genetic changes, which are indistinguishable from changes that could be found in nature or created through conventional breeding methods. Compared with first-generation techniques, all second-generation GEd tools allow for targeted changes and greater control over where these changes occur in the genome. This improved precision suggests that new GEd tools are now safer, as they pose less risk of generating unintended changes in the genome[13] – and therefore are likely to require less regulatory scrutiny.

The enabling environment for GEd applications continues to evolve. Regulatory processes have focused only on gene-editing tools as yet, while the situation for other genome-editing applications remains unclear everywhere. To date, no GEd crop products have been commercially released in LMICs and very few in high-income countries.[14] This may change soon, as several GEd products (specifically those based on gene-editing tools) in the R&D and regulatory pipeline are poised to enter the marketplace (Table 1).

PRIVATE AND PUBLIC SECTOR ENGAGEMENT

The public sector has taken a strong leadership role in developing GEd applications. Of the approximately 1,400 families of GEd-related patents and patent applications in agriculture, most have been submitted by public sector entities, primarily state-sponsored research institutes in China and public universities and research centers in the United States.[15] In LMICs, public sector entities such as CGIAR and its Centers are showing increasing interest in GEd applications (Table 1), and regional and national research organizations are now conducting R&D for GEd applications, including research on maize, sorghum, rice, beans, cassava, and fonio.

Public sector initiatives increasingly aim to develop public–private and international collaborations that enable the management of R&D, technology transfer, and product stewardship capacities necessary to deliver GEd applications to producers. Recent advances have improved access to and availability of funds that prioritize regional or multilateral collaboration and integration of private or public–private entities. For example, in Latin America, initiatives from FONTAGRO and the Inter-American Development Bank (IDB) are funding capacity building in GEd for public research institutions, with South–South support from public research institutes in Argentina, Brazil, and Chile.[16] However, the donor landscape is mixed. Some donors have supported GMO application development, including Australia's CSIRO, Japan's JIRCA, the Bill & Melinda Gates Foundation, and the US Agency for International Development, providing limited investments in GEd. Other donors, especially in the European Union, have either decreased investments in GMO applications or have not invested in GEd development at all.[17]

In the private sector, nimble start-ups are taking the lead in advancing GEd applications.[18] Multinational corporations have attempted to keep up with the rapid growth of GEd applications and to overcome lags in capacity building by gradually acquiring start-ups that have developed commercially viable applications. The market potential of such acquisitions is attractive to multinational corporations, but their market access may be hampered by slow-moving regulatory processes, high costs of regulatory compliance, and their lack of experience in negotiating complex GEd-related regulatory processes across different jurisdictions.[19] Where time and cost considerations are a serious constraint, multinationals and other private sector developers are likely to push forward only those GEd applications with potential to become commercial blockbusters.

Given the focus of GEd investments on commercially profitable applications, shifting attention to agro-climatic- and region-specific varieties of LMIC-appropriate crops is a challenge for public sector farmer-led plant improvement communities and for some smaller private sector developers.[20] Private financing for such R&D will likely remain limited because varieties tailored to LMIC agro-environments often lack the economies of scale needed to generate an attractive return for private investors. The emergence of local start-ups using GEd technologies for LMIC crops also continues to lag due to the high initial investments required in specialized R&D infrastructure and other capacities. As a result, the development process in LMICs typically depends upon public systems that also face significant financial and resource constraints for deployment acceleration, marketing, and product stewardship.[21]

TABLE 1 CGIAR genome-editing bio-innovations to increase resilience to climate change impacts

TRAIT	In the R&D pipeline	Potential projects
Disease and insect resistance	bananas, cassava, rice, maize, wheat, potato	cassava
Enhanced heat tolerance		potatoes
Enhanced input use and reduced GHG emissions	rice	
Enhanced nutrition and quality and safety traits	cassava, cacao	beans, wheat, maize
Weed resistance		sorghum
Reduced postharvest loss		wheat

Source: Based on K. V. Pixley, J.B. Falck-Zepeda, R. Paarlberg, P.B. Phillips, I. Slamet-Loedin, K. Dhugga, H. Campos, and N. Gutterson, "Genome Edited Crops for Improved Food Security of Smallholder Farmers," Nature Genetics, forthcoming.

Note: Potential projects refer to those that are feasible with the current state of knowledge, application advancement, demand, and an appropriate funding level.

POLICY, REGULATORY, AND SOCIAL LICENSE FRAMEWORKS

Science-based regulations and transparent regulatory processes offer investment security for companies and public sector entities developing GEd technologies and can help deliver valuable and safe technologies to producers in LMICs. GEd crops are likely to face some of the same issues experienced by genetically modified (GM) crops in the approval, transfer, and adoption processes, and in securing "social license," that is, public acceptance of these processes and products.

Although safety and economic assessments and regulatory decisions on GM crops have demonstrated a history of safe and productive use for society,[22] existing biosafety regulatory processes continue to pursue a strict interpretation of the precautionary approach embedded in the 2003 Cartagena Protocol on Biosafety and the 1992 Convention on Biological Diversity. This approach makes these processes both costly and time-consuming. Such financial and time costs may be prohibitive for small and medium enterprises (SMEs) and the public sector.[23] In sum, these national measures would likely hamper the deployment of GEd technologies and crops.[24]

Regulatory frameworks for GEd crops are gradually being developed through a mix of approaches.[25] Countries with ample R&D and regulatory experience with GM technologies – including Argentina, Brazil, Chile, Colombia, Honduras, Paraguay, Uruguay, Australia, Canada, and the United States – will likely regulate GEd crops that have no permanent presence of foreign DNA (that is, DNA coming from other species) in the same way they regulate conventional crops. China's position on regulation remains a bit unclear, although it has invested heavily in developing GEd applications[26] and has recently announced new GEd regulatory guidelines.[27] The United Kingdom and Japan have signaled their intent to consider regulating GEd crops with different safety assessment processes than those required for GM crops under the 2003 Cartagena Protocol. The European Union and New Zealand have indicated in principle that GEd applications will be regulated as GM. This regulatory landscape is in flux and can be expected to change over time.

The success of GEd applications depends not only on science, R&D, and regulatory processes, but also on societal "buy-in" by a broad set of actors.[28] Establishing social license will require securing political support for innovation, ensuring and enhancing public participation and transparency, and making communication and outreach an integral part of the decision-making process. Some concerns can be resolved by responding to consumer and special interest groups' questions about new technologies, but certain segments of society may always remain opposed to GEd products.

Nevertheless, building transparency, using the best available evidence to address concerns, and communicating complex scientific concepts clearly to the public can help build the credibility of regulatory and decision-making processes and systems. Several comprehensive studies and reviews look at willingness to pay or consume GM products, and others discuss the role of science communications and actors.[29] This literature can help identify avenues to address and secure social license. It should be noted, however, that the ability of scientific and regulatory communities to respond to some societal concerns in a robust manner may be constrained by the limitations of available assessment methods.[30]

USING GEd TECHNOLOGIES TO ADDRESS CLIMATE CHANGE AND FOOD INSECURITY

A portfolio of policy actions has the potential to create an enabling environment for the development and deployment of safe, effective GEd applications that benefit producers and consumers in LMICs. Providing impetus for GEd applications will require: 1) enhancing innovation through capacity building, partnerships and networking, and improved regulatory processes, and 2) enhancing participation and inclusion through farm-level adoption, the use of evidence and transparency to address concerns, and the consideration of women's knowledge and needs in designing programs and selecting crops and traits.

ENHANCING INNOVATION AND TECHNOLOGY DEVELOPMENT

INCREASE CAPACITY FOR AGRICULTURAL R&D AND ENTREPRENEURSHIP. Current R&D efforts must expand beyond crop productivity improvements to include the development of stress-tolerant and climate-resilient crops, which are critical for LMIC farmers, including smallholders and women farmers, to safeguard and improve food and nutrition security. Both public and private sectors have a role to play in accomplishing this goal, but capacity building and an enabling environment will also be required. By working together, public and private sector actors can support the efficiency, inclusiveness, innovation, resilience, and sustainability of GEd applications.

Investing in **human and infrastructure capacities** can expand access to GEd applications while also addressing the economies-of-scale effects that have constrained work on LMIC crops.[31] To ensure that high regulatory costs do not impede the work of public organizations or private SMEs and start-ups, support should be targeted to developing partnerships between public and private institutions and entities with existing technical, legal, management, and regulatory skills to address GEd issues. Partnerships that foster broader public–private sector engagements can also contribute to an enabling environment that broadly promotes the development, transfer, and adoption of GEd crops.

The process of "democratizing" GEd tools – that is, making them available to SMEs and small organizations to create products that will be delivered to producers – could be accelerated through the development of "bio-foundries" and technology **incubators** that provide infrastructure and support in the early stages of product development.[32] By reducing initial investment requirements and accelerating R&D processes, these incubators could launch a new generation of bio-entrepreneurs offering differentiated and tailored products to targeted LMIC markets.

Capacity building or strengthening is also needed in both the public and private sectors to support the development of **strategic approaches** and policy design of regulatory, intellectual property, and product stewardship frameworks. Capacity building can facilitate commercial success by reducing barriers and thus securing economic benefits, especially for new companies and the public sector.[33]

Developing policies and management procedures for **intellectual property** rights and benefit sharing for public and small private research institutions is also critical, as well as investment in capacities to improve all aspects of intellectual property management. Building new partnerships can help to ensure firms and organizations have the capacity to negotiate intellectual property rights and licenses for operating and deploying technologies to farmers.[34]

PROMOTE INNOVATION THROUGH PARTNERSHIPS AND R&D NETWORKING. The development and deployment of GEd applications will benefit from an R&D ecosystem that fosters inclusive and innovative R&D approaches, linkages, and networking, as well as inter-country regulatory coordination and convergence in LMICs. For example, start-ups are increasingly focusing their R&D efforts on fruit and vegetable products of interest to LMICs.[35] To ensure GEd applications are approved in the countries where they are needed, developers – especially start-ups and the public sector – require support for regulatory compliance. These actors can use GEd networks to draw on the substantial experience accumulated by organizations and countries related to regulatory issues and compliance, intellectual property, licensing, and preparation of data necessary for assessment. For example, GEd developers have been attracted to regulatory processes in countries like Colombia, Guatemala, and Honduras that allow them to conduct safety assessment activities, such as field and performance trials. The data generated can then be used in submissions to regulatory authorities in other countries, which may open new markets in these countries for GEd applications.

SUPPORT STREAMLINED AND INNOVATIVE REGULATORY PROCESSES. Functional and streamlined biosafety review processes have been proposed and used in several countries for GEd applications. These allow the regulatory authority to undertake an initial technical review of a proposed gene-edited product to determine how it should be evaluated. If a full biosafety regulatory assessment is not deemed necessary based on the lack of a permanent presence of foreign DNA, then the product proceeds through the standard national regulatory and registration processes for conventional varieties.[36] If the initial technical review determines that the product is subject to a full assessment, then it follows GMO regulations.

This feasible and streamlined regulatory approach has led to significant progress in R&D and deployment of GEd applications. In some countries, clear and precise regulations have allowed researchers from universities and public research centers to develop GEd applications. Since Argentina approved the use of a feasible and streamlined regulatory approach in 2015, crops generated by the public sector and research centers using gene-editing have represented 59 percent of all applications using GEd tools, while only 8 percent have been for older approaches such as GM.[37]

The case of bananas resistant to Fusarium tropical race 4 (TR4), an important fungal disease, demonstrates the need for streamlined approaches and for capacity building and collaboration. The Fusarium TR4 fungus is most common in LMICs that have little capacity to develop or evaluate GEd applications to address the disease. Bananas resistant to TR4 have been produced in developed countries where streamlined approval processes have been adopted, but where the disease is not present. Finding ways to connect capacity and needs across countries and regions is the logical next step for protecting banana crops and for introducing other new GEd crops in LMICs.

ENHANCING PARTICIPATION AND INCLUSION

PROMOTE FARM-LEVEL ADOPTION OF GEd PRODUCTS. Public policies, regulations, standards, and investments must support improved availability, access, and affordability of high-quality seeds and traits for smallholder farmers, especially women farmers, given the role they play in guaranteeing household food security.[38] Creating inclusive seed systems will require pragmatic reforms and investments in seed policy harmonization, common standards, and certification requirements to ensure their suitability to local social, economic, and environmental contexts. Seed system reforms and investments must also consider strategic interventions that support scalable, climate-smart practices to achieve climate resilience and increase productivity, as well as gender and social equity. Women farmers, for instance, often have limited access to quality seeds and planting material. Thus, a gender lens is needed to understand women's often informal access to seeds and their preferences for different crop traits. Women tend to adopt improved varieties of crops that are central to household food security in quality and quantity, whereas men tend to favor cash crops directed to the market. However, women often cannot access these crops and planting materials due to their lack of purchasing power and access to information. This behavior has been shown for different crops in many regions, including improved cassava in the Caribbean and maize in Central America and Mexico, among other places.[39]

ADDRESS CONSUMER AND OTHER STAKEHOLDER CONCERNS. Achieving social license, or "buy-in," for GEd applications from relevant stakeholders including producers, consumers, decision-makers, and other stakeholders will require both better evidence and more strategic communication. More robust knowledge is needed about GEd applications themselves and their potential societal impacts, as well as evidence on "what works" regarding policies, incentives, property rights, and governance. To generate this evidence, **integrated assessments** will be needed across multiple scales, including local, national, regional, and global levels, and gaps in the impact assessment capability of LMICs must be addressed. Versatile, robust, and easy-to-use analytical tools and big data can help identify constraints to adopting new technologies and distributing benefits. These tools can also improve the understanding and inclusion of broader societal concerns such as gender, institutions, and rights in technical and social change processes.

Strategic **outreach and communication** efforts are also necessary for local and national buy-in, both to address consumer and other stakeholders' concerns and interests and to develop implementation and delivery capacities to support country-led participatory and collaborative efforts for use of GEd applications.[40] As part of this, decision-makers must consider the diverse interests of various stakeholders in their assessments of regulatory frameworks for GEd applications.[41] Effective communication also depends on the transparency of R&D, regulatory frameworks, and technology transfer processes. This **increased transparency** can be achieved by helping build up actors' empowerment and knowledge as relevant stakeholders in broader efforts for building value chains. Furthermore, capacity for assessing evidence on GEd applications must be increased. Both transparency and capacity will help with strategic efforts to share knowledge and communicate through multiple channels and media, which are crucial to facilitating GEd uptake.[42]

PROMOTE INCLUSION. Improving technologies to meet the needs of smallholders, including women farmers and other vulnerable groups, and increasing their access to GEd applications will build resilience to climate change impacts and contribute to food security and improved nutrition. All farmers should have access to better GEd crop **information and extension** that is tailored to their needs. Extension activities need to consider local conditions, be accessible to differing levels of literacy (for example, by using pictures), use local languages, and emphasize farmers' local knowledge. Messages should be realistic and present concrete examples of how GEd applications can perform and solve real on-farm problems that are not easily addressed through conventional plant breeding or pest control methods, such as integrated pest management.

Participatory methodologies will be essential to making GEd applications accessible and contextually appropriate for smallholder farmers.[43] GEd applications developed using participatory methods, especially those that seek to maximize agro-biodiversity and match valuable seed use to specific contexts, can support the diversification of poor people's livelihoods. This approach is important to pursue in LMICs, as seed suitability must be ensured for highly diverse preferences and supply systems. Well-known approaches such as farmer field schools and farmer-to-farmer initiatives that include a variety of stakeholders can also help breeders learn about drivers of adoption and trait selection.[44] The choice of GEd applications for specific contexts will need to be carefully considered and developed through robust research evaluations. Using local knowledge and networks will allow farmers to access appropriate strategies that meet their needs and thus help ensure socially beneficial outcomes.

INCLUDE WOMEN'S VOICES. Extension for GEd applications can facilitate the inclusion of women farmers. Women are more likely to farm on marginal plots and have less access to inputs such as fertilizers, quality seeds, and irrigation schemes. They are also more likely to lack financing and information, which constrains their adoption of GEd crops. For example, providing packages with small quantities of seeds has been shown to stimulate the adoption of improved varieties by women farmers in Africa[45] and could be effective in encouraging women to adopt GEd seeds. This strategy may be attractive to the public sector, local seed companies, and nongovernmental organizations to stimulate demand for GEd seeds and to increase farmers' access by making the technology affordable.

Women's voices should also be heard in the **selection of GEd applications**, that is crops and crop traits. Women are most likely responsible for raising crops for home consumption and sale in local markets. Often, women's time and labor constraints prevent them from choosing the most profitable crop options, such as cash crops including grains.[46] Instead, the crops that women often prioritize as quality home food sources are otherwise neglected and underutilized. Enhancing the capacity of public and private research institutions to focus on these crops can increase access to nutritionally rich food sources, foster dietary diversity,[47] and reduce micronutrient deficiencies, especially among lactating women and young children.[48]

Women farmers must also be able to **participate in adopting GEd applications**. To ensure their inclusion, it will be important to reflect on women farmers' leadership and issues that enhance or hinder rural women's access to new technologies. Using a gender lens can improve the "menu" of crops to be improved and contribute to a more inclusive redesign of the commercial agenda for both public and private sectors. To improve understanding of how GEd crops can boost food security and incomes, women must be empowered to access information about these crops through advisory services, experiential learning strategies, and women-centered extension education strategies. More broadly, there is a need to focus on the development and adoption of gender-responsive GEd applications. To do this, women must also be included in exploring how genome editing is developed and regulated within research institutions and political structures.

CONCLUDING REMARKS

As GEd applications become more routine, we can expect increased standardization of R&D tools and procedures, expanded use of core development tools, and growth in crop-specific development platforms to introduce valuable traits, even in crops with little commercial promise. These developments, coupled with an acceleration of gene discovery processes and improvement in genomic mapping techniques, should speed the identification and deployment of valuable crop traits. Expanding the use of GEd tools will increase the need to adopt practical, effective assessment and regulatory frameworks, foster social acceptance of GEd technologies, and provide appropriate crop traits, extension, and support for all farmers. Putting GEd applications to work will also require broader economic, legal, and policy reforms for agrifood value chains, many of which are discussed in other chapters of this report. Most importantly, both public and private sectors must work together to develop an enabling environment that will allow bio-innovation to flourish and contribute to well-being, resilience, and climate change mitigation and adaptation. Systematic GEd landscape evaluations anticipate an increased interest in GEd crops that are resistant to intractable diseases (fungi, bacteria, viruses), and less interest in management of insect pests and weeds, especially in Latin America. Therefore, during this transition period, existing technologies and production practices that can address environmental and food security concerns related to pest and weeds will still play a role. This includes first-generation bio-innovations, such as GMOs, and other products that have a proven record of safety and value when developed and used responsibly.

CHAPTER 11

Food Value Chains
Increasing Productivity, Sustainability, and Resilience to Climate Change

ALAN DE BRAUW AND GRAZIA PACILLO

Alan de Brauw is a senior research fellow, Markets, Trade, and Institutions Division, International Food Policy Research Institute, Washington, DC. **Grazia Pacillo** is a senior economist, Alliance for Bioversity International and CIAT, Rome.

KEY MESSAGES

- Key nodes along food value chains, from crop production patterns to consumption, will have to adapt in response to climate change.

- Higher temperatures and humidity resulting from climate change will lower on-farm productivity and increase food spoilage and contamination along food value chains, with implications for food prices and nutrition.

- Consumer demand for sustainably produced products can create incentives for upstream change in value chains, but can also jeopardize livelihoods of poor farmers.

- Climate change is a threat multiplier. Resource scarcity and food insecurity can trigger grievances and conflict, and further disrupt value chains, especially amid widespread inequality.

Three action-ready solutions can begin to address climate change impacts in food value chains:

- Monitor the impacts of climate change, especially for vulnerable populations. Governments must monitor consumption, with particular attention to ensuring poverty does not increase and diets do not deteriorate.

- Create an enabling environment for cold chain development. In the value chain midstream, cold chains can reduce food loss and waste. However, growth of private sector investment will depend on government provision of adequate infrastructure.

- Support simple, low-cost options to reduce aflatoxins. At the local level, appropriate technologies to reduce aflatoxin contamination are available for all farmers and aggregators. Farmers will need government or NGO assistance to understand their options for reducing aflatoxin risks.

Climate change will drive responses and adaptations throughout agrifood systems. Changes in growing conditions for many crops will alter agricultural production patterns. Along with these shifts in crop production, rising temperatures, changes in humidity levels, and increased extreme weather will also affect the value chains through which agricultural products are traded, aggregated, processed, and sold to consumers. This chapter illustrates how incentives for producers and other value chain actors will change as climate change reduces the effectiveness of inputs, such as herbicides and pesticides, increases the risks of spoilage faced by middlemen and retailers, and potentially leads to increases in transaction costs. Whole value chains may be affected from farmer to consumer; for example, if international shipping costs rise with increasing fuel costs, export-oriented chains for select products in some countries may become unprofitable and even disappear. Although research has largely neglected the impacts of climate change on value chains beyond the farm, one thing is clear — many value chain actors along with farmers will need to adapt to new realities, as they showed they were capable of in the face of disruptions from the COVID-19 pandemic.[1]

IMPLICATIONS OF CLIMATE CHANGE FOR AGRIFOOD VALUE CHAINS

The potential impacts of climate change on key elements of agrifood value chains are illustrated in Figure 1. Climate change can be expected to reshape these value chains in three ways: through gradual changes; through increased likelihood of shocks; and through increased potential for conflict. While crop production is most obviously affected by climate change, risks of postharvest losses will increase and incentives for finance and insurance providers will also change. Threats to livelihoods and food security increase the risk of civil strife and conflict, which can disrupt whole value chains.[2] Consumers may add to the pressures for change across entire value chains not only through changes in diets but also through demand for sustainably produced products. All these changes have implications for value chain actors from smallholders to urban consumers.

FIGURE 1 Potential impacts of climate change on an agrifood value chains

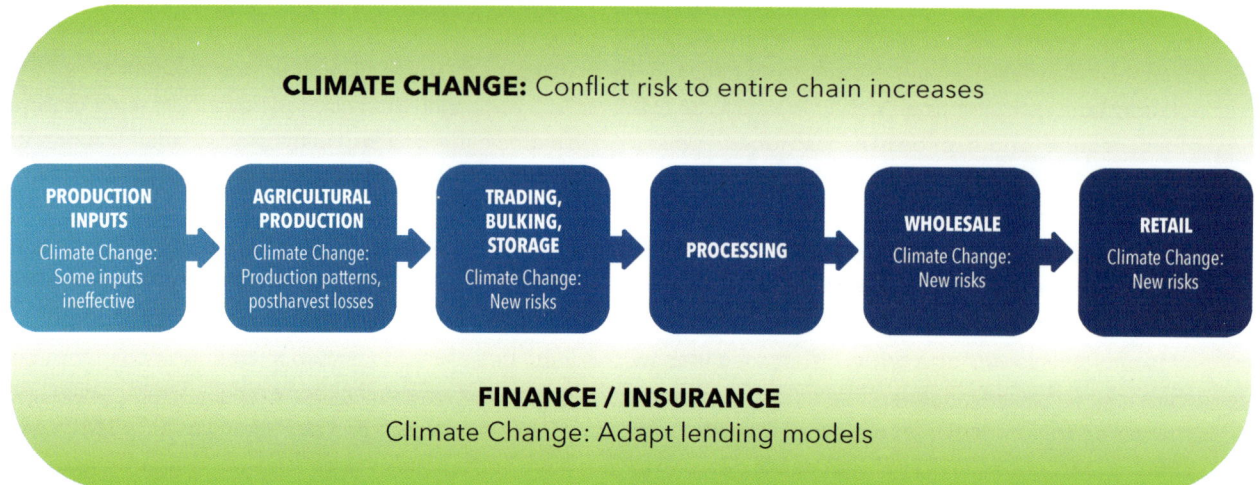

Source: Authors' illustration.

PRODUCTION CHANGES

Gradual changes in precipitation patterns, temperatures, and humidity levels in low- and middle-income countries (LMICs) are expected to increase stress on agricultural production systems.[3] New climate patterns will also affect weed growth, disease prevalence, and pest populations and potentially reduce the efficacy of herbicides, pesticides, and integrated pest management. In addition to changes in rainfall and humidity, the risk of extreme weather events increases with climate change. These events can have even larger effects, as they can affect production in future years.[4] Together, these factors can reduce the maximum potential yields for crops, affecting their economic viability. Depending upon resulting yields, market conditions, and transaction costs within value chains, crop production patterns may change dramatically.[5] As a result, the downstream value chains that follow crops, whether for domestic use or exports, will need to adapt. For example, if it becomes too dry to grow peanuts in Senegal's "peanut basin," farmers may switch completely to growing millet and sorghum, meaning peanut traders would need to adapt their purchases and find buyers for these grains.

FOOD WASTE AND LOSS AND RELATED NUTRITION IMPACTS

Higher temperatures and humidity levels will increase the risk of postharvest losses. For grains, greater humidity could lengthen drying times, increasing the likelihood that they will be stored before properly dry and thus raising their susceptibility to pests and contamination with aflatoxins or other molds. Fruits, vegetables, and animal-source foods are usually stored for shorter periods of time, but these perishable products begin to spoil more rapidly, and higher temperatures will accelerate that process. Increased loss of perishable products is particularly concerning because these foods are the source of critical micronutrients that are already insufficient in the diets of many LMIC populations (see Chapter 8).[6] The effects of climate change on food spoilage could make perishables and their associated micronutrients even scarcer.[7] While perishables can be dried or otherwise processed to slow or stop spoilage, these techniques can also lead to nutrient loss. In addition to increasing scarcity of these foods, foodborne pathogens, like salmonella in animal-source products, will likely become more prevalent.[8]

Recent work suggests that postharvest losses already average around 14 percent of total potential harvests, but vary substantially by both crop and region.[9] As most of these estimates are for nonperishable crops, among perishable crops these losses may be higher; few estimates are either survey-based or account for reduced food quality.[10] While spoilage can occur in any of the value chain nodes beyond the farm, risks are highest for traders and aggregators who deliver crops in bulk to processors. For fresh foods, evidence suggests that losses are concentrated on the farm and at the retail level.[11] At the retail level, climate change could increase losses as informal markets often lack infrastructure to cool perishables.[12]

FINANCE, INSURANCE, AND SERVICES

Financial service providers in agrifood systems can be expected to change their behavior gradually in response to changes in perceived risks both on the farm and further along the value chain. While finance and insurance companies may be able to adapt to growing risks by simply adjusting their lending or insurance terms, both aggregate investment funding and insurance available locally could decline. Other firms along value chains may also change their investment strategies, whether or not they depend on availability of finance, which could likewise cause cascading changes.[13] For example, storage can be affected; in Nigeria and other humid climates, maize is susceptible to mold when stored before it is dry, and with increasing humidity, traders are becoming more averse to storing maize at all.[14] In turn, financial instruments dependent upon storage (that is, warrantage) become riskier, potentially reducing farmers' access to finance. And in value chains in which farmers depend on finance for inputs, such as sugarcane or coffee chains, reduced inputs would reduce yields and total harvests in a region, thus reducing the returns to trading and processing those crops as well, which can lead value chain actors to exit the market.

SMALLHOLDERS AND VULNERABLE POPULATIONS

Smallholders are likely to bear the brunt of climate change impacts on value chains. Because it is inherently more complex to help a hundred small farms adapt than one large farm, it will be more difficult and costly to help smallholders adapt to gradual changes. Even if governments are willing to help, remote smallholders are hard to reach and may be resistant to adopting new practices, if these are perceived as risky. In addition, smallholder access to credit or formal sources of insurance will be the first to be reduced as uncertainty increases. This is particularly true for female and minority value chain actors, whose access to finance is already more constrained.[15] Moreover, if the risks of storing grain or other food products increase, traders may also reduce the amount they are willing to pay farmers.

CIVIL STRIFE AND CONFLICT

In places where the climate crisis sparks civil unrest and conflict, further cascading effects on the functioning of value chains are likely. Evidence suggests that natural resource scarcity and food insecurity can lead to violence when exacerbated by climate impacts and associated with other insecurities, such as poverty, inequality, and overall sociopolitical fragility.[16] Lower agricultural productivity and natural resource scarcity due to climate extremes and variability could reduce food supply and quality, and lead to higher food prices that would compound food and nutrition insecurities among the poor and vulnerable, potentially sparking protests, riots, or armed conflict in already fragile contexts.

In addition, economic shocks resulting from climate change may lead to an increase in criminal activity or other disruptive behavior that escalates challenges for agrifood value chain actors, for example, by increasing risks of theft when transporting foods to, through, or from conflict-affected areas.[17] Climate-induced tensions and conflicts may be most frequent where there are large structural inequalities, characterized by social and political marginalization and existing vulnerability.[18] Such conflict risks can affect whole agrifood value chains, not just individual nodes. Conflict can thus intensify the damage to value chains. In response, value chains must both adapt to climate change and do so in a conflict-sensitive manner.

CONSUMER DEMAND

Consumer demand may shift as a result of climate change concerns, as some well-informed consumers will likely increase their demand for more sustainably produced foods. If these shifts are large enough and drive an increase in prices for these products, consumer demand can create incentives for producers and processors to shift toward more environmentally sustainable crops and technologies; however, there are several challenges. First, consumers will want assurance that the products they purchase are sustainably produced, thus increasing demand for traceability and "process standards." Process standards certify the way that foods were produced; existing standards include certification for organically grown products and other measures of environmental and social sustainability. At present, the products most commonly certified under such standards are coffee and cocoa.[19] If demand increases substantially, organizations that provide certification, such as the Rainforest Alliance, will need to increase capacity. Second, increased demand for foods with sustainability certification can lead to increased inequality by reducing market options for poorer farmers who lack the resources to meet the new standards.[20] And if farmers expand their production of certified products too quickly, it is likely that they will need to sell a portion of the certified food into uncertified markets at lower prices, affecting their incentive to participate.[21]

ADAPTING AGRIFOOD VALUE CHAINS

Clearly, agrifood value chains must *adapt* to climate change. Value chains offer less potential to help *mitigate* climate change, however. Despite the growing complexity of some value chains, evidence on greenhouse gas emissions suggests that the value chain steps between production and consumption – including processing and transporting agricultural products to end markets – only account for 18–29 percent of total emissions from agrifood systems, even for products traded over long distances.[22] Since this range represents a total over a wide range of products and levels of value chain complexity, there are no easy fixes for reducing those emissions. For example, research suggests that "buy local" movements will not materially reduce emissions, and instead might increase them, as there are returns to scale in moving bulky agricultural products.[23] Even effective interventions to reduce emissions between farms and retailers may have little overall effect.

Though reorganizing downstream value chains to reduce emissions may not be a cost-effective way to mitigate climate change, other interventions could help avoid deterioration of other important outcomes, such as nutrition-related outcomes. Solutions are clearly needed to assist smallholders, and particularly women, in adapting to climate change and to changes in the value chains into which they sell their crops.

In this section, we consider two potential solutions that can be initiated now, both of which require government intervention. First, we consider how governments and other stakeholders can act to prevent climate-related food waste and loss that reduce food security and nutrition. We focus on storage technologies that can reduce both postharvest losses and greenhouse gas emissions, even if those reductions cannot be measured at the macro level. Second, we suggest how better monitoring can be used to help government and the private sector identify problems along value chains, adapt value chains to climate change, and ensure stability.

PREVENTING FOOD WASTE AND LOSS

Possible solutions to the problem of food spoilage differ by crop type. For perishable, micronutrient-rich foods, greater availability of cold storage and cold chains can maintain or even improve access to these fresh foods by reducing spoilage. For a cold chain to work properly, cold storage must be available at or near the farm soon after harvest, refrigerated trucks must be available to transport produce to larger cold storage facilities or to retailers, and people who handle the food must be trained in proper handling procedures. Cold chains are dependent on complementary infrastructure, particularly good roads and reliable sources of electricity (see Chapter 9). Some emerging cold chain technologies adaptable to settings with little existing infrastructure include solar-powered or electrically efficient cold rooms that can be used in villages.[27] However, prices for these technologies will remain high so long as demand is low, and demand is likely to remain low as long as producers do not perceive large income effects from using such technologies.

Further along the chain, there is a need for companies that are able to invest in refrigerated trucks, which in turn depends on adequate road infrastructure to ensure sufficient returns on private investment. Without public investments in this necessary supporting infrastructure, private sector entrepreneurs will not find investment in cold chain technologies attractive. However, given that the social benefits of such investments extend well beyond the food system and are likely to be quite high, government has an important role to play in providing the infrastructure to foster cold chain development. The benefits of these investments are evident in the growing economies of

Southeast Asia, where strong road networks and rural electrification preceded cold chain development that has expanded the availability of perishable food products in markets.[28]

For grains and legumes, the major concern from a health perspective is the growth of aflatoxins. More widespread use of aflatoxin-reduction technologies could reduce losses and prevent an increase in contamination. Several technologies, from simple to complex, can reduce aflatoxins; the key is to use those that are context appropriate, cost-effective, and sustainable. For example, simple technologies, like spreading tarps underneath crops that are drying in fields, have been shown to cost effectively reduce aflatoxin levels.[24] Hermetic bags are a second solution, though trials have shown that while farmers are willing to use the bags if free, value chains for bags are not well developed, so farmers have difficulty obtaining them post-trial, and therefore tend to stop using them.[25] Improved sales outlets for bags could help, but it is not clear whether farmers would be willing to pay for them. Solutions such as Aflasafe, a biocontrol product developed by the International Institute for Tropical Agriculture, are also effective at reducing aflatoxins but are expensive, and would likely require specialized value chains to ensure the existence of buyers for aflatoxin-safe grains before they could be cost-effective.[26] For smallholders, then, promoting the use of tarps spread underneath drying crops would seem to be a cost-effective and already available solution.

ADDRESSING INSTABILITY AND CONFLICT

For these solutions to work, as well as many other efforts to address climate change throughout the food system, agrifood value chains and the wider economy require stability and security. An integrated approach to maintaining stability in the face of gradual change and shocks requires both technical solutions to challenges, like those proposed above, and "restorative" and "sustainable" solutions.[29] Restorative solutions enable stability by creating a common platform for dialogue. These collaborative dialogues aim to understand the main causes of discontent arising from climate impacts and to facilitate cooperation by building trust and legitimacy.[30] Sustainable solutions address root causes of conflict and grievances through collective-action approaches for institutionalizing joint management systems. Such systems link local communities on an equal footing with public and private decision-makers and work across multiple levels and sectors (see Chapter 7 on landscape management).

The effectiveness of such solutions will depend heavily on the context, on the structural drivers of grievances, and on how effective monitoring systems are in detecting disruptions occurring at different levels of the value chain. To help meet these needs, CGIAR is developing a "Climate Security Observatory" – a decision-support tool that will provide real-time or almost real-time scientific evidence on how climate exacerbates existing social, economic, and political risks and insecurities, including the potential for conflict.

CONCLUSION

As our climate changes, agrifood value chains must adapt to new cropping patterns and changes in investment and input needs. Governments must safeguard against the risk of increasing food and nutrition insecurity, and agrifood value chains must be transformed to address climate security concerns. In the short term, policymakers can focus on ways to reduce food loss and waste in value chains, particularly for perishables, to yield more food from their agrifood systems and potentially to alleviate the local environmental stress associated with food systems development. In the medium term, investments in climate-smart infrastructure, including new roads and electrification to support development of cold chains, will be important to safeguard food and nutrition security. To ensure that civil strife and conflict are not fostered by climate change, investments will be needed not only in monitoring but also in ensuring that smallholders and other vulnerable value chain actors can adapt, and that both diets and livelihoods are protected and improved.

CHAPTER 12

Digital Innovations
Using Data and Technology for Sustainable Food Systems

JAWOO KOO, BERBER KRAMER, SIMON LANGAN, ANIRUDDHA GHOSH, ANDREA GARDEAZABAL MONSALUE, AND TOBIAS LUNT

Jawoo Koo is a senior research fellow in the Environment and Production Technology Division, International Food Policy Research Institute (IFPRI), Washington, DC. **Berber Kramer** is senior research fellow in the Markets, Trade, and Institutions Division, IFPRI. **Simon Langan** is digital innovation director, International Water Management Institute (IWMI), Colombo. **Aniruddha Ghosh** is a scientist, Alliance of Bioversity and CIAT, Rome. **Andrea Gardeazabal Monsalue** is a monitoring and evaluation manager, International Maize and Wheat Improvement Center (CIMMYT), El Batán, Mexico. **Tobias Lunt** is chief data scientist, Development Data Lab.

KEY MESSAGES

- Digital innovations offer unprecedented potential for managing climate risks across the entire agrifood system — from producers to markets and value-chain services to policymakers.

- Farmers can benefit from localized weather information services, digital extension services, and weather index-based insurance schemes. Along food value chains, internet-connected sensors can monitor food quality and safety risks, while digital innovations in insurance, credit, and banking can increase access to risk-reducing services for all food system actors.

- Innovations in weather and climate forecasting can improve early warning systems and public and private sector decision-making; climate information services have great potential to save lives and reduce damages from extreme weather events.

- However, rural food production areas are underserved by digital infrastructure. Hundreds of millions of small farmers, especially in sub-Saharan Africa, do not have mobile network coverage or internet access, and more than 300 million cannot access digital climate advisory services. Women, in particular, have limited access to digital services.

Critical steps to take now include:

- Invest to bridge the digital divide. Both private and public investment are needed to address this gap. Given low returns on investments in connectivity in rural areas, policy incentives and public-private partnerships should promote private investments that benefit vulnerable populations and are inclusive of women.

- Strengthen agrifood information systems. Decision-makers often lack timely, reliable, and actionable information. Research can help governments

identify where better data can best contribute to reducing climate impacts. Digital technologies can provide cost-effective real-time monitoring for forecasting; and expansion of weather stations can provide localized weather data for farmers.

- Cultivate digital capabilities to manage climate risks. Strategic investments in "soft" infrastructure – digital climate services, advisory services, actionable information for producers, private-public partnerships for data production, and equal access to financial services – can all boost capacity to identify, manage, and respond to climate risks.

Climate change and associated extreme weather events directly impact the functioning and sustainability of food systems.[1] The increasingly erratic onset of seasonal rainfall and prolonged heat stress during growing seasons are already causing crop losses.[2] As of late 2021, for example, Madagascar's three successive seasonal droughts had put 1.35 million people at risk of the world's first climate-change-induced famine.[3] In the United States, the number of days between billion-dollar weather-related disasters has fallen from more than 80 in the 1980s to just 18 in recent years.[4] Without adequate preparation, these weather hazards disrupt food supply chains by interrupting production and cause problems farther along these chains by raising costs and prices of processing, storage, transport, retail, and consumption and reducing business revenues.[5]

POTENTIAL OF DIGITAL INNOVATIONS TO MANAGE CLIMATE RISKS

While global warming is a threat to food systems, there are unprecedented opportunities for technological solutions to contribute to climate change mitigation and adaptation in food production. Digital tools already provide food producers with timely insights and services that support improvements in agricultural productivity and profitability. Across low- and middle-income countries (LMICs), the number of digital agricultural services (such as digital advisory services, digital procurement, e-commerce, digital finance, and smart farming services) increased rapidly during the past decade, from 53 reported in 2009 to 713 in 2019, with beneficial impacts.[6] In sub-Saharan Africa, for example, meta-analysis studies show that producers who adopted digital extension and financial services increased their incomes by 20 to 40 percent[7] and their adoption of recommended agrochemical inputs by 22 percent.[8] In Ethiopia, video-mediated digital extension services reached wider audiences and were more effective in increasing adoption of improved technologies than conventional extension approaches.[9]

Digital technologies can be a particularly powerful, innovative tool for managing climate risks across food systems — from producers to markets and value-chain services to policymakers. As weather patterns become more variable at the farm level,[10] producers can use digital technologies to access localized weather and climate information services in order to optimize farm management decisions, such as irrigation scheduling and crop variety selection.[11] Digital extension services help small-scale producers to communicate with experts, who can diagnose farm-specific problems and prescribe best-bet climate-smart practices.[12] Weather index-based crop and livestock insurance schemes, which assess climate-induced losses remotely and provide payouts through digital financial services, offer producers an increasingly important option for managing climate risk.[13] By using data from mobile phones[14] and satellite remote-sensing,[15] these digitally enabled insurance products can estimate agricultural losses faster and at a lower cost and can make timely payouts for losses.

All along food supply chains, the postharvest use of internet-connected smart sensors, such as time temperature indicators (TTIs) and tech-enabled traceability devices, allow value-chain actors to detect potential food safety and quality issues (such as the prolonged exposure of dairy products to high temperatures during transportation) and thus reduce health risks and postharvest losses.[16] Digital innovations for value-chain services — financial services, credit, and insurance — can reduce transaction costs, address information asymmetries in traditional markets, and make crop insurance and digital finance more affordable and inclusive.[17] For public sector agencies and government entities, real-time food systems that monitor data collected, analyzed, and disseminated using digital technologies can improve early warning systems and help policymakers to make informed decisions to prepare for and mitigate risks.[18]

Digital innovations in forecasting can support climate science with information on weather, climate variability, and climate change, and can play a vital role in supporting food system resilience to worsening weather extremes. For example, every month, the European Centre for Medium-Range Weather Forecasts releases a seven-month global forecast of temperature and precipitation.[19] These forecasts use advances in Earth-system monitoring and prediction modeling capabilities to extend conventional short-term weather forecasts (up to two weeks) to seasonal climate predictions.[20] Enhanced weather and climate forecasting skills improve early detection and warning systems for floods and droughts. With this information, policymakers can prepare for disasters and reduce damages by, for example, declaring emergencies early and getting resources where they are likely to be needed in advance.[21] Between 2000 and 2017, advances in flood early-warning systems were estimated to have already helped to reduce global flood-related human casualties by 45 percent and the number of people affected by floods by 24 percent.[22] Conservative estimates based on a meta-analysis of global studies suggest that the benefit-cost ratio for reliable climate information services is about 10 to 1, with potential global benefits as high as US$30 billion per year in increased agricultural productivity and $2 billion per year in reduced asset losses.[23]

THE DIGITAL DIVIDE

The potential of digital innovation is clear, yet its reach is far from universal. Evidence shows that farmers who own mobile phones, where available, are able to reduce climate risk by communicating with community members to learn about potential climate shocks early, by sharing information on how to manage them, and by promoting social learning that influences technology adoption.[24] Yet rural food production areas, especially in the global South, are underserved by the enabling digital infrastructure that is key to connecting all agrifood system actors. More than 600 million people still live outside of areas covered by mobile networks, two-thirds of whom are in sub-Saharan Africa.[25] Globally, fewer than 40 percent of small farms are covered by mobile internet.[26] Rural areas in Africa are particularly underserved; only 6 percent of rural households have internet access, compared to 28 percent of urban households, and in sub-Saharan Africa, only 13 percent of small farmers have ever accessed a digital service.[27] Overall, mobile internet connectivity tends to perform poorly in LMICs where the agriculture sector drives the economy (Figure 1a) or the majority of the population lives in rural areas (Figure 1b). This creates a serious challenge for initiatives to use digital innovations for managing climate risks in agriculture. Slow progress on electrification in LMICs further limits the affordability and coverage of digital technologies;[28] moreover, in sub-Saharan Africa, the number of people without access to electricity increased in 2020 because of the COVID-19-related economic slowdown. More than 300 million small-scale producers worldwide lack access to digital climate advisory services because of this gap in digital infrastructure.[29] As a result, unmanaged risks hinder producers' adoption of other improved technologies.[30]

Beyond the coverage gap of infrastructure, there are also complex socio-technical dimensions of the digital divide, particularly related to gender.[31] Social norms in many cultural contexts may hinder women's access to and use of technology, including mobile phones.[32] Across LMICs, women are 7 percent less likely than men to own a mobile phone and 15 percent less likely to use mobile internet.[33] Moreover, evidence suggests that the gender digital divide widens as technologies become more expensive and sophisticated.[34] This inequality in connectivity can further undermine

FIGURE 1 Digital connection, agricultural GDP, and rural populations

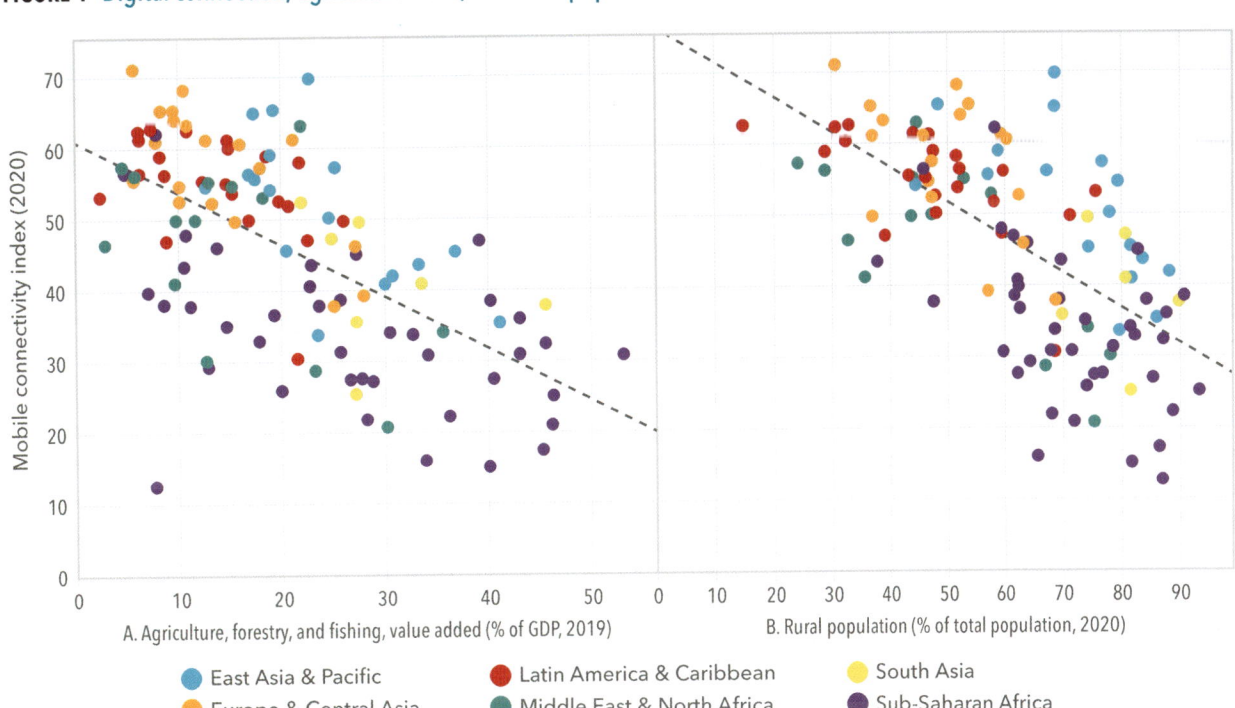

Source: Authors' analysis using the Mobile Connectivity Index, GSM Association, *The State of Mobile Internet Connectivity 2021* (London: 2021) and the World Development Indicators (World Bank, 2022).

Note: Data include 199 low- and middle-income countries.

women's access to digital solutions for managing climate risks in agrifood systems.

POLICY RECOMMENDATIONS

To unlock the potential of digital innovations equitably at scale, three critical areas require policymakers' action and investment.

INVEST TO BRIDGE THE DIGITAL DIVIDE

The promise of digital technologies to help manage climate-related risks for food systems cannot be fully realized if inequalities in access to digital infrastructure, digital literacy, and the gender digital divide are not addressed for all food systems actors. Recognizing the critical role of digital infrastructure investment for achieving economic growth, social development, and climate action, the UN Sustainable Development Goals (SDGs) included a target for the least-developed countries to achieve universal and affordable internet access by 2020 (SDG 9.C). However, this ambitious goal with its accelerated timeline was not met. As of 2019, 21 percent of the population in the least-developed countries was still not covered by mobile networks.[35]

Enabling policies that, for example, incentivize people to adopt new technologies (such as public awareness campaigns, subsidies, and training and education) and private sector actors to invest in infrastructure (such as tax benefits) are needed to address the digital divide.[36] As research continues to generate robust and granular evidence of digital innovations' positive impacts across agrifood systems, policymakers should support public investment and create incentives for private sector entities to invest in the expansion of digital infrastructure in rural areas, prioritizing food production zones. The cost of connecting all of Africa to the internet is estimated at US$100 billion,[37] which will only be feasible with strong private sector involvement. However, the private sector is unlikely to invest in rural infrastructure without incentives because the cost of deploying and maintaining infrastructure in rural areas can be two to five times the cost in urban areas, but generates only a tenth of the revenue.[38] Innovative financing models are being explored to bridge connectivity gaps, including governments agreeing to connect public sites (such as schools and hospitals) and to provide citizens with subsidized "connectivity vouchers" to ensure sufficient revenues for private investors.[39] In the United States, taxing the private sector actors that stand to benefit the most from connectivity, such as digital advertisers, is being discussed as a potential source of funds to close the digital divide.[40] In addition, research is needed to assess the economywide benefits of simultaneous investment in digital infrastructure and electrification and their potential synergistic impact on achieving SDGs.

Policymakers should also prioritize regulations on responsible data management (including who owns the data and how they are used) and enabling environments that ensure equitable access to technologies and services — while being mindful of the potential unintended consequences of investing in digital innovations for women and other vulnerable groups. Women can make important contributions toward climate-resilient food systems, adopting climate-smart practices in their livelihood and household roles (for example, climate-sensitive livestock feeding practices, improved grain storage, and better food processing practices) if they have the same means and access to information as male farmers.[41] For example, a study in Africa showed that female farmers can contribute more to lowering aflatoxin risks, which are increasing with climate change.[42] To bridge the digital gender gap, the U.S. Agency for International Development's WomenConnect Challenge (WCC) program invested in 12 digital solutions designed specifically to improve women's access to digital technologies. These include a safe virtual space for peer-to-peer education, an online information delivery service specifically for women, and a digital literacy training program.[43] To promote women's equitable access to credit, another WCC pilot in the Dominican Republic developed a new credit-scoring model designed for women, and found that 93 percent of women secured more credit with this model than the conventional model.[44]

STRENGTHEN AGRIFOOD INFORMATION SYSTEMS FOR TIMELY DECISION-MAKING

The concept of a food system emphasizes the complex connectivity and interdependence of activities, actors, and institutions across food, ecology, economy, and society required to achieve a sustainable state of food security.[45] Similarly, the concept of climate-resilient agrifood systems encompasses the dynamic interactions among different policy areas — connecting agricultural policies with environmental, health, and socioeconomic outcomes under a changing climate. Due to this complexity, there are substantial information gaps across agrifood systems, and decision-makers often lack access to timely, reliable, and actionable information for making informed decisions.[46] Without appropriate data and information readily accessible when and where needed, decision-makers cannot detect potential risks early (such as climate variability and associated pest infestations and food price hikes in markets) or implement preparatory measures to manage risks systematically, leaving countries vulnerable to climate shocks. Reliable baseline data is equally important for understanding how agrifood systems perform and where the priority areas are; without this information, estimates of climate impacts before and after shocks can be unreliable.

Weak information systems can lead to misguided investments and wasted budgets,[47] slow economic growth,[48] and cause other unintended consequences to agrifood systems. In Zambia during the 2015/16 El Niño event, for example, the government introduced a precautionary export ban during the maize growing season out of concern that drought would limit the food supply. However, the late onset of rains actually increased average yields in Zambia's northern region. This information did not reach the policymakers in a timely way, so the export ban was left in place, resulting in lower domestic maize prices and a substantial increase in poverty among net maize sellers.[49]

Identifying the most critical information gaps is important, given the complexity of developing and maintaining holistic information systems that cover all aspects of agrifood systems, a challenge compounded by the urgency of threats from climate change. Rather than striving to fill all the data gaps across agrifood systems, digital innovations and research can help governments to identify where better data can best contribute to reducing climate impacts and strategize to address these priority areas. For example, Johns Hopkins University and the Global Alliance for Improved Nutrition (GAIN) recently launched the Food Systems Dashboard.[50] The Dashboard is designed to provide food-system-wide baseline indicators globally in a coherent data visualization framework that emphasizes the interconnections across multiple food system sectors. It also highlights where data are limited or outdated, notably on diets, food consumption, and subnational disaggregated data. However, the dashboard does not yet capture interactions among key components of the food system.

Especially in the context of climate adaptation, digital technologies can be cost-effective for capturing timely data, analyzing complex interactions, and generating insights on the functioning of food systems. Digital platforms can support the development of improved statistical systems that capture the dynamics of interactions among components of the food system as well as shocks threatening its resilience.[51] Real-time monitoring of food, water, and land systems is increasingly possible — at a lower cost and with greater accuracy than ever before — using satellite remote-sensing, advanced analytics, and communication technologies.[52] Digital development partnerships between public and private sector innovators (supported by safeguarding laws, enabling policies, and institutional arrangements) can generate synergistic impacts across the economy while strengthening agrifood information systems.[53] Such partnerships can incentivize the private sector to release data assets publicly in order to fill gaps in ground-truthing data, as in the case of the Lacuna Fund and ACRE Africa (Box 1). Such data are sought by agricultural scientists developing artificial intelligence (AI) and machine-learning-based applications that can analyze large datasets and aid decision-making processes.

The expansion of weather stations, complemented by satellite remote-sensing-based estimates, to improve the provision of reliable, localized weather data is another priority investment for increasing climate resilience in agrifood systems. The utility of weather and climate services depends on the availability of good quality baseline weather observation data, without which changes in growing conditions are difficult to monitor and predict. Despite the fundamental importance of these data, the World Meteorological Organization's latest assessment of the state of climate services shows a concerning increase in the divide in data collection, particularly affecting the least developed countries.[54] In Africa, the density of weather stations is only one-eighth of the minimum

> **BOX 1 UNLOCKING THE VALUE OF UNDERUTILIZED PRIVATE SECTOR DATA**
>
> The Lacuna Fund is a global coalition providing support to low- and middle-income countries to produce datasets for use in machine-learning applications. Lacuna has awarded funding to ACRE Africa, a Kenya-based for-profit agricultural insurance broker, to incentivize the publishing of ACRE's unique data in the public domain. The focus is on data that will be highly relevant for agricultural data scientists working to train machine-learning algorithms and develop artificial intelligence (AI)-based applications (such as automated estimation of crop production losses from climate shocks). The new dataset includes georeferenced crop images along with labels on input use, crop management, phenology, crop damage, and yields from 11 counties in Kenya, giving scientists around the world "eyes on the ground" to develop AI for agriculture.[a] To avoid privacy violations, the georeferenced data has been published in a usable format that does not reveal the exact locations of farmers.
>
> As demonstrated by this partnership example, the private sector more broadly has a treasure trove of underutilized data that could advance agricultural data science and help to localize AI applications. A key question for policymakers is how to bridge critical data gaps by incentivizing and facilitating the sharing of these data, while also protecting the privacy of food systems actors, responsibly managing data, and sharing benefits.

recommended for timely weather forecasting and climate monitoring for early warning systems; moreover, 35 percent of the existing stations appeared to be non-operational in 2019. As a result, only 44 percent of Africa's population is estimated to be covered by early warning systems, compared to 70 percent in Asia and 60 percent in South America. Public-private partnerships can help to fill this gap. For example, the Trans-African Hydro-Meteorological Observatory (TAHMO) operates 115 low-cost automated weather stations across rural Kenya, and its data are shared with the Kenya Meteorological Department.[55] Crowdsourcing through citizen science projects can also be an effective way to fill data gaps, engage with digital-savvy youth, raise citizens' awareness and capabilities, and better communicate policies.

CULTIVATE DIGITAL CAPABILITIES TO MANAGE CLIMATE RISKS

For these investments in digital infrastructure and information systems to be effective in managing climate risks, "soft" infrastructure must also be cultivated.[56] Strategic investments are needed to increase producers' capabilities to find location-specific climate information, interpret information to understand the hazards, and make decisions. This process will require multisectoral coordination among digital climate services, research communities, and food systems actors.

First, digital climate services should reach all vulnerable populations in order to support a broad range of benefits, including improvements in yields and incomes and reductions in risks. Policies can promote the inclusive planning and co-design of climate services to help increase resilience of vulnerable agricultural communities.[57] Laws and regulations are needed to govern open access to climate data, ensure data quality in compliance with international climate data quality standards, and enforce equity components (for example, gender equity) of services.[58] To guide policymakers and investors in helping to connect the 300 million small-scale agricultural producers who lack access by 2030, the Global Commission on Adaptation (GCA) consulted with 57 organizations, including CGIAR, to formulate six principles of successful digital climate services: ensuring data quality, equity, stakeholder engagement, accountability and transparency, financial sustainability, and design for scaling.[59]

Second, digital climate services need to communicate timely, localized, and actionable climate information. While an increasing number of climate data sources are available, communicating climate science and research findings with the public is often challenging due to their inherent uncertainties.[60] Collaboration with design research communities (such as human-centered design), co-development with multisectoral expertise (for example on nutrients and biodiversity), and embracing alternative scientific approaches (including local indigenous knowledge) in the development and provision of climate information can help digital climate services find the most effective way of communicating complex information to ensure it is relevant for local agrifood system contexts.[61]

Third, because early warnings are shown to be effective only when people know what actions to take in response,[62] researchers can add value to climate information by providing more actionable insights.[63] For example, a climate forecast of "50 percent chance of crop failure due to potential drought in the upcoming season" can be enriched by crop modeling analyses that predict the effect of alternative management practices, such as, "Crop failure risk can be reduced to 20 percent if drought-tolerant varieties are planted." Collaborative research and response capacity-building campaigns are needed to determine what actions should be undertaken when early warnings are released.[64]

Finally, food system actors, especially women and other vulnerable groups, need support to take advantage of digital financial services for investing in climate-smart farming practices and managing climate risks. In Kenya, for example, the majority of farmers (82 percent as of 2019) already use mobile financial services, but only 1 percent use these services to manage climate risks in agriculture through crop or livestock insurance or loans.[65] While more research is needed, the specific practices of mobile financial services appear to be unsuitable for agriculture in terms of the transaction amounts, frequency of use, repayment periods, and interest rates. To be more gender-responsive, digital financial services should consider women's needs, for example, the role of female farmers in managing frequent household emergencies.[66] The active participation of food systems actors in the design and provision of digital financial services could create more tailored, empowering digital finance products for women and men to manage climate risks. Policy options for advancing women's digital finance inclusion include, for example, making official identity systems universally accessible to all women and girls, facilitating women's universal ownership of mobile phones, supporting consumer protections that address women's needs and ensure data privacy and security, and requiring financial institutions to provide sex-disaggregated data in reporting and make them public.[67]

REGIONAL DEVELOPMENTS

CLIMATE CHANGE IS A TRULY GLOBAL THREAT, BUT ITS IMPACTS DIFFER AROUND the world. Regions and countries urgently need to identify and implement policy responses that reflect local needs and opportunities. This section examines the effects of climate change on national and regional food systems in Africa, the Middle East and North Africa, Central Asia, South Asia, East and Southeast Asia, and Latin America and the Caribbean. For each major region, promising innovations and policy directions to promote the resilience and sustainability of food systems are considered:

- Scaling up social protection programs in Africa south of the Sahara
- Strengthening the focus on climate adaptation in Africa
- Rethinking water use in the Middle East and North Africa
- Promoting climate-smart practices and crop diversification in Central Asia
- Reforming agricultural support policies in South Asia
- Improving financing for climate change mitigation and adaptation in East and Southeast Asia
- Supporting global food security and sustainability in Latin America and the Caribbean

AFRICA	116
MIDDLE EAST AND NORTH AFRICA	120
CENTRAL ASIA	124
SOUTH ASIA	128
EAST AND SOUTHEAST ASIA	133
LATIN AMERICA AND THE CARIBBEAN	137

AFRICA

JEMIMAH NJUKI, SAMUEL BENIN, WIM MARIVOET, JOHN ULIMWENGU, AND CAROLINE MWONGERA

Jemimah Njuki is former director for Africa, **Samuel Benin** is deputy division director, **Wim Marivoet** is a research fellow, and **John Ulimwengu** is a senior research fellow, Africa office, International Food Policy Research Institute. **Caroline Mwongera** is a senior scientist with the Alliance of Bioversity International and CIAT.

Africa's food systems are evolving rapidly, driven by the rise of an African middle class, growing urbanization, shifts in the labor force from farming to nonfarm jobs, and increased availability of digital technologies. As a result, the entire food system is changing, marked by rising food import bills as the gap between Africa's food production and consumption widens, by growing consumer demand for more diverse, higher-quality and safer foods, and by new preferences regarding packaging, shopping outlets, and financial and electronic payment services.[1] Despite growing demand and competition for African farmland, low-input, rainfed production systems with low yields[2] still predominate, and the entire food value chain is led by smallholders and small and medium enterprises (SMEs),[3] with women playing critical roles in the production, processing, retailing, preparation, and waste management of food.[4]

With more than 50 percent of Africa's population depending on rainfed agrifood systems for their livelihoods, the impact of climate change on Africa's food security and other outcomes will be enormous. However, differences across the continent in food systems and their transformation will shape those outcomes. For example, although northern and southern Africa are projected to experience the largest increases in temperature and decreases in rainfall, the negative impact of rising temperatures on GDP is projected to be greatest in western and eastern Africa.[5] For the coastal and island countries, the projected rise in sea levels, leading to coastal degradation and erosion, will compound the other negative effects of climate change.[6] Similarly, other crises such as the recent COVID-19 pandemic and locust and fall armyworm infestations that have exposed the fragility of Africa's food systems have also had different effects across the continent. For example, COVID-19's impact has varied across the continent, with Egypt, Morocco, South Africa, and Tunisia alone accounting for about 68 percent of Africa's reported deaths.[7] And while the locust infestation was concentrated in eastern Africa,[8] the fall armyworm has invaded most of sub-Saharan Africa.[9] In addition, although food insecurity, hunger, and undernourishment have worsened all across Africa, the prevalence of undernourishment, for example, is highest in eastern (28.1 percent) and central (31.8 percent) Africa, and lower in the western (18.7 percent), southern (10.1 percent), and northern (7.1 percent) regions.[10] The COVID-19 pandemic has further exacerbated the situation, resulting in production losses, lower household incomes, and declining nutrition levels among the most vulnerable.[11] As with these other crises, the most vulnerable populations disproportionately bear the burden of climate change, as shown in a recent assessment of the climate resilience of pastoralists and agro-pastoralists in sub-Saharan Africa.[12]

COMMITMENT TO ADAPTATION AND BUILDING RESILIENCE TO CLIMATE CHANGE

Responses to climate change in Africa must focus on adaptation, given the region's vulnerability to impacts in agriculture and throughout its food systems, and the region's relatively small contribution to global greenhouse gas (GHG) emissions.

AFRICAN COMMITMENTS AND PROGRESS

As part of the 2014 Malabo Declaration, African heads of state and governments committed to building a climate-resilient African food system, including preparedness to respond to present and future climate variabilities and shocks, and providing social protection provisions for rural and vulnerable groups.[13] The specific commitments and targets to reach by 2025 are: 1) improving the resilience capacity of at least 30 percent of farmer, pastoral, and fishing households by equipping them to mitigate, adapt to, and recover from shocks and stresses; 2) having 30 percent of agricultural land under sustainable land and water management or climate-smart agriculture practices; and 3) creating investment and an enabling environment for resilience initiatives, especially for disaster preparedness plans, early warning and response systems, social safety nets, and weather-based index insurance.[14]

According to the second biennial review of progress on the Comprehensive Africa Agriculture Development Programme (CAADP), conducted in 2019, 11 African countries (Burundi, Cabo Verde, Ghana, Ethiopia, Mali, Mauritania, Morocco, Rwanda, Seychelles, Tunisia, and Uganda) are on track to meet these commitments,[15,16] up from just 7 countries in 2017.[17] With respect to individual indicators, only 25 countries were on track for households' resilience to climate-related shocks, 12 for share of agricultural land under sustainable management practices, and just 1 for existence of government budget spending lines dedicated to resilience-building initiatives. However, the lack of good quality data hampers accurate assessment of countries' progress. For example, 30 countries lack data on households' climate resilience, and 22 countries lack data on use of sustainable management practices in agriculture.

To meet international commitments, many countries are working to keep their national GHG emissions within the required range through reporting their Nationally Determined Contributions (NDCs) and developing and implementing their National Adaptation Plans (NAPs) in line with the UN Framework Convention for Climate Change (UNFCCC). Fifty-three countries have submitted their first NDCs and are now submitting revised versions.[18] However, integrating these goals with the overall national development priorities of individual countries and the continent will be challenging as the bulk of the NDC targets set by African countries are conditional on external support.

FINANCING FOR ADAPTATION AND RESILIENCE

While funding needs for adaptation are large, tracking development finance going to Africa reveals that adaptation funding is only half the amount of the overall mitigation budget, that disbursement ratios are lower for adaptation (46 percent) than for mitigation-related projects (56 percent), and that most adaptation programs offer loans as opposed to grants (Figure 1).[19] Given Africa's relatively low contribution to GHG emissions compared to other global regions (about 3 percent),[20] it seems that the investments in Africa for mitigation are disproportionate relative to its GHG contribution, and that the focus should be on adaptation. Furthermore, although developed countries pledged to mobilize US$100 billion annually to address climate change in developing countries through 2025, that commitment has not been met (see Chapter 5).

Nevertheless, important investments are being made at the country and continental levels. In Kenya, for example, CGIAR is supporting the development and implementation of the Kenya Climate Smart Agriculture Implementation Framework (KCSAIF), with monitoring and evaluation frameworks at the county and subnational levels for measuring progress in increasing agricultural productivity and building the climate change resilience of agrifood systems and value chain actors.[21] At the continental level, the African Development Bank, under the African Financial Alliance on Climate Change, has committed to mobilize $25 billion for low-income African countries. It has also established the Africa NDC Hub to serve as a resource pool for its member countries, with a focus on fostering long-term climate action, mobilizing resources for implementation, and coordinating other NDC-support activities on the continent.[22]

PROMISING INNOVATIONS

Many innovations for building resilience of food systems (discussed in other chapters of this report) hold great promise for Africa, but are largely unproven. These include innovations in agricultural insurance, precision agriculture, renewable energy, and monitoring and evaluation of climate policy. Social

FIGURE 1 African and global CO_2 emissions (1951–2017) and disbursements of public adaptation and mitigation funds (2014–2018)

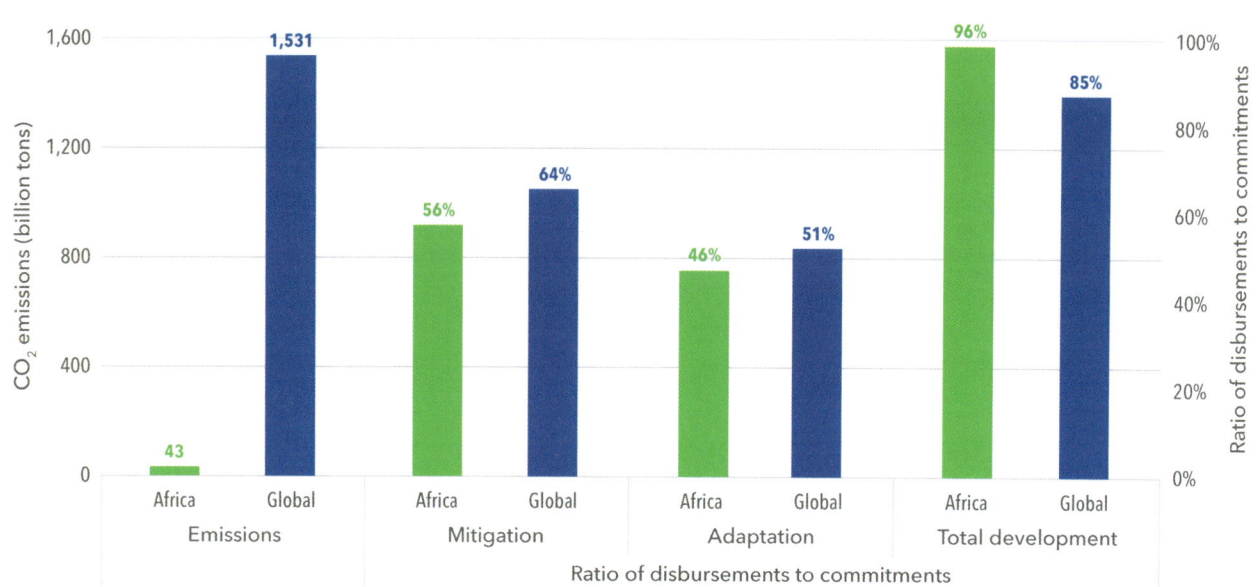

Source: Based on data from: H. Ritchie, "Who Has Contributed Most to Global CO2 Emissions?" *Our World in Data*, October 1, 2019; G. Savvidou, A. Atteridge, K. Omari-Motsumi, and C.H. Trisos, "Quantifying International Public Finance for Climate Change Adaptation in Africa," *Climate Policy* 21, 8 (2021): 1020-1036. Reproduced with permission of the publisher.

protection, discussed in last year's Global Food Policy Report, must also play a critical role.[23] Availability and increased use of digital technologies and information can potentially help catalyze adoption of all these innovations (see Chapter 12).

Use of agricultural insurance remains low at the continental level, with only 3 to 6 percent of farmers being covered.[24] Since the early 2000s, various index-based agricultural insurance products, seen as more promising than traditional insurance products that rely on actual crop loss assessments, have been tested across the continent. Coverage of these, which ranges from a few hundred farmers in a single country to about 400,000 farmers across multiple countries, depends on the primary company and collaborating organizations, type of commodities covered, and number of products offered, among other things.[25] A major challenge hindering wide implementation of these products is reducing the basis risk, that is, the mismatch between the index triggering the insurance payout and the actual damage experienced by the farmers. Innovations to reduce spatial basis risk involve indexes that use remote-sensing data (such as soil moisture and vegetation) to better proxy rainfed yield,[26] such as index-based livestock insurance that is based on a normalized-difference vegetation index,[27] the Africa Risk Capacity drought insurance that is based on soil water availability, or a water requirements satisfaction index.[28] However, it may be difficult for farmers to understand how such remote-sensing data are correlated with weather variability and thus trigger payouts. Innovations toward this include ACRE-Africa's hybrid index and bundle for comprehensive coverage, which includes both different shocks (drought, storms, pests, and diseases) and different stages from planting to harvest.[29] These combine index-based insurance products with other technologies such as picture-based tools to assess actual damages, or bundle insurance products with credit for climate-smart practices and technologies.[30]

Efforts to address climate change must tap into renewable energy sources to build the resilience of African food systems. Over half of Africa's population has no access to modern electricity services.[31] Lack of access to reliable, affordable, and sustainable energy contributes to food loss and limits the efficient use and growth of food supply chains, which in turn affects food availability and access.[32] Moreover, 78 percent of the population relies on traditional biomass for cooking and heating, with implications not only for climate change but also for women's health and labor.[33] Many African countries have committed to transitioning to green energy which, together with increasing sustainable agricultural practices, is prioritized in more than 70 percent of the African NDCs and NAPs.[34] Several countries including South Africa, Ghana, and Nigeria are investing in electricity grids and off-grid technologies (especially mini-grids) to expand rural electrification and reduce dependence on forest products (see Chapter 9),[35] which has reduced the share of people in sub-Saharan Africa without access to electricity from 69 percent in 2010 to 57 percent in 2018.[36] A key challenge is the low demand for electricity, especially in rural areas, due primarily to poverty as well as unreliability of existing service provision, high costs, and a lack of policy support for clean energy sources.

KEY RECOMMENDATIONS

Innovations in Africa should prioritize climate change adaptation, and seek to capture "sweet spots" where adaptations can also deliver co-benefits to mitigation.[37] This includes exploring sustainable intensification as a climate-smart option to build resilience and increase crop yields while also contributing to mitigation through reduction of GHG emissions.[38] In addition, innovations should be seen as immediate and complementary means to accelerate economic recovery and inclusive development more broadly, as defined by national and sectoral masterplans.[39,40] As communities become healthier, wealthier, and more educated, they will be better placed to adapt to the negative consequences of climate change.

LOCALIZE INNOVATIONS. Because of the immense diversity across the continent, innovations that work well in one country or region might not work well in another. Any promising options for transforming African food systems should be locally assessed and based on local projected climate vulnerability and prevailing biophysical and socioeconomic conditions. More contextualized research, data, and information are needed to inform climate change adaptation. Research and analysis should consider the expected impacts on food system components and actors all along value chains.[41] It should support better assessment of trade-offs among competing options, in terms of both effectiveness and cost-efficiency, in meeting climate change goals along with other development goals (see Chapters 11 and 12). Although several platforms, models, and datasets have been developed in this direction, including the meta-dataset Evidence for Resilient Agriculture (ERA),[42] none of these initiatives focus on cost-efficiency at the continental level, which is critical information for resource-constrained African policymakers.

INCREASE AND IMPROVE INVESTMENT FOR RESILIENCE. Taking resilience-building innovations to scale will require increased finance flows to Africa. Public finance and foreign direct investment (FDI) for food systems development and climate change adaptation in Africa must be substantially increased by scaling up both commitments and disbursement ratios, as well as the quality of investments (see Chapter 5). To mobilize more resources that can speed up the implementation of Africa's national adaptation plans, procedures to access climate finance need to be simplified, harmonized, and mainstreamed across donors and funding vehicles.[43,44] Each national government should also involve and empower local and subnational actors and institutions in the design, monitoring, and evaluation of climate change adaptation projects, employing their local expertise.[45] This will require continued efforts to increase the participation, representation, and legitimacy of vulnerable and marginalized groups within national processes to address climate change along with other long-term challenges facing the continent.

While Africa's food systems are evolving rapidly in response to multiple factors, including changing demographics, growing urbanization, shifts in labor supply, and increased availability of digital technologies, more than 50 percent of the population depends on rainfed agrifood systems for their livelihoods. As a result, the impact of climate change on the continent's food security and other outcomes will be enormous. Considering the diversity in food systems and other factors across the continent, how African countries respond to climate change will be critical for food systems. Overall, the focus must be on adaptation and on investing in innovations, such as those in agricultural insurance, precision agriculture, renewable energy, and social protection, that can build the resilience of Africa's food systems.

MIDDLE EAST AND NORTH AFRICA

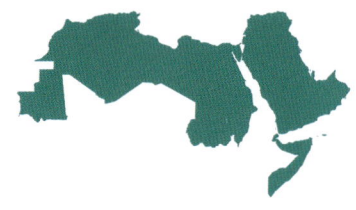

CLEMENS BREISINGER, AMGAD ELMAHDI, YUMNA KASSIM, AND NICOSTRATO PEREZ

Clemens Breisinger is a senior research fellow and program leader, Development Strategy and Governance Division, International Food Policy Research Institute (IFPRI), Nairobi. **Amgad Elmahdi** is water sector lead, Green Climate Fund, Songdo, Republic of Korea, and former director for MENA, International Water Management Institute (IWMI). **Yumna Kassim** is a senior research associate, Development Strategy and Governance Division, IFPRI, Cairo. **Nicostrato Perez** is a research fellow, Environment and Production Technology Division, IFPRI, Washington, DC.

In many countries of the Middle East and North Africa region (MENA), the COVID-19 pandemic has increased budget deficits, reduced economic growth, and raised poverty and unemployment levels.[1] Ongoing conflicts, political turbulence, and fragility in Sudan, Syria, Yemen, and elsewhere have hindered both the response to the pandemic and long-term economic reform and development processes. The Russian invasion of Ukraine has further exacerbated the recent rise in global food prices and created additional pressure for economies and populations in MENA countries that are net food and fuel importers, such as Morocco, Tunisia, and Egypt. For example, in Egypt and Yemen, wheat and wheat products represent between 35 and 46 percent of caloric intake per person. Egypt is the world's largest importer of wheat, with imports accounting for about 60 percent of total wheat use in the country, and Yemen's cereal import dependence ratio is estimated to be 97 percent. As such, the Russia-Ukraine war poses a serious food security threat for Egypt and threatens to further exacerbate undernourishment and reliance on external assistance in Yemen.[2]

Beyond these immediate crises, climate change, along with water scarcity, poses a short- and long-term challenge for food and water security, sustainability, and development. Climate change and variability threaten to destabilize agricultural production and further reduce water availability and accessibility in the region, with implications for the broader food system. Yields are projected to decline for most crops in MENA, while the global impacts of climate change may lead to substantial increases in consumer prices for the region's major food imports (Table 1). Consumption, nutrition, and food security are all likely to suffer,

TABLE 1 Impact of climate change on food security in MENA region by 2050 (percent change)

Indicators	Cereals	Fruits & Vegetables	Oilseed Crops	Pulses	Roots & Tubers	Meat Products
Yield	-4.18	-1.78	-6.86	-17.20	-0.17	–
Production	-5.67	1.53	0.78	-26.55	9.06	-1.22
Consumer prices	12.84	13.89	20.82	10.56	35.17	8.72
Consumption	-1.93	-2.23	-7.28	0.54	-9.84	-1.24

Source: IMPACT simulations.
Note: Values are percentage changes from a no-climate-change assumption. Climate change scenario is based on HadGEM general circulation model, SSP2 and RCP 8.5. "–" means not significant changes.

putting more people at risk of reduced water availability, hunger, and malnutrition.

In addition to these longer-term impacts of climate change, extreme climate events such as droughts, floods, and storms are already becoming more frequent and intense, costing lives and millions of dollars in lost revenue and damages.[3] In Egypt, for example, the costs of climate change impacts to the agriculture sector alone are estimated at US$1.84 billion per year over the next 30 years.[4] For the region as a whole, GDP could shrink by 6–14 percent by 2050 due to climate change and water scarcity.[5]

Conflicts over water are expected to become more frequent and severe with climate change, as rising temperatures and more volatile rainfall reduce water quality and quantity. Notably, about 60 percent of the region's water resources are transboundary water bodies, with every country sharing at least an aquifer with a neighboring country.[6] Increased competition over this scarce resource will be compounded by rapid population growth, economic development, and conflict-related migration in several MENA countries.[7] In addition, sea-level rise, coastal degradation, and water variability driven by climate change will threaten the viability of low-lying areas (particularly the Nile Delta in Egypt), harbors, and other critical food system infrastructure, and will potentially lead to loss of fertile land, social disruption, displacement, and migration.[8]

IMPLEMENTING "SHOVEL-READY" POLICIES, REFORMS, AND INNOVATIONS

To turn these threats into opportunities for development, existing "shovel-ready" policies and investments for climate change adaptation and mitigation need to be implemented by public institutions and the private sector, working in tandem, and capacities need to be built to adapt interventions to local contexts.

To mitigate the impacts of the war in Ukraine and address the vulnerabilities arising from high and increasing food import dependency, MENA countries should diversify their food import sources, utilize the potential for agricultural production remaining amid water shortages and climate change, endeavor to reduce food waste, diversify diets away from imported cereals toward locally grown staples, and increase the use of targeted cash transfers for the poor.

PARADIGM SHIFT ON WATER MANAGEMENT. A shift in approaches to water use that embraces the need to conserve water resources and develop new water sources has the potential to increase resilience to climate change. Better demand management — including greater water conservation, water efficiency, and water re-use and recycling — can reduce pressure on water supplies. This shift can also promote the use of new water supply sources, such as rainwater and desalinated water. Innovative pricing mechanisms, such as those established in Morocco and Tunisia for cost recovery for water, would allow the region's agricultural water sector to become financially sustainable and to meet its maintenance and operational costs while also promoting more efficient water use by end-users. In addition, these water pricing policies can deliver on financial goals (achieving full-cost recovery), social goals (ensuring access for all), and environmental goals (incentivizing sustainable use). Knowledge building, including data on existing water and uses, hydrology of regional water flows, and ecosystem services, as well as engagement with communities and the private sector for knowledge sharing can support these changes. Projects to promote water-saving technologies and improve irrigation management, such as those implemented in Jordan and Egypt, often benefit from co-designed interventions that involve partnerships among all key private and public market players rather than targeting end-users directly.[9] At the farm level, a number of technologies and services can improve water management and resilience.

INVESTING IN PROMISING FARM TECHNOLOGIES AND SERVICES. Investments in R&D and scaling-up of promising technologies should foremost include those focused on greater efficiency of water use, including irrigation and water management technologies. Solar photovoltaic water pumping stations along with water-saving drip irrigation kits can help to increase resilience of irrigated systems, reduce operational and maintenance costs, increase yields, and promote crop diversification.[10] Moreover, judicious use of groundwater and/or treated wastewater could reclaim and transform desert lands for agricultural use, which could further boost agriculture's resilience to climate change.

Mobile phone applications can provide farmers with geo-specific information customized to their plots, weather conditions, and crop types. These apps can make irrigation recommendations based on the crop type, irrigation system, farm size, planting time, types of water pumps, energy sources, and soil type (Figure 1). The IRWI APP, for delta areas in Egypt, and LARI-LEB, for Bekka Valley in Lebanon, have reduced the time and amount of irrigation by 30 percent on average.[11] Other promising technologies include improved seeds, greenhouses, hydroponics, optimized fertilizer application, improved groundwater management, and digital agricultural tools and services, particularly for extension and market performance (see Chapter 12).[12]

CREATING AN ENABLING ENVIRONMENT FOR CLIMATE-FRIENDLY FOOD SYSTEMS

Adopting these policy and technology changes will require an enabling environment of coherent policies, incentives, investments, and capacity building across all food system sectors. Creating an enabling business environment, fostering climate-smart trade, and expanding international cooperation will help MENA to meet country-level 2030 development agendas and international goals, including the UN Sustainable Development Goals.

ENABLING BUSINESS ENVIRONMENTS. More business-friendly environments with clear regulatory and enforcement frameworks will support all food system actors and make domestic markets more competitive. Examples include investments in food system infrastructure, promotion of food quality standards, frameworks to stimulate more efficient water use in the agriculture sector, disincentives for food and water waste, and encouragement of more sustainable practices. When combined with targeted, sector-specific public investments and social safety nets, macroeconomic and fiscal policies that phase out unsustainable energy and fertilizer subsidies and incentivize climate-smart investments have proven successful in countries like Egypt.[13] The private sector can help close the financial gap through public–private partnerships. These partnerships are new and rapidly evolving tools for funding in MENA, from the agro-industrial sector to food trade. In fragile and post-conflict areas, such as Iraq, public–private partnerships may also offer a strategy to resume construction, maintenance, and operation of abandoned projects.[14]

CLIMATE-SMART TRADE. Climate-smart trade policies and openness to trade will be critical for MENA, even though some policymakers may be inclined to promote self-sufficiency goals as a result of

FIGURE 1 Mobile applications that support agricultural production

Source: IRWI App (Irrigation Water Information Application) and LARI-LEB App (Lebanese Agricultural Research Institute).

the Russia-Ukraine war and related uncertainties around wheat imports. But net food imports are likely to further increase due to rising demand and limited potential for domestic wheat production. Vulnerabilities to production shocks from climate change and extreme climate events, as well as prolonged health emergencies and pandemics, are also likely to grow. Export crops will remain important as a source of income and foreign exchange, particularly where MENA countries have a comparative advantage, as in fruits and vegetables. Rather than focusing on unrealistic self-sufficiency goals for wheat, efforts to phase out water-intensive crops could facilitate further specialization in high-value agricultural products while addressing countries' food security concerns. Incentives to promote trade in "virtual water" — that is, water embodied in crops — could expand regional production of less water-intensive, more heat-tolerant, and high-value export crops like fruits and vegetables and increase imports of water-intensive crops like rice and wheat, with positive impacts on sustainability and resilience amid climate change.

INTERNATIONAL COOPERATION. Regional and international cooperation in climate change adaptation efforts can also improve production, lower food prices, and facilitate the import of strategic crops or food items.[15] Areas for cooperation may include R&D, financing, trade, technologies, innovations, digitalization, and management support, as well as capacity building. Regional collaboration and cooperation are also essential for conflict-resolution and problem-solving around water resource management. For example, the establishment of water allocation mechanisms where water resources are shared across borders can support more sustainable water use and reduce conflict.

LESSONS FROM OTHER SHOCKS FOR CLIMATE CHANGE

The pandemic and the invasion of Ukraine, in addition to imminent threats from climate change–induced extreme weather events, have clearly shown MENA countries that a well-functioning, resilient food system is critical and that "climate-proofing" the regional food system must be a priority. In several MENA countries, the negative impacts of COVID-19 were cushioned by keeping domestic and international food markets open; diversifying food imports; ensuring that food producers, traders, and retailers could operate within sensible enabling environments (including soft and hard institutions); and providing safety nets for the most vulnerable. These responses will also be crucial to dealing with the impacts from the war in Ukraine. Some of the same principles — coupled with a push to "green" MENA economies and investment in a new water paradigm, as well as tools and technologies for more sustainable agriculture — could support adaption to climate change and pave the way for greater prosperity.

CENTRAL ASIA

KAMILJON AKRAMOV, KAHRAMON DJUMABOEV, AND ROMAN ROMASHKIN

Kamiljon Akramov is a senior research fellow, Development Strategy and Governance Division, International Food Policy Research Institute, Washington, DC. **Kahramon Djumaboev** is a researcher in Water Management, International Water Management Institute, Central Asia Regional Office, Tashkent. **Roman Romashkin** is deputy director of the Eurasian Center for Food Security at Lomonosov Moscow State University, Moscow.

Increasing evidence shows that shifts in Central Asia's climate are already occurring at an accelerating rate, and the region's aridity is expected to increase, with consequences for agricultural production.[1] While this warming trend applies to all areas and seasons, regional and seasonal trends vary. Lowland areas have seen a greater warming trend than mid-altitude and upland regions. The average annual temperature in Tajikistan, where mountains cover 93 percent of the land area, has increased by 0.3°C–1.2°C since the 1950s. In contrast, in Turkmenistan, where 80 percent of the land area is flat desert, the average annual temperature increased by 1.1°C–2.4°C during the same period.[2] In Uzbekistan, the average minimum temperature rose by 2.0°C and average maximum temperatures by 1.6°C between 1950 and 2013. The warming was steepest in spring (0.39°C per decade) and more modest in winter (0.13°C per decade).[3] In contrast, Kazakhstan has seen the sharpest rise in winter temperatures (0.35°C per decade), with less warming in summer months (0.18°C per decade).[4] Unlike the clear warming trends, precipitation changes have shown no clear trend, with significant variations observed across the region.

The frequency and intensity of extreme climate events, notably heatwaves, droughts, and dust storms, have increased during recent decades throughout Central Asia. The probability of heatwaves is expected to rise further, especially in Tajikistan, Turkmenistan, and Uzbekistan, and Central Asia's arid climate and steadily high summer temperatures make the region particularly susceptible to drought.[5]

CLIMATE CHANGE IMPACTS ON FOOD SYSTEMS

Climate change is already affecting food systems in Central Asia. Rising temperatures and extreme climate events have contributed to reductions in land and soil quality, availability of water resources for irrigation and human use, and crop yields. Land degradation is a major challenge, resulting from the combination of rising temperatures and aridity, unsustainable agricultural and irrigation practices, and overgrazing of pastureland.[6] About a quarter of the population of both Kazakhstan and Uzbekistan already live on degraded land.[7] Moreover, with increased probability of drought, the region is vulnerable to dryland expansion and desertification. These impacts have high costs. Land degradation reduces agricultural incomes by lowering crop yields and livestock productivity and by necessitating increased input use.[8] The annual cost of land degradation in Central Asia between 2001 and 2009 has been estimated at about US$5.85 billion, equivalent to 3 percent of GDP in Kazakhstan and Uzbekistan, 4 percent of GDP in Turkmenistan, and 10–11 percent of GDP in Kyrgyzstan and Tajikistan.[9]

Renewable water resources in Central Asia are limited. These resources, including critical seasonal runoff (meltwater), will be affected by a combination of changes in precipitation, increasing glacier melt, earlier snowmelt, and higher evapotranspiration that will aggravate regional water shortages and reduce water available for irrigation in the summer months.[10] Already as a result of rising temperatures, nearly a third of the glacial area in the Tien Shan and Pamir mountains has disappeared since 1930.[11]

Water scarcity can be physical (lack of water of sufficient quality) or economic (lack of adequate infrastructure plus technological and institutional constraints).[12] Currently, only Uzbekistan is considered physically water-stressed, with water availability averaging just 1,505 m³ per capita per year.[13] Turkmenistan also has relatively low renewable water resources per capita (2,407 m³ per year). However, over the past three decades, all Central Asian countries have experienced a steady decrease in the per capita availability of renewable water resources due to the rising population (Figure 1). Turkmenistan and Uzbekistan are severely water-scarce countries in economic terms, with withdrawal levels for available renewable water resources reaching more than 100 percent[14]; Kyrgyzstan and Tajikistan, which withdraw 33 and 45 percent of their renewable water resources, respectively, are considered water-stressed countries (Figure 2). Most of the region is expected to face a significant increase in water stress[15] due to climate change.[16] This is a major concern for the sustainability of Central Asia's agrifood systems, as irrigation is essential for agriculture in most of the region, and

FIGURE 1 Total renewable water resources per capita, Central Asia, 1988–2022

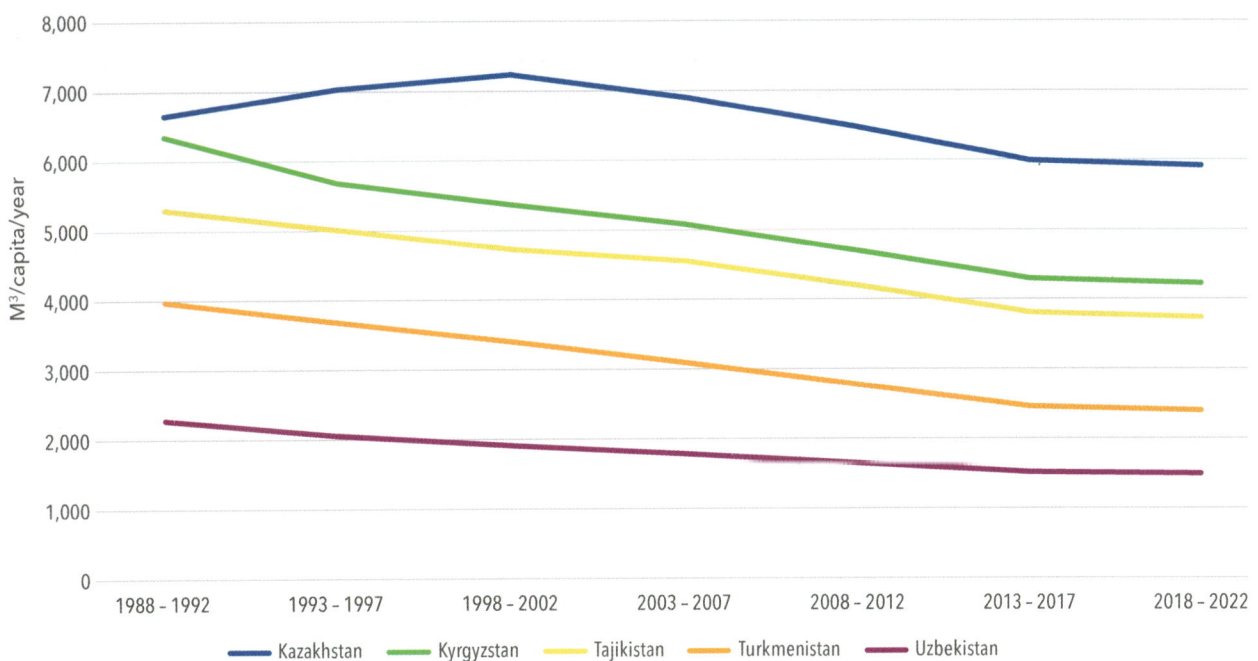

Source: FAO Aquastat 2021.

FIGURE 2 Proportion of renewable water resources withdrawn, Central Asia, 2018–2020

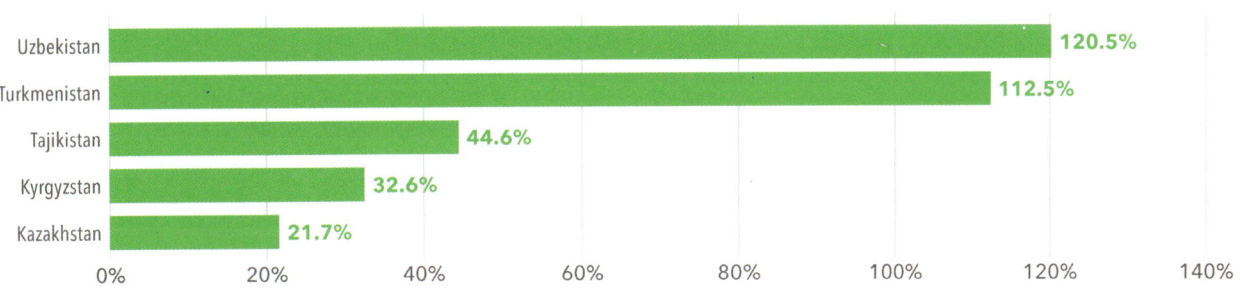

Source: FAO Aquastat 2022.

Note: Abundant water resources = up to 25%; Water stressed = 25% to 60%; Water scarcity = 60% to 75%; Severe water scarcity = above 75%.

CENTRAL ASIA 125

competition for water is increasing, leading to intraregional conflicts.

Increased aridity and water stress levels, together with the greater frequency of extreme weather events, have already caused sizable declines in the region's crop yields. For example, crop yields in Tajikistan dropped by 30–40 percent in 2000/01 and again in 2007/08 as result of droughts. Combined with the global food crisis, this has led to more than 2 million people being undernourished.[17] Nevertheless, Central Asian countries achieved significant progress in reducing child malnutrition during the past two decades. Child stunting rates in Kazakhstan declined from 17.7 percent in 2000 to 6.7 percent in 2020, in Kyrgyzstan from 29.9 percent to 11.4 percent, in Tajikistan from 41.6 percent to 15.3 percent, and in Uzbekistan from 29.5 percent to 9.9 percent.[18] However, the risk of malnutrition in the region is likely to increase, especially among children and other vulnerable groups, as the adverse effects of climate change may become a significant challenge for agriculture, especially for crop yields. In addition to indirect effects related to land degradation, water availability, and extreme weather events, climate change affects crop yields directly through changes in precipitation, temperatures, and carbon dioxide availability. Projections from IFPRI's IMPACT and DSSAT models show that Kazakhstan will experience modest losses in cereal yields but significant losses in potato yields in the period up to 2050, while Tajikistan and Uzbekistan will likely face higher yield losses for major crops.[19] The projections for Tajikistan indicate that adverse effects of climate change may lead to a significantly higher (4.6 percent) number of malnourished children than there would be without climate change.[20]

POLICY RESPONSES

Policy responses to climate change have been limited in Central Asia until recently. Climate policies predominantly aim to reduce the adverse effects of weather extremes and focus on production of staple crops rather than building resilience.[21] However, at least two potential interventions could support adaptation.

CROP DIVERSIFICATION

Agricultural crop diversification is a potentially feasible and cost-effective means to increase the resilience of Central Asia's agricultural systems. Greater crop diversity can improve resilience by reducing threats from pest outbreaks and pathogens (which are expected to worsen under climate change) and buffering crop production from the effects of growing climate variability and extreme weather events. The production structures and economic incentives adopted in the 1990s encouraged the production of just a few crops. Together with the belief that specialized agricultural systems are more productive than diversified systems, these policies have hindered past attempts to promote diversification.[22] As a result, reforms that support agricultural diversification have only recently emerged in Central Asia. The region's governments have now begun to adopt crop diversification policies, with support from major bilateral and multilateral development partners.[23] For example, the governments of Uzbekistan and Tajikistan are encouraging farmers to grow horticultural crops, and the Kazakh government is encouraging farmers to allocate more land to oilseed crops.

Crop diversification can be implemented in various forms and at a variety of scales, allowing farmers to choose a context-appropriate strategy that both increases resilience and provides economic benefits. For example, the cultivation of more winter crops may provide good market opportunities for farmers. However, efforts may be needed to promote the adoption of these crops, including providing information to farmers and ensuring availability and access to seeds, fertilizers, and plant protection products. Another potential opportunity, arising as a result of climate change, is the shift of farming to some areas that were previously too cold for optimal farming of summer crops.[24]

WATER-SAVING TECHNOLOGIES

Implementation of climate-smart agricultural (CSA) technologies — such as conservation technologies, water-saving irrigation technologies, drought-resistant crops, and climate-resilient varieties — remains limited despite their potential benefits. Of these, water-efficient technologies are critical, given the vital role of irrigation for agricultural production in Central

Asia. Such technologies, including drip irrigation, sprinklers, and smartsticks, can support adaptation to climate change and boost productivity, contributing to improvements in farmers' livelihoods and food security.[25] However, high initial and maintenance costs for this equipment and lack of necessary technical capacity are major constraints to large-scale implementation in the region.

In this regard, a significant policy shift recently took place in Uzbekistan, where in 2020 the government adopted a program that redirects subsidies from the energy sector to support water-saving technologies. The evidence suggests that improved irrigation practices, such as drip irrigation, can save approximately 30 percent of the water used compared to traditional irrigation practices.[26] These technologies would also reduce energy use and carbon emissions and improve crop yields in pump irrigated areas. Research by the International Water Management Institute (IWMI) in the Karshi region of Uzbekistan has shown that optimized irrigation could save half a trillion liters of water, spare 259 gigawatt-hours of electricity, and cut 122,000 tons of CO_2 emissions.[27] Considering that the irrigation sector accounts for about 21 percent of total electricity use in Uzbekistan, the adoption of water-saving technologies should lead to significant energy and water savings. This could be a promising path forward for other countries in Central Asia, where, overall, 18 percent of irrigated areas are under pump irrigation, covering about 2.84 million hectares.[28]

LOOKING FORWARD: OBSTACLES AND LESSONS

Demand for nutritious food in Central Asia is rising as a result of population and income growth.[29] At the same time, the food supply faces significant constraints due to increasing land degradation and water scarcity. While policies that encourage water-saving technologies and crop diversification can help mitigate these problems in the short and medium term, additional innovations are needed.

First, digital technologies (such as precision agriculture, weather-station-based irrigation advisory systems, and unified digital market platforms) enabled by internet and mobile phone connectivity could help farmers efficiently and sustainably use limited land and water resources, improve decision-making, better manage risks and variability, and enhance efficiency (see Chapter 12).[30] However, Central Asia lacks the necessary infrastructure for a digital transformation of the food system, a problem that has been highlighted by the COVID-19 pandemic.[31] Regional governments need to invest in digital connectivity infrastructure and encourage public-private partnerships to accelerate the digital transformation of the agrifood sector.

Second, regional policy coordination is needed, given that Central Asian countries share common water and other essential resources for agriculture. Poor coordination of policies has been a serious obstacle. For example, during the pandemic, some national governments moved to impose export restrictions, which led to higher food prices and exacerbated food insecurity issues already worsened by the crisis. Moreover, Central Asia historically has a complicated relationship with water, and the "water tensions" between upstream (Kyrgyzstan and Tajikistan) and downstream (Kazakhstan, Turkmenistan, and Uzbekistan) countries are well known.[32] This underscores the importance of policy coordination and transboundary water cooperation across the region to make the agrifood system inclusive and resilient to climate-change-related shocks. In this regard, regional intergovernmental bodies such as the International Fund for Saving the Aral Sea (IFAS) and the Interstate Commission for Water Coordination (ICWC) need to improve their efficiency and promote evidence-based decision-making.

Finally, one of the biggest challenges in Central Asia is the lack of reliable data on climate change and food systems, including land, labor, and water productivity, as well as the costs and benefits of water-saving technologies. This problem is compounded by inadequate analytical capacity in the region. Developing a robust and comprehensive framework on how climate change will impact the region and its consequences for water and food security and nutrition will require innovative data solutions.

SOUTH ASIA

ADITI MUKHERJI, AVINASH KISHORE, AND SHAHIDUR RASHID

Aditi Mukherji is a principal researcher, International Water Management Institute (IWMI), New Delhi.
Avinash Kishore is a research fellow, South Asia office, International Food Policy Research Institute (IFPRI), New Delhi, and **Shahidur Rashid** is director for South Asia, IFPRI, New Delhi.

South Asia is a climate change hotspot, with many climate-induced risks compounded by significant existing vulnerabilities. All available indicators — glacier melt in the Himalayas, sea-level rise, extreme weather events, precipitation patterns, and the frequency of natural hazards turning into disasters — have worsened since 1950. Rapid economic growth in recent decades, which is expected to continue, will likely exacerbate these alarming trends, and the high economic costs of climate change in the region are projected to increase further.

Policy actions vary across countries, but all governments in the region have signed the Paris Agreement on climate change. However, all South Asian countries are lagging in undertaking some critical actions[1] that would directly contribute to both climate change adaptation and mitigation in the region, and COVID-19 has led to a significant reduction in climate-related investments.[2]

In South Asia, as in much of the developing world, total greenhouse gas (GHG) emissions have been rising since the 1990s.[3] Agriculture is both a major source of emissions and the sector most affected by climate change. Globally, agriculture and associated land use and land-use change activities account for roughly a quarter of all GHG emissions. Three-quarters of these emissions can be attributed to developing countries, and South Asia is a major contributor. With these broad trends in mind, this chapter focuses on 1) observed and projected climate impact drivers (also called climate hazards) that are driving climate change and its societal impacts, 2) observed and projected impacts on agrifood systems, and 3) policy solutions that can be implemented in the near term to adapt to and mitigate climate change in the region.

CLIMATE CHANGE IN SOUTH ASIA

Anthropogenic GHG emissions, including CO_2 and other GHGs, have already caused a temperature rise of about 1.1°C globally.[4] In South Asia, hot extremes (both the frequency and intensity of daily maximum temperatures), heavy precipitation events, and other climate shocks have increased since 1950. These changes are expected to accelerate with every increment of global temperature increase.[5] South Asia's pre-existing vulnerabilities — high levels of poverty, governance challenges, and limited access to basic services and resources — amplify the region's climate risks, with potentially devastating effects if warming continues at this pace.[6]

MEAN AND EXTREME TEMPERATURES HAVE RISEN AND ARE PROJECTED TO RISE FURTHER. The temperature rise in South Asia over the past century is well documented, and the projections are dire. However, temperatures have not risen evenly across the region. For example, while the average temperature rise was 0.7°C between 1901 and 2018, a larger increase was observed over the Himalayas (1.3°C). In Bangladesh, average annual temperatures have increased in the last six decades, with warming accelerating from 2001 onward.[7] South Asia has experienced a lower-than-global-average temperature rise mainly due to the cooling effect of aerosols, including short-lived climate pollutants, which have their own negative health and agricultural consequences.[8] For the region as a whole, annual mean temperatures are projected to increase by between 1.2°C and 4.3°C (under low- and high-emissions scenarios) by the end of the century.[9] The average temperature across India is projected to rise by between 2.4°C and 4.4°C by 2100.[10] Similarly, summer heat waves are projected to triple or quadruple by 2100 in India,[11] and increase at the rate of 0.71 days per decade in Pakistan.[12]

HIMALAYAN GLACIERS ARE MELTING AT UNPRECEDENTED RATES. These glaciers, which are an important source of the region's rivers, have lost more mass since 2000 than in the entire 20th century.[13] With temperature rise over the Hindu Kush Himalayas in the northwestern area of South Asia potentially reaching around 6°C under high emissions scenarios, glaciers could lose up to two-thirds of their volume by 2100, with severe consequences for people living in downstream river basins — including reduced water availability for agriculture, hydropower, and domestic use.[14] Melting of Himalayan glaciers also affects vulnerable mountain populations upstream who directly depend on glacier meltwater for their livelihoods. Projected intensification of monsoon precipitation is expected to increase annual river flows, but seasonal flows reduced by lower baseflow due to early melting will affect both irrigation and hydropower generation, while extreme precipitation events are likely to make flood events more intense and impactful.

WARMING-INDUCED INCREASE IN MONSOON RAINFALL. Mean precipitation in South Asia has increased since the 1950s due to higher GHG emissions, although the cooling effect of aerosols has kept the rate of precipitation increase below the global average.[15] Both low and high extremes have increased, especially since the 1980s, with more frequent dry spells and more intense wet spells.[16] Notably in India, summer monsoon precipitation has shown declining trends over the last few decades, with larger decreases over the main breadbasket region of the Indo-Gangetic belt;[17] and in Bangladesh, a decline in precipitation of about 84 mm per decade was observed from 1981 to 2010.[18] Projections show increases in both mean and extreme precipitation for the entire region by the end of this century, with some variation across countries, ranging from about a 17 percent increase in mean precipitation in Bangladesh to more than 27 percent in India under a high emissions scenario.[19] Studies also project that extreme precipitation events will be 1.7 times more likely in Bangladesh by 2050 than they are now.[20]

CLIMATE HAZARDS ARE PROJECTED TO WORSEN. Over 750 million people in South Asia are currently exposed to climate hazards, primarily floods and droughts.[21] In India, reduced summer monsoon rainfall has increased the frequency and spatial extent of droughts from the 1950s onward.[22] Similarly, in Pakistan, the intensity and severity of droughts have increased over the last century.[23] Greater incidence of localized extreme precipitation events has increased flood risks in India, and projections show increases in flood intensity and occurrence in the Indus, Ganges, and Brahmaputra River basins.[24] On the other hand, the tropical cyclones over the Northern Indian Ocean region have become less common, although very severe cyclonic storms have become more frequent in the post-monsoon season.

Event attribution studies are relatively rare in South Asia and remain an important knowledge gap. The 2017 Bangladesh flood is one rare case. In 2017, flash floods damaged about 220,000 hectares of nearly harvestable *boro* paddy in Bangladesh. Several attribution studies show that anthropogenic climate change increased the likelihood of the extreme rainfall events that led to this flood.[25]

OBSERVED AND PROJECTED IMPACTS ON AGRIFOOD SYSTEMS

The impacts of climate change on South Asia's agrifood systems include decreases in yields, depleted natural resources, and associated income losses. Globally, total factor productivity (TFP, a measure of technological innovation and efficiency in production) in agriculture has fallen by about 21 percent since 1961 due to climate change, and warmer areas like South Asia have seen the largest declines.[26] At the same time, the likelihood of climate shocks affecting crops including wheat, soybeans, and maize increased in most global breadbaskets, and especially in South Asia.[27] Increased water demand for agriculture is draining aquifers and could worsen food and nutrition insecurity in the region. Negative impacts on agricultural GDP and trade outcomes have also been projected at higher levels of warming.

DECLINING YIELDS IN ALL MAJOR AGRICULTURAL SUB-SECTORS WILL CONTINUE. The unprecedented suite of climatic changes has caused crop yield declines and production losses in the region, with a few exceptions for certain crops and subregions. For instance, in Bangladesh, temperature and rainfall explain 12 percent of crop production variability,[28] and rising temperatures are linked to decreasing wheat yields.[29] However, productivity of some crops has benefited from temperature and humidity trends, notably rainy season paddy (*aus* and *aman*), though not irrigated *boro* paddy. In Pakistan, declines in rice and wheat yields are observed with climate change, although the use of heat-tolerant varieties has provided some resilience and forestalled greater impacts.[30] In the country's mountainous Khyber Pakhtunkhwa region, however, increased average precipitation improved crop yields between 1985 and 2016.[31]

In India, agricultural production data (1967–2016) for several crops show that average land productivity decreases as average temperatures increase,[32] and this impact accelerates at higher levels of warming. Projections show yields of India's crops falling by 1.8 to 6.6 percent by mid-century (2041–2060) and by 7.2 to 23.6 percent by end-century (2061–2080) under a middle-of-the-road scenario for climate change.[33] Yield losses are projected to be higher in rainfed conditions,[34] and yield losses in wheat and maize are projected to be higher than yield losses in rice. In Nepal, flood-induced damages to the area under paddy are already significant and projected to rise further, reaching 50 percent by the end of the century under a scenario with little emissions mitigation.[35]

In India and Bangladesh, subsistence fisheries provide up to 60 percent of animal protein in people's diets,[36] but increasing salinity intrusions in inland aquaculture ponds have resulted in fish mortalities.[37] Globally, fisheries productivity is projected to fall in tropical and subtropical regions like South Asia.[38]

GROUNDWATER DEPLETION IS ACCELERATING RAPIDLY. South Asia accounts for a quarter of global groundwater use. Groundwater extraction for agriculture contributes to food security, livelihood support, and poverty alleviation.[39] Yet groundwater depletion – at rates exceeding 2 cm per year in the Indo-Gangetic aquifer[40] – is straining the limits of these resources, leading to lower agricultural production and loss of related benefits,[41] including effects on the adaptive capacity of communities.[42] Episodic recharge during extreme rainfall events has been recorded in India's semi-arid tropics,[43] but the future impact of climate change on groundwater recharge is uncertain.

EXISTING VULNERABILITIES TO FOOD AND NUTRITION SECURITY WILL GROW. The impacts of climate change in South Asia, compounded by the impacts of COVID-19, will make it very difficult for the region to achieve the goal of zero hunger (SDG2) by 2030.[44] Climate change shocks have direct and long-term impacts on food and nutrition security. For example, a district-level analysis from India concluded that the odds of children suffering from stunting, wasting, and being underweight or anemic increases by 30 to 60 percent in districts with high climate vulnerability compared to districts with low vulnerability.[45]

A global study that accounts for extreme weather events estimates that, by 2050, the number of people at risk from hunger will increase by 11 to 20 percent, with South Asia (along with sub-Saharan Africa) at greatest risk, and estimates that South Asia will need three times its current food reserves to offset the impacts of such events.[46] Similarly, an earlier study estimated that food shortages caused by climate change could lead to a significant increase in the number of malnourished children in South Asia.[47] In Bangladesh, near-term projections estimate a reduction of up to 17 percent in total calorie consumption by 2030 due to climate change.[48]

NEGATIVE IMPACTS ON AGRICULTURAL GDP AND TRADE ARE PROJECTED. Climate change is projected to reduce agricultural GDP through declining crop yields, and increase consumer prices, with greater losses associated with higher warming levels. For example, the Pakistan floods of 2010 caused economic losses of US$4.5 billion,[49] while the losses caused by the 2014 drought in India are estimated at $30 billion.[50] Floods in Bangladesh in 2017 caused a 30 percent year-on-year rise in paddy prices. Economic projections for Bangladesh for the short term estimate a modest decline of 0.11 percent of GDP by 2030, and a 1.23 percent fall in agricultural GDP.[51] Another study projects loss of ecosystem services as a result of climate change in the range of $18 to 20 million by 2050 in Bangladesh under low- and high-emissions scenarios.[52]

Water scarcity is also projected to worsen with climate change. Of the world's five basins where water scarcity-led GDP losses are projected to be highest, three (Indus, Sabarmati, and Ganges-Brahmaputra) are in South Asia. In the Indus Basin alone, GDP losses are expected to exceed $5,000 billion by 2100.[53]

Negative impacts on agricultural trade are also projected. In Pakistan, empirical results show a decline in agricultural export trade (1975–2017) attributable to climate change.[54] Moreover, recent evidence suggests that the welfare losses caused by decreases in food production, increases in food prices, and decreases in food consumption cannot be adequately compensated through trade and fiscal policies.

AGRICULTURAL GHG EMISSIONS

In South Asia, GHG emissions from agriculture are largely attributable to 1) methane emissions from rice cultivation, 2) application of urea, 3) animal husbandry, 4) burning of crop residues, and 5) use of fossil energy for irrigation.[55] These practices are supported by government policies that promote their continued use. Some of the most productive and intensively cultivated areas, like the Punjab and Haryana regions in the northwest and Bangladesh and West Bengal in the east, are locked into rice–wheat or rice–rice monocropping by policies. Both state and national governments provide incentives for farmers to produce these staples through public procurement, price guarantees, preferential provision of subsidized inputs, and disproportionately large investments in the development of improved varieties of these two crops (see Chapter 2, Box 2). The use of urea, a major emitter of nitrous oxide, is heavily subsidized across the region, especially in irrigated areas, resulting in its widespread overuse. South Asia has the world's largest livestock population, which is another major methane source, particularly the region's large bovine population. India alone is home to 30 percent of the total global bovine population. Moreover, the share of animal products in the total value of regional agricultural output is rising; in fact, the livestock sector is benefiting from its relative resilience to drought. In addition, subsidized or free fossil-fuel-based electricity is used for pumping water in large areas of India and Pakistan and parts of northwest Bangladesh, making groundwater irrigation both unsustainable and carbon intensive (see Chapter 9). Currently, more than 260 km³ of groundwater are extracted every year, largely using fossil fuels.[56]

ADAPTATION AND MITIGATION ACTIONS

Some proven strategies can make agriculture more resilient to climate change *and* mitigate GHG emissions from food production, including changes in practices and crops, and most importantly in South Asia, reforming agricultural policies that promote unsustainable, GHG-intensive practices.

TECHNOLOGIES AND CROP DIVERSIFICATION

IRRIGATION AND SOIL MOISTURE CONSERVATION are perhaps the most widely used and most effective adaptive strategies at the farm level for building resilience to droughts, dry spells, and high degree days. However, increases in water requirements for crops and the demand for irrigation in a warming world may worsen water scarcity.[57] Rationalization of electricity subsidies and use of grid-connected solar pumps can support increased use of irrigation while reducing GHG emissions. In India, for example, solar irrigation pumps could reduce the country's CO_2 emissions from agriculture by 8–11 percent.

DIVERSIFICATION TO CROPS more resilient to higher temperatures (such as maize) and moisture stress (coarse cereals, legumes) and allied activities (animal husbandry, poultry, and fisheries) can also support farm-level adaptation. Switching from rice to other crops in the water-scarce northwest and peninsular India will reduce GHG emissions, save scarce water and energy used for irrigation, and reduce crop residue burning. Moreover, the area under rice could be reduced in the region without threatening food security.[58]

AGRICULTURAL MECHANIZATION is reducing farmers' reliance on cattle for draught power and allowing animal husbandry to specialize in milk and meat production, thus leading to a gradual reduction in the cattle population while increasing its economic value. However, in India, recent changes in government policies may have slowed the reduction of unproductive cattle.

POLICY REFORMS

Rationalizing food, fuel, and fertilizer policies can support both adaptation to climate change and mitigation of GHG emissions. Switching to crop-neutral food policies, including allocation of resources for R&D and other input and output subsidies, can help reduce

GHG emissions from food production and make agriculture more resilient to climate change. Four major reforms are worth considering.

INCREASE INVESTMENT IN R&D. Countries in South Asia underinvest in agricultural R&D. As of 2009, the region invested $0.21 to $0.40 in agricultural research for every $100 of agricultural output, well below the $0.61 invested by sub-Saharan African countries and $1.14 by Latin American and Caribbean countries.[59] Governments in the region should increase budget allocations for agricultural research and direct a larger share of this budget to non-cereal crops and allied sectors (livestock and fisheries), given their rising share in agricultural GDP (see Chapter 4).

REFORM FERTILIZER SUBSIDIES. Lowering the subsidy on urea could help reduce emissions of nitrous oxide. Urea is heavily subsidized in all large South Asian countries, and its excessive application is common. Governments hesitate to decontrol the price of urea, fearing strident opposition from farmers.[60] For example, when implementing its nutrient-based subsidy in 2010, India's government partially decontrolled the prices of other fertilizers but left the subsidy for urea untouched. A sharp increase in the price of urea may be politically unfeasible, but a gradual reduction in the subsidy may be possible. India used this strategy successfully to phase out its diesel fuel subsidy. Switching to non-distortionary direct cash transfers for fertilizer subsidies is another option that South Asian governments have tried or are seriously considering.[61] However, successful implementation of a direct cash transfer system requires significant investments in data and monitoring systems.[62]

REFORM ENERGY POLICIES. Electricity use for irrigation is heavily subsidized or free in India, resulting in wasteful use of energy and water and high GHG emissions. Proposals to raise electricity rates are deemed politically unviable, but governments are exploring other ways to reduce use of grid electricity for irrigation. One such scheme in India, PM-KUSUM, aims to convert a million irrigation pumps to solar power, which will decarbonize groundwater irrigation. Connecting this solar power production to the electricity grid will allow farmers to use the power to irrigate their land and/or sell it to the electricity company at a pre-fixed price. This option may create incentive for farmers to use less water and energy for crops.[63] Pilot tests of business models for implementing direct benefit transfers to subsidize electricity for irrigation — which in essence pay farmers for saving energy — are also being tried in some Indian states, including Gujarat, Karnataka, Punjab, and Rajasthan.[64] India's government proposed switching to a direct transfer subsidy in the 2021 Electricity Amendments Bill, but dropped this option in response to strong objections from farmers and other stakeholders. The rollback suggests that such reforms should be undertaken by the state governments, which are responsible for distribution and supply of power to rural and urban consumers.

REFORM AGRICULTURAL SUPPORT. Guaranteed prices for two main cereals (rice and wheat) in much of South Asia have led to overproduction of these crops and disincentivized crop diversification. Given high methane emissions from paddy, rolling back the output price subsidy could help reduce GHG emissions from agriculture. However, resistance from the special interest groups that benefit from these policies may hamper reform efforts. In one reform effort, the state government in Haryana, India, is offering a cash incentive of INR 7,000 (US$92) per acre to farmers who replace paddy with crops like maize and pulses in water-scarce areas, but the impact of this incentive scheme has not yet been assessed.

Climate change presents immediate and long-term challenges for South Asia. Glacier melt, sea-level rise, groundwater depletion, extreme weather events, and frequency of natural hazards are evident and likely to worsen in the coming decades. Solutions to these challenges are complex, encompassing human activities in all sectors of the economy at the country, regional, and global levels. But policy actions within agriculture can be taken now to support adaptation and mitigation. Governments in the region can undertake several critical actions, including scaling up climate-smart technologies, crop diversification, and elimination of policy incentives that encourage unsustainable use of water, fossil fuels, fertilizers, and land resources. The political economy of such reforms will be difficult, as is evident from recent farmer protests in India, but the cost of inaction for future generations would be too high.

EAST AND SOUTHEAST ASIA

KEVIN CHEN AND YUE ZHAN

Kevin Chen is a senior research fellow, International Food Policy Research Institute (IFPRI), Beijing, and a chair professor, Zhejiang University, Hangzhou, China. **Yue Zhan** is a research analyst, IFPRI, Beijing.

The East and Southeast Asia region is among the most vulnerable to natural hazards and climate change, due to the high concentration of population and economic activity along its extensive coastlines and its heavy reliance on agriculture and other natural resources to providing livelihoods and income.[1] Economies in the region are slowly diversifying away from agriculture and other natural resource-based activities, but most countries still depend substantially on these sectors. Agriculture is the largest economic sector for Cambodia and Lao PDR,[2] and remains one of the top five sectors across the region (with the exceptions of Singapore and Brunei). Thus, East and Southeast Asian countries, and especially the agriculture-dependent communities within them, will suffer climate change impacts, including threats to the region's food security. Moreover, because the region is a major contributor to global grain exports, most notably rice, the impacts of climate change in East and Southeast Asia will also affect food security globally.

CLIMATE CHANGE IMPACTS AND CHALLENGES

On the production side, East and Southeast Asia are already experiencing growing intensity and frequency of extreme weather events, including droughts, floods, typhoons, and heatwaves, as well as sea level rise, that are causing significant economic, environmental, and social damage.[3] The Global Climate Risk Index ranked Myanmar, Philippines, Bangladesh, and Thailand among the top 10 countries globally most affected by extreme weather events (Figure 1), and overuse of chemical inputs and water also endanger long-term sustainability of agriculture in the region. These impacts threaten the region's role as a major producer of grain and industrial crops (of which the most important are rubber, palm oil, and sugarcane), with many countries projected to experience a relative decline in agricultural productivity as climate change progresses (compared to a scenario with no climate change).[4]

On the consumption side, rising household incomes and urbanization have shifted food preferences from cereals toward animal products. Animal-source foods require more resource-intensive production and produce more greenhouse gas (GHG) emissions than plant-based products. Beef generates the most GHG emissions per kilocalorie consumed and has the largest water footprint. However, high average consumption of rice in the region also leads to significant GHG emissions and water use per capita from rice production.[5] This has important implications for land and water use, as well as for climate change.

Projections from IFPRI's IMPACT model indicate that with climate change, average calorie consumption in the region in 2050, compared with a no-climate change scenario, will be about 3 percent lower and cereal consumption 5 percent lower. In a no-climate change scenario, the number of people at risk of hunger in the region is projected to fall from 268 to 92 million between 2000 and 2050. Climate change will reduce this progress, and the number of people at risk of hunger will only fall to 105 million in 2050 under a middle-of-the-road scenario, leaving 13 million more people at risk of hunger than the no-climate change scenario.

The region's agriculture sector will have to adapt and adjust to climate change impacts over the medium and longer term; it will also have to contribute to climate change mitigation. Globally, agriculture currently generates 19-29 percent of total greenhouse gas (GHG) emissions.[6] Emissions from agriculture

FIGURE 1 Map of the Global Climate Risk Index 2000 – 2019

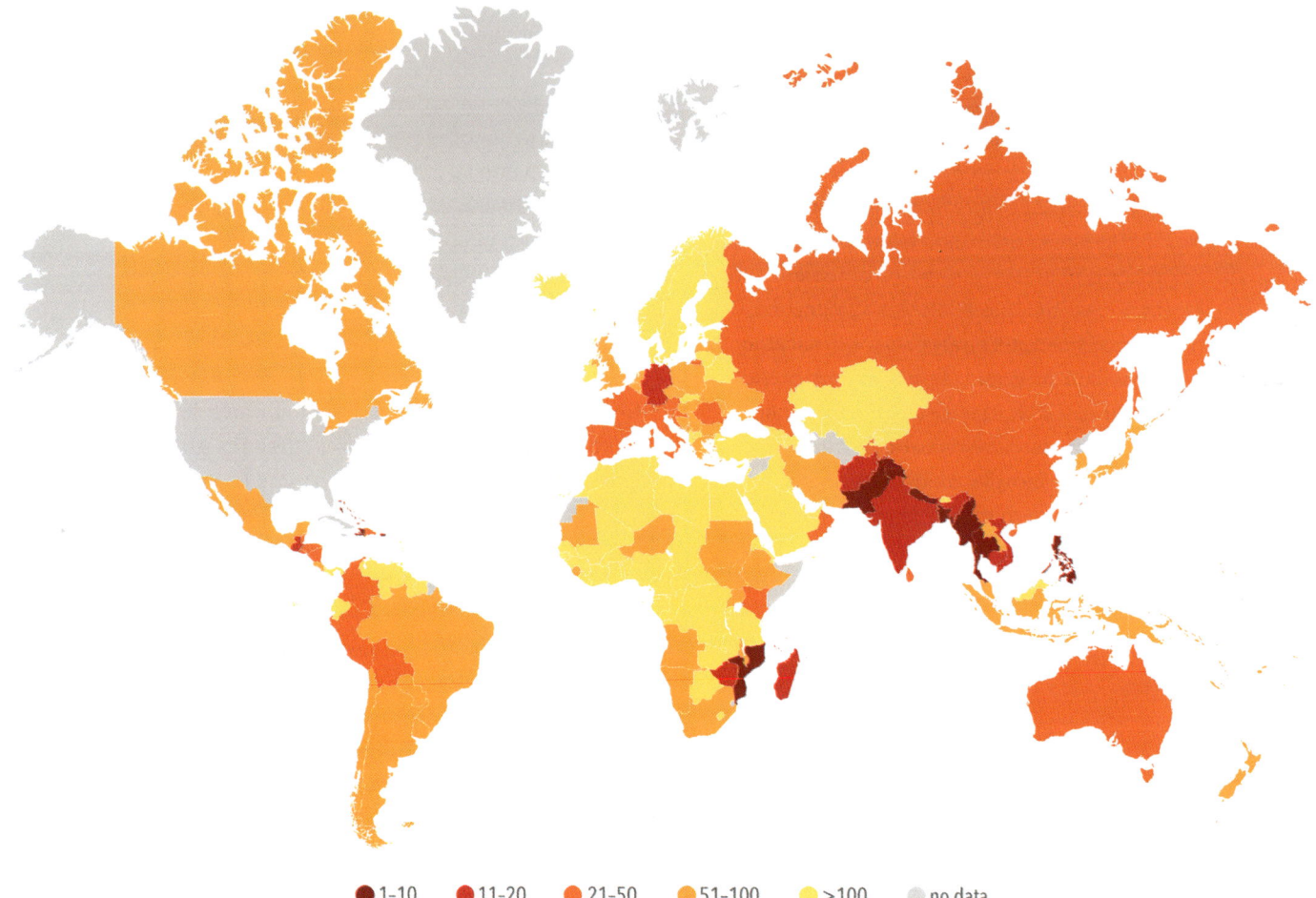

● 1–10 ● 11–20 ● 21–50 ● 51–100 ● >100 ● no data

Source: Germanwatch, *Global Climate Risk Index 2021* (Bonn: Germanwatch, 2021). www.germanwatch.org/en/cri.

have been highest in Asia over the past 30 years. In 2019, the equivalent of 4.1 billion metric tons of CO_2 was produced on agricultural lands in Asia. Among the regions, South Asia and East Asia have the highest agricultural production-related emissions, followed by Southeast Asia.[7] These emissions are mainly from deforestation, livestock production, and soil and nutrient management.[8] Therefore, allied to the pressure created by climate change is the need to better manage the region's natural resources and find ways to produce food sustainably. Curbing emissions in the region is critical to advancing the global climate change agenda, and the agriculture sector must play a role.

POLICIES AND RESPONSES

REGIONAL AND NATIONAL STRATEGIES

Countries in the region have made substantial progress in recognizing climate change issues in their planning processes, either by explicitly mentioning these issues or identifying means to address them in their development plans, as identified in various official communications, including the National Communications, Nationally Determined Contributions, Adaptation Plans, and Adaptation Strategies. The People's Republic of China (PRC), Japan, and the Republic of Korea have announced the ambitious target of carbon neutrality by mid-century. Common climate change priorities for agriculture and natural resources are found in East and Southeast

Asian countries: increasing forest cover is a specified target for Cambodia, Lao PDR, Malaysia, and Myanmar. Sustainable forest management is among commonly prioritized activities in the region, which is a major deforestation hotspot due to logging and clear-cutting for food production, cash crops, and other agriculture.[9] The promotion of climate-resilient agriculture and farming systems (irrigation, cultivation, crop rotation, livestock/pasture management) are also priorities.[10] At the regional level, under the broad umbrella of ASEAN regional cooperation on climate change adaptation, several activities have been implemented, including capacity building workshops, information exchange workshops, forums, and implementation of specific projects. An expansion of regional cooperation from ASEAN countries to countries outside ASEAN will provide opportunities to enhance adaptation actions in the region.

CLIMATE CHANGE ADAPTATION PRACTICES

The region has a demonstrated commitment to climate-smart agriculture (CSA) to support adaptation.[11] The most common adaptation techniques in the region are changes in cropping patterns and cropping calendars and improved farm management, for example, a switch from rice-rice to rice–maize cropping patterns and use of drought-resistant and heat-resilient varieties.[12] Climate-smart technology is also an active area of research and implementation in the region. Countries that have made progress in climate change risk assessments, adaptation planning, and CSA are well placed to share this expertise. For example, the International Rice Research Institute (IRRI) has developed the Rice Crop Manager (RCM), a web-based smartphone application currently deployed in many parts of the Philippines and Viet Nam.[13] The RCM app enables agricultural extension staff to support sustainable productivity gains for poor rice farmers through cost-effective crop management.

Starting in 2015, the CGIAR Research Program on Climate Change, Agriculture and Food Security (CCAFS) began scaling up CSA practices using the climate-smart village approach in Southeast Asia, bringing tailored CSA interventions to target communities through continuous participatory research and evaluation. With the efforts of various partners, the initial 7 climate-smart villages have grown to 90. Based on this experience in Cambodia, Lao PDR, Myanmar, the Philippines, and Viet Nam, the pathways for scaling up can be followed through knowledge transfer, policy incidence, and commercialization.[14]

CLIMATE CHANGE MITIGATION PRACTICES

Adaptation can make a substantial contribution to reducing the damages expected from climate change, but would not be sufficient on its own. Mitigation practices can generate significant synergies with adaptation by improving ecosystem functioning, increasing water availability, and improving resilience to drought, pests, and other threats. A range of existing technologies and management actions could be deployed now, including improved crop and grazing land management; site-specific nutrient management based on crop needs; restoration of degraded lands and soils; livestock management; manure and bio-solid management; and sustainable and bio-energy use.[15] Progress in implementing these practices in the region has been slow and efforts are needed to accelerate their adoption. Projections for China suggest that the combination of improving agricultural technologies, reducing food loss and waste, and shifting dietary patterns could reduce GHG emissions from the food system by 47 percent as of 2060.[16] Therefore, to address climate change, a comprehensive emissions reduction strategy should be formulated for the whole agrifood system – from farm to consumer.

FINANCING CLIMATE CHANGE MITIGATION AND ADAPTATION

New ways of financing climate change adaptation and mitigation must be found. Estimates of financing needs for the region vary widely, reflecting the uncertainties associated with potential climate change scenarios and their likely impact. However, emerging estimates of the additional investment needed indicate a financial gap of hundreds of billions of dollars every year for several decades to come (see Chapters 2 and 5).[17]

Financial support to smallholder farmers is required to address the multiyear gap between investments and financial benefits if they are to adopt climate-smart practices. In particular, transition financing involving the use of loans wherein repayment terms are pushed back to accommodate multiyear return gaps should be considered (see Chapter 9). Access to transition funds would help farmers to cover upfront costs for seed and new equipment that farmers face when adopting climate-smart practices.

Where upfront costs are not entailed, farmers can benefit from payment schemes for environmental services. Governments can create incentives through public payments (or charges) to private resource users for the enhancement (or degradation) of ecosystem services.[18] For example, payment for environmental services (PES) can be designed to compensate resource managers for the maintenance or restoration of environmental services such as clean water. China has conducted numerous national, provincial, and local experiments over the past decade with different forms of eco-compensation, focused primarily on water resources management.

LOOKING FORWARD

With climate change expected to put downward pressure on yield growth of many crops, the role of agricultural research and development (R&D) and innovation systems will become increasingly important for agricultural development. For East and Southeast Asian countries, compared with other countries at a similar level of development, there appears to be significant scope to increase investments in R&D and innovation systems to help safeguard future levels of productivity growth and to mitigate some of the expected effects of climate change.[19]

For some countries, reform of distorting trade policies and agricultural input and output price support policies is needed (see Chapters 2 and 3). For example, in the Philippines, support policies for rice work against incentives for adaptation and can increase producers' exposure to climate risks.[20] Similarly, for the region as a whole, current trade distortions and war-fueled reemergence of food inflation have the potential to magnify the global price effects from climate change.

LATIN AMERICA AND THE CARIBBEAN

EUGENIO DÍAZ-BONILLA, CAROLINA NAVARRETE-FRIAS, AND VALERIA PIÑEIRO

Eugenio Díaz-Bonilla is a special advisor, Inter-American Institute for Cooperation on Agriculture, and senior visiting research fellow, International Food Policy Research Institute (IFPRI), Washington, DC. **Carolina Navarrete-Frias** is a senior advisor on Environmental Policy and Biodiversity, Alliance of Bioversity and CIAT, Cali. **Valeria Piñeiro** is a senior research coordinator, Markets, Trade, and Institutions Division, IFPRI, Washington, DC.

Food systems in Latin America and the Caribbean (LAC) play a vital role in the region's economies, in global food security, and in the global response to climate change. They operate, however, within a region that has suffered from economic stagnation in recent years. While LAC enjoyed annual growth of 2.1 percent in per capita incomes between 2000 and 2011 as the commodity cycle was trending upward (even with the 2008–2009 global crisis), when the cycle turned downward, per capita growth fell to a meager 0.2 percent per year until 2019. The region's economies experienced the worst eight years since the 1980s, and the stagnation affected political stability in LAC.[1] Then COVID-19 hit. Despite having only about 8 percent of the world's population, the region recorded more than 30 percent of global COVID-related deaths. Several factors have made LAC particularly vulnerable to the pandemic, including its high level of urbanization, significant income inequality (which also limits access to high-quality health services), informality of labor markets, prevalence of obesity, and the ongoing economic stagnation.[2]

In 2020, due to the pandemic, LAC experienced the deepest economic recession among developing regions, and recovery is expected to be slow (Figure 1). Governments in LAC countries reacted to the pandemic with lockdowns and various health initiatives, while also increasing economic and social interventions to mitigate the negative impacts of lost income and employment. Most countries have suffered significant health and economic setbacks, despite implementing diverse policies in regard to movement restrictions and the allocation of public funding.[3] Human capital has been affected by the interruption in education; nutritional problems associated with insufficient and less healthy diets; and the strain on job skills and abilities due to long unemployment periods. Currently, much is still unknown about the dynamics of the pandemic, both in LAC and globally, but COVID-19 appears likely to become an endemic condition with long-term implications for societies, economies, and human health.

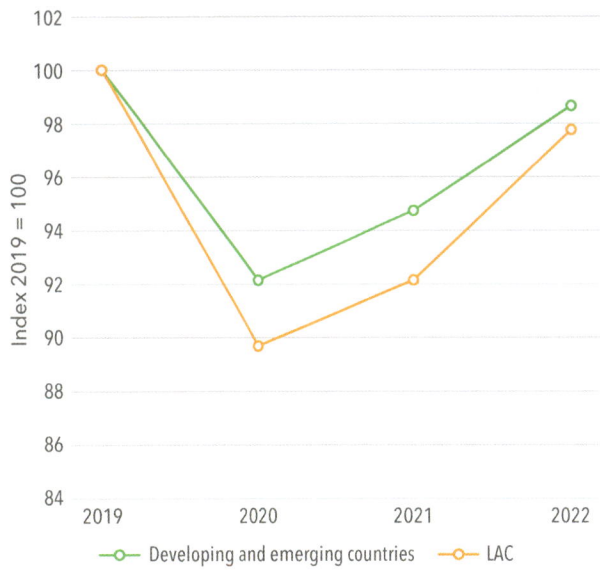

FIGURE 1 GDP per capita

Source: Based on data from International Monetary Fund (IMF), *World Economic Outlook*, October 2021.

In the very short term, it is necessary to increase vaccination coverage, reinforce prevention policies, and strengthen access to and provision of healthcare in LAC. As of mid-October 2021, large disparities in vaccination rates existed in the region (Figure 2). Vaccination coverage in some countries is above the average for developed nations, while in others, vaccination rates are below those of lower-income developing countries.

For the longer term, LAC countries, like all developing countries, require integrated recovery programs that are coordinated at the highest governmental level. These are needed to alleviate the economic and human costs of the pandemic and to contend with preexisting economic and social problems, while also tackling the current and future challenges of climate change.

These programs will require strong support from the international community, including debt restructuring and write-offs, and from multilateral development banks operating in the region (see Chapter 5). The Special Drawing Rights (SDRs) recently allocated by the International Monetary Fund will need to be used more creatively to set up a trust fund that guarantees the issuance of bonds to finance pandemic recovery programs and a climate-positive transformation of economies, including food systems.[4]

LAC'S ROLE IN GLOBAL FOOD SECURITY AND SUSTAINABILITY

Agrifood systems in the LAC region are crucial to food security at both the national and global levels. LAC is the world's largest *net* food- and agriculture-exporting region (exceeding the combined value of the United States, Canada, Australia, and New Zealand), making it pivotal to global food security. The region is also critical to global environmental sustainability and biodiversity: it holds about 36 percent of the CO_2eq stock kept in all forests worldwide and more than a third of the total global volume of renewable water resources. Of the world's top ten most biodiverse countries,

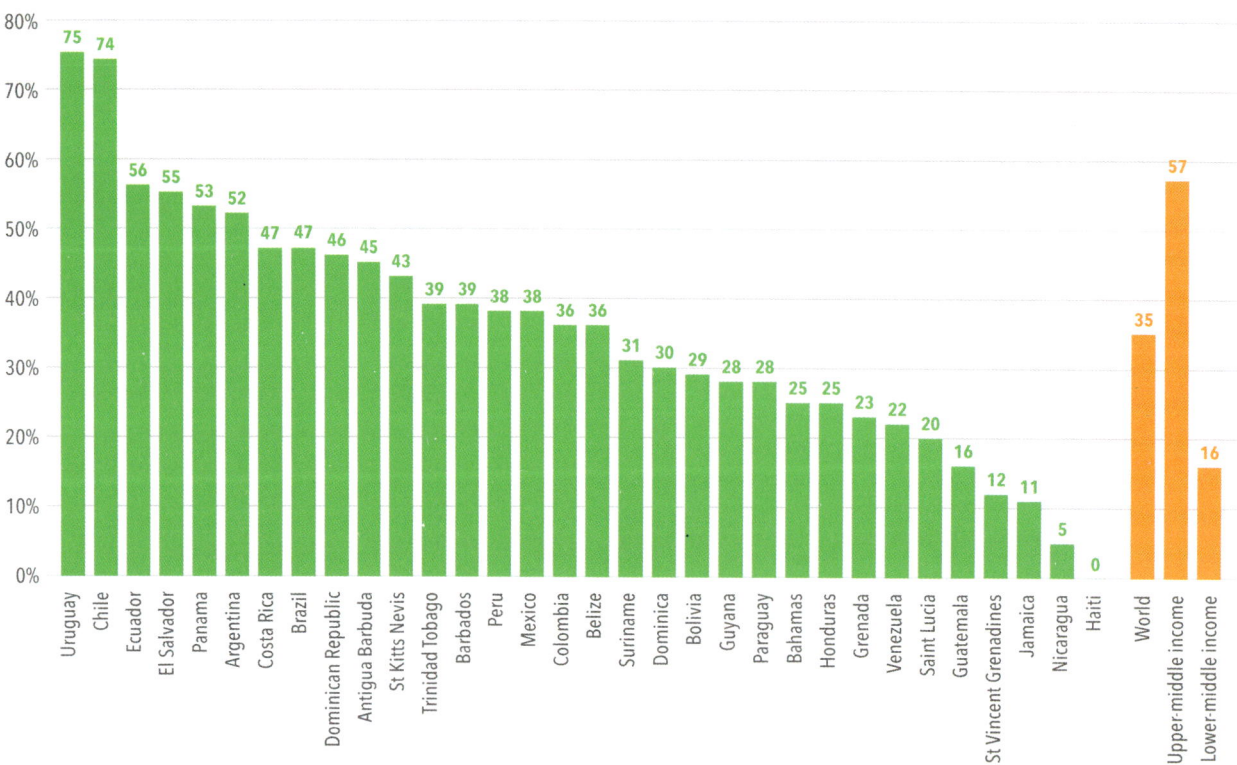

FIGURE 2 Population fully vaccinated in LAC countries, compared to global rates of vaccination

Source: Compiled from Our World in Data (https://ourworldindata.org/covid-vaccinations).

six are located in the region, including the two most biodiverse countries.[5] The LAC region also contains about 23 percent of the world's forested areas. Environmental and biodiversity functions are closely interlinked with forest cover and ecosystem integrity. However, deforestation continues to increase, with the region's forested area declining from 1.07 billion hectares in the early 1990s to 932 million hectares in 2020.[6] Deforestation has led to increases in greenhouse gas emissions, and reduced positive contributions to the oxygen and water cycle, carbon sequestration, and the preservation of biodiversity.[7]

Several international agreements and partnerships aim to reduce deforestation and forest degradation. Policy implementation research could inform choices and guide the development of solutions for the underlying problems of forest governance. However, the success of these efforts depends on the integration of lessons learned from implementing previous international conventions, forest laws, and other environmental policies.[8] The Glasgow Leaders' Declaration on Forests and Land Use [9] represents a new opportunity to reverse current trends; this declaration was signed by 141 countries, including Brazil and most LAC countries with large forested areas. Changes in overall incentives (such as through adequate pricing of carbon) and supportive financing will also be necessary (see Chapter 5).

Two key priorities of the 2030 Agenda for Sustainable Development are preserving the planet and providing food that is sufficient, safe, affordable, and nutritious. These priorities are not easily reconciled, however.[10] In addition to the LAC region's dual roles for global food security and environmental sustainability, agrifood systems are very important for individual countries' economic development, poverty alleviation, food security, population retention, and crime reduction, particularly in rural areas. Achieving these multiple goals will require major efforts to scale up innovations and investments in the intensification of sustainable agriculture and to transform food systems to make them carbon neutral.

OVERVIEW OF INNOVATIONS, POLICY OPTIONS, AND OBSTACLES[11]

LAC countries have long been experimenting with different innovations and policy approaches, in many cases with support from CGIAR centers. Here we focus on the most recent ones.

CLIMATE-SMART AGRICULTURE TECHNOLOGIES. The most promising climate-smart agriculture (CSA) technology clusters in the region include intercropping, green manures/cover crops, organic inputs, silvopasture, conservation agriculture, mulching, improved pastures, stress-tolerant crops, and adequate management of grazing, fertilizer, and water use. Local strategies are key to tailoring CSA innovations to farmers' needs, including preserving or improving genetic diversity to strengthen resilience and applying alternate wetting and drying (AWD) for water-use efficiency and emissions reductions. Those innovations generate co-benefits and "multiple wins" that are key to strengthening resilience and attaining global goals such as the Sustainable Development Goals and the Paris Agreement on mitigation and adaptation. As these technology clusters are scaled up, context-specific opportunities and limitations within LAC must be considered.

WATER QUALITY AND EFFICIENCY. The development of water-smart strategies requires addressing information gaps on water quality and sources of contamination, as well as implementing policies to address sources of water pollution in the long term. Innovations in water-efficient irrigation infrastructure, drip irrigation, and supportive policies are crucial, given that agriculture uses 70 percent of water in LAC and many farmers still depend on rainfed systems.

RESTORATION OF DEGRADED LANDSCAPES. LAC has large degraded landscape areas with low agricultural productivity. Restoring these degraded landscapes presents an opportunity to enhance habitat connectivity, protect biodiversity, and increase ecosystem services, while also fostering community-based activities and supporting livelihoods through sustainable business models.

SUSTAINABLE LIVESTOCK SYSTEMS. LAC produces approximately 26 percent of the world's bovine meat and exports nearly 20 percent of its own production, representing about 28 percent of global bovine meat exports (average 2015–2019).[12] The livestock sector is an important source of employment in the region, particularly for small farmers,[13] but it poses environmental challenges. To address these issues, the LAC region is pursuing various innovations to increase profitability, reduce emissions, and limit land degradation and land use conversion, including silvopastoral systems of varying intensity, site-specific agriculture for pastures (which adjusts practices and inputs to specific locations), and plot division and rotational grazing.

BIODIVERSITY IN AGRICULTURAL LANDSCAPES. Agroecological knowledge must be incorporated into agricultural practices to reduce pests and disease, mitigate climate change, promote crop pollination, and restore soils, landscapes, and ecosystem services. These goals align with the priorities set by the UN's Post-2020 Global Biodiversity Framework. Agroecological approaches can be combined with other production applications to advance the technology frontier and offer flexibility in the selection of the appropriate mix of technologies.

ANCESTRAL PRACTICES. LAC benefits from ancestral knowledge that could be combined with CSA practices to strengthen resilience to climate change. This traditional knowledge includes bio-indicators for predicting climate variability, wild genetic diversity (such as potato landraces in Peru), and platforms, terraces, canals, and ponds that can conserve soil and retain water. Furthermore, the strong social base of local communities can support governance mechanisms to reduce land use change.

INSTITUTIONAL INNOVATIONS. The promotion and scaling up of technological solutions will require strong and sound institutions, cross-sector coordination, and appropriate policies. The region's Local Technical Agroclimatic Committees (LTAC) are one example of a relevant institutional innovation in Latin America. These committees, which are set up by governments in close collaboration with local actors, help smallholder farmers to prepare for climate change. LTACs develop local and regional forecasts showing potential effects on crops, and they introduce ways to reduce vulnerability and negative impacts (such as alternative dates for sowing and irrigation). These committees are now present in Chile, Colombia, El Salvador, Guatemala, Honduras, Mexico, Nicaragua, Panama, Paraguay, and Peru. LTACs can identify and promote bundles of technologies and services that are appropriate to local conditions and needs, and share these through South–South technical exchanges to strengthen the management of climate risk at a regional level.

DIGITAL AGRICULTURE FOR CLIMATE ACTION. In LAC, investments in digital agricultural technology are still low.[14] CSA strategies should make use of the digitalization of agriculture to innovate with agroclimatic services and virtual roundtables for farmers and entrepreneurs; promote knowledge-sharing, data-driven agronomy, and digital extension through the internet of things; and employ simulation modeling and peer-to-peer learning (see Chapter 12).

BARRIERS TO CSA ADOPTION AND POSSIBLE SOLUTIONS. Common barriers to CSA adoption include limited access to training and information and to credit and markets, limited availability of required inputs (such as improved seeds), and high costs of equipment (such as irrigation facilities). Furthermore, the global crisis driven by the pandemic and, more recently, the conflict in Ukraine is increasing inflation in LAC and causing key agricultural inputs, such as fertilizers, to become scarce and expensive. Measures to counter the effects of these crises include lowering trade tariffs for agricultural inputs and producing alternative inputs on site (such as compost). Other solutions include further expanding LTACs, designing country-based CSA investment plans, and promoting financial alternatives for scaling up CSA best practices. The LAC region has great potential to implement practices that increase carbon sequestration (using appropriate soil and vegetation maps), leverage carbon markets and credits, and take advantage of COP26 agreements on the operationalization of these tools.

BUILDING BACK BETTER

Countries in the LAC region have been experimenting with and scaling up a variety of innovations, in many cases with support from CGIAR centers. These innovations relate to modern crop and livestock production and management techniques, promote low-emissions agricultural practices and technologies, strengthen national research organizations to deliver more nutritious and climate-resilient varieties, use local knowledge and practices, expand climate information services, and experiment with new finance partnerships. The private sector, including small and medium enterprises and entrepreneurs, could play a bigger role in promoting the adoption of environmentally friendly technologies, and private-public partnerships could help support the availability and financing of these technologies as well as the monitoring of policy schemes with environmental objectives (for example, through certification and accountability).

LAC is still suffering from the crisis caused by the pandemic and compounded recently by the war in Ukraine, which may affect its key role for both global food security and environmental sustainability. Countries in the region need to develop and effectively implement comprehensive public programs to "build back better" from the pandemic and transform food systems, with a strong application of science, technology, and innovation, while taking advantage of ancestral knowledge. All this will require significant financing and support from international stakeholders, as well as financial and political commitment from the governments in LAC (see Chapter 5). Failure to do so will not only affect the region, but the whole world.

PROJECTIONS FROM IFPRI'S IMPACT MODEL: CLIMATE CHANGE AND FOOD SYSTEMS

POLICYMAKERS, ANALYSTS, AND CIVIL SOCIETY FACE INCREASING CHALLENGES TO reducing hunger and sustainably improving food security. Modeling alternative future scenarios and assessing their outcomes can help inform policy choices. The International Food Policy Research Institute's IMPACT model is an integrated system of linked economic, climate, water, and crop models that allows for the exploration of such scenarios.

METHODOLOGY

The IMPACT model was used to evaluate impacts of climate change on aggregate food production, food consumption (kcal per person per day), net trade of major food commodity groups, and the population at risk of hunger. At IMPACT's core is a partial equilibrium, multimarket economic model that simulates national and international agricultural markets. Links to climate, water, and crop models support the integrated study of changing environmental, biophysical, and socioeconomic trends, allowing for in-depth analysis of a variety of critical issues of interest to policymakers at national, regional, and global levels. IMPACT benefits from close interactions with scientists across CGIAR and other leading global economic modeling efforts around the world through the Agricultural Model Intercomparison and Improvement Project (AgMIP).

KEY FINDINGS FROM GLOBAL AND REGIONAL AGGREGATIONS

The following tables summarize results from the latest IMPACT projections to 2030 and 2050, for a scenario that includes the impacts of climate change and a "baseline" scenario that assumes no climate change (for comparison). These results update previous projections by showing aggregations to six regions: Central and West Asia and North Africa; Eastern and Southern Africa; Latin America and the Caribbean; South Asia; Southeast Asia; West and Central Africa; and the rest of the world. The baseline projections indicate that global food production will grow by about 60 percent over 2010 levels by 2050 in the context of climate change — 8 percentage points less than would be the case without climate change (Table 1). Production and demand are projected to grow more rapidly in developing countries, particularly in Africa, than in developed countries, due to projected growth in population and incomes. However, the long-term impacts of COVID-19 and other current geopolitical factors have not yet been incorporated in these projections. Diets are also shifting toward higher-value foods, including more fruits and vegetables, processed foods, and animal-source foods, outside of high-income countries. Meat production is projected to double in South Asia and West and Central Africa by 2030 and triple by 2050 (Table 2). Despite this growth, per capita consumption levels in developing countries will remain less than half of those in developed countries. The demand for processed foods also shows up in the growing production of oil crops; by 2050 production is expected to more than double in Southeast Asia and West and Central Africa. Production of fruit and vegetables is projected to more than double in most regions (Central and West Asia and North Africa; East and Southern Africa; and West and Central Africa) by 2050. By the same year, average dietary energy consumption is projected to increase by about 10 percent globally to more than 3,000 kcal per capita per day. However, regional differences in access to food mean that nearly 500 million people are projected to remain at risk of hunger. Globally, about 70 million more people will be at risk of hunger because of climate change, including more than 28 million in East and Southern Africa.

Notes following the tables provide additional details on the data. For more information about IMPACT, see https://www.ifpri.org/project/ifpri-impact-model.

TABLE 1A Aggregate food production (index, 2010 = 1.00)

Region	Without CC 2010	Without CC 2030	Without CC 2050	With CC 2030	With CC 2050
CENTRAL & WEST ASIA & NORTH AFRICA	1.000	1.487	1.954	1.468	1.919
Afghanistan	1.000	1.335	1.728	1.337	1.758
Algeria	1.000	1.544	2.023	1.423	1.717
Armenia	1.000	1.278	1.421	1.445	1.785
Azerbaijan	1.000	1.296	1.490	1.226	1.318
Cyprus	1.000	1.399	1.856	1.156	1.266
Egypt	1.000	1.469	1.962	1.430	1.912
Georgia	1.000	1.348	1.581	1.330	1.509
Iran	1.000	1.483	1.962	1.520	2.055
Iraq	1.000	1.770	3.155	1.750	3.100
Israel	1.000	1.695	2.688	1.658	2.509
Jordan	1.000	1.728	2.515	1.819	2.737
Kazakhstan	1.000	1.340	1.590	1.298	1.493
Kyrgyzstan	1.000	1.407	1.833	1.668	2.502
Lebanon	1.000	1.512	1.956	1.548	2.032
Libya	1.000	1.649	2.280	1.620	2.191
Morocco	1.000	1.613	2.274	1.428	1.828
Palestine	1.000	1.769	2.591	1.640	2.247
Rest of Arabia	1.000	1.933	3.133	2.037	3.450
Saudi Arabia	1.000	1.762	2.740	1.755	2.709
Sudan	1.000	1.744	2.469	1.446	1.761
Syria	1.000	1.560	2.163	1.671	2.545
Tajikistan	1.000	1.254	1.427	1.218	1.329
Tunisia	1.000	1.657	2.498	1.428	1.857
Turkey	1.000	1.401	1.601	1.441	1.701
Turkmenistan	1.000	1.410	1.681	1.373	1.616
Uzbekistan	1.000	1.281	1.491	1.266	1.455
Yemen	1.000	1.809	3.065	1.932	3.290
EAST ASIA & PACIFIC	1.000	1.293	1.563	1.281	1.543
Australia	1.000	1.356	1.662	1.273	1.480
Japan	1.000	1.239	1.518	1.302	1.690
New Zealand	1.000	1.284	1.558	1.365	1.739
South Korea (ROK)	1.000	1.246	1.434	1.252	1.436
EAST & SOUTHERN AFRICA	1.000	1.646	2.390	1.574	2.203
Botswana	1.000	1.617	2.134	1.712	2.383
Burundi	1.000	1.664	2.266	1.613	2.131
Djibouti	1.000	1.911	3.636	1.945	3.769
Eritrea	1.000	1.722	2.754	1.620	2.523
Ethiopia	1.000	1.652	2.452	1.657	2.477
Kenya	1.000	1.756	3.121	1.794	3.134
Lesotho	1.000	1.617	2.085	1.699	2.264
Madagascar	1.000	1.563	2.248	1.604	2.397
Malawi	1.000	1.598	2.118	1.301	1.429
Mozambique	1.000	1.264	1.388	1.178	1.216
Namibia	1.000	1.379	1.696	1.364	1.673
Rwanda	1.000	1.683	2.253	1.270	1.311
Somalia	1.000	2.086	3.898	1.947	3.468
South Africa	1.000	1.495	1.869	1.477	1.793
Swaziland	1.000	1.537	1.849	1.465	1.678
Tanzania, United Rep.	1.000	1.640	2.424	1.563	2.227
Uganda	1.000	1.894	3.055	1.778	2.712
Zambia	1.000	1.535	2.037	1.467	1.855
Zimbabwe	1.000	2.141	3.186	2.058	2.942
EUROPE	1.000	1.146	1.276	1.116	1.218
Albania	1.000	1.344	1.686	1.317	1.594
Austria	1.000	1.182	1.391	1.248	1.545
Baltic States	1.000	1.197	1.286	1.279	1.455
Belgium-Luxembourg	1.000	1.125	1.232	1.097	1.161
Bulgaria	1.000	1.208	1.340	1.139	1.202
Croatia	1.000	1.153	1.258	1.011	0.965
Czech Republic	1.000	1.183	1.300	1.269	1.517
Denmark	1.000	1.080	1.161	1.128	1.267
Finland	1.000	1.043	1.092	1.222	1.466
France	1.000	1.075	1.155	1.013	1.032
Germany	1.000	1.091	1.205	1.101	1.222
Greece	1.000	1.300	1.618	1.211	1.382
Hungary	1.000	1.138	1.237	1.043	1.043
Iceland	1.000	1.298	1.559	1.305	1.560
Ireland	1.000	1.048	1.054	1.086	1.143
Italy	1.000	1.179	1.360	1.077	1.132
Netherlands	1.000	1.157	1.296	1.190	1.338
Norway	1.000	1.132	1.268	1.281	1.602
Other Balkans	1.000	1.201	1.331	1.081	1.076
Poland	1.000	1.229	1.372	1.253	1.495
Portugal	1.000	1.191	1.394	1.186	1.377
Romania	1.000	1.181	1.289	1.080	1.096
Slovakia	1.000	1.155	1.247	1.183	1.324
Slovenia	1.000	1.195	1.321	1.100	1.135
Spain	1.000	1.196	1.409	1.150	1.300
Sweden	1.000	1.055	1.113	1.184	1.401
Switzerland	1.000	1.226	1.487	1.277	1.608
United Kingdom	1.000	1.032	1.038	1.053	1.107
EURplus (EUR+FSU)	1.000	1.174	1.309	1.135	1.248
FORMER SOVIET UNION	1.000	1.240	1.386	1.178	1.316
Belarus	1.000	1.229	1.305	1.069	0.998
Moldova	1.000	1.226	1.380	1.214	1.399

TABLE 1A Aggregate food production (index, 2010 = 1.00)

Region	2010	Without CC 2030	Without CC 2050	With CC 2030	With CC 2050
Russia	1.000	1.258	1.438	1.227	1.430
Ukraine	1.000	1.212	1.313	1.114	1.183
LATIN AMERICA & CARIBBEAN	1.000	1.455	1.831	1.411	1.716
Argentina	1.000	1.417	1.748	1.418	1.741
Belize	1.000	1.136	1.235	0.959	0.857
Bolivia	1.000	1.529	2.028	1.584	2.201
Brazil	1.000	1.520	1.946	1.412	1.657
Chile	1.000	1.476	1.881	1.584	2.154
Colombia	1.000	1.402	1.753	1.495	1.989
Costa Rica	1.000	1.276	1.486	1.238	1.410
Cuba	1.000	1.313	1.539	1.084	1.062
Dominican Republic	1.000	1.420	1.789	1.473	1.911
Ecuador	1.000	1.475	1.898	1.581	2.191
El Salvador	1.000	1.394	1.772	1.392	1.748
Guatemala	1.000	1.384	1.775	1.226	1.379
Guyanas	1.000	1.374	1.679	1.385	1.704
Haiti	1.000	1.508	2.069	1.518	2.090
Honduras	1.000	1.376	1.747	1.234	1.421
Jamaica	1.000	1.206	1.415	1.148	1.279
Mexico	1.000	1.351	1.619	1.311	1.534
Nicaragua	1.000	1.505	2.108	1.388	1.808
Other Caribbean	1.000	1.273	1.535	1.232	1.439
Panama	1.000	1.465	1.833	1.452	1.800
Paraguay	1.000	1.572	2.170	1.488	1.961
Peru	1.000	1.456	1.780	1.699	2.430
Uruguay	1.000	1.418	1.678	1.447	1.750
Venezuela	1.000	1.397	1.737	1.286	1.477
NORTH AMERICA	1.000	1.294	1.582	1.159	1.295
Canada	1.000	1.284	1.666	1.312	1.766
Greenland	1.000	1.042	1.045	1.042	1.045
USA	1.000	1.295	1.571	1.138	1.231
SOUTH ASIA	1.000	1.571	2.053	1.499	1.902
Bangladesh	1.000	1.406	1.632	1.330	1.458
Bhutan	1.000	1.395	1.621	1.485	1.832
India	1.000	1.627	2.164	1.549	2.003
Nepal	1.000	1.327	1.601	1.373	1.705
Pakistan	1.000	1.329	1.630	1.262	1.493
Sri Lanka	1.000	1.398	1.732	1.391	1.712
SOUTHEAST ASIA	1.000	1.238	1.367	1.249	1.396
Cambodia	1.000	1.176	1.302	1.102	1.155
China, PR	1.000	1.232	1.342	1.255	1.399
Fiji	1.000	1.158	1.341	1.090	1.176
Indonesia	1.000	1.262	1.466	1.251	1.442
Lao PDR	1.000	1.311	1.553	1.266	1.463
Malaysia	1.000	1.231	1.393	1.224	1.370
Mongolia	1.000	1.659	2.265	1.748	2.446
Myanmar	1.000	1.347	1.552	1.335	1.525
North Korea	1.000	1.173	1.293	1.222	1.399
Other Indian Ocean	1.000	1.251	1.520	1.250	1.516
Other Pacific Ocean	1.000	1.229	1.452	1.190	1.356
Other Southeast Asia	1.000	1.513	2.010	1.495	1.980
Papua New Guinea	1.000	1.432	1.879	1.464	1.970
Philippines	1.000	1.325	1.676	1.313	1.654
Solomon Islands	1.000	1.295	1.607	1.230	1.454
Thailand	1.000	1.150	1.240	1.091	1.114
Timor l'Este	1.000	1.366	1.748	1.352	1.715
Vanuatu	1.000	1.263	1.578	1.226	1.477
Viet Nam	1.000	1.247	1.360	1.193	1.236
WEST & CENTRAL AFRICA	1.000	1.631	2.367	1.562	2.179
Angola	1.000	1.385	1.632	1.247	1.318
Benin	1.000	1.694	2.465	1.569	2.129
Burkina Faso	1.000	1.627	2.369	1.526	2.104
Cameroon	1.000	1.851	2.845	1.689	2.381
Central African Rep.	1.000	1.813	2.790	1.723	2.531
Chad	1.000	1.606	2.350	1.512	2.102
Congo, Rep. of	1.000	1.644	2.295	1.562	2.071
DRC	1.000	1.718	2.492	1.678	2.389
Equatorial Guinea	1.000	1.569	2.002	1.523	1.896
Gabon	1.000	1.558	2.025	1.518	1.911
Gambia	1.000	1.763	2.763	1.593	2.273
Ghana	1.000	1.654	2.307	1.591	2.137
Guinea	1.000	1.874	2.760	1.839	2.666
Guinea-Bissau	1.000	1.606	2.309	1.603	2.242
Côte d'Ivoire	1.000	1.627	2.153	1.623	2.142
Liberia	1.000	1.690	2.410	1.672	2.372
Mali	1.000	1.716	2.879	1.608	2.544
Mauritania	1.000	2.059	3.484	1.999	3.222
Niger	1.000	1.831	3.074	1.663	2.554
Nigeria	1.000	1.585	2.305	1.530	2.159
Other Atlantic	1.000	1.409	1.785	1.561	2.062
Senegal	1.000	1.784	2.993	1.621	2.371
Sierra Leone	1.000	1.773	2.580	1.703	2.430
Togo	1.000	1.818	2.796	1.744	2.612
WORLD	1.000	1.346	1.641	1.305	1.554

See note at the end of this section for details on this data.

TABLE 1B Per capita food consumption (kcal per capita per day)

Region	Without CC 2010	Without CC 2030	Without CC 2050	With CC 2030	With CC 2050	Region	Without CC 2010	Without CC 2030	Without CC 2050	With CC 2030	With CC 2050
CENTRAL & WEST ASIA & NORTH AFRICA	2,963	3,089	3,217	3,049	3,118	Mozambique	2,149	2,425	2,627	2,348	2,480
Afghanistan	2,149	2,239	2,452	2,206	2,349	Namibia	2,261	2,515	2,810	2,449	2,666
Algeria	2,977	3,098	3,163	3,061	3,071	Rwanda	2,064	2,536	2,943	2,454	2,775
Armenia	2,713	3,005	3,228	2,973	3,138	Somalia	1,691	1,887	2,389	1,864	2,339
Azerbaijan	3,210	3,245	3,264	3,209	3,163	South Africa	2,962	3,229	3,397	3,156	3,257
Cyprus	2,554	2,489	2,551	2,466	2,499	Swaziland	2,376	2,389	2,769	2,304	2,592
Egypt	3,395	3,580	3,783	3,520	3,645	Tanzania, United Rep.	2,178	2,398	2,603	2,307	2,434
Georgia	3,091	3,554	3,855	3,521	3,763	Uganda	2,391	2,585	2,795	2,518	2,662
Iran	3,079	3,109	3,228	3,067	3,126	Zambia	1,924	2,255	2,646	2,144	2,427
Iraq	2,342	2,651	2,773	2,618	2,685	Zimbabwe	2,052	2,269	2,702	2,168	2,510
Israel	3,639	3,810	3,998	3,759	3,872	**EUROPE**	3,371	3,426	3,526	3,397	3,451
Jordan	3,181	3,469	3,872	3,429	3,767	Albania	2,921	3,066	3,229	3,043	3,160
Kazakhstan	3,111	3,494	3,536	3,457	3,441	Austria	3,711	3,767	3,874	3,733	3,790
Kyrgyzstan	2,762	3,054	3,302	3,014	3,205	Baltic States	3,253	3,384	3,444	3,357	3,380
Lebanon	3,312	3,582	3,719	3,531	3,615	Belgium-Luxembourg	3,699	3,702	3,772	3,671	3,696
Libya	3,215	3,422	3,498	3,383	3,403	Bulgaria	2,784	3,018	3,142	2,992	3,076
Morocco	3,287	3,592	3,856	3,553	3,755	Croatia	2,990	3,063	3,156	3,039	3,096
Palestine	2,271	2,807	3,231	2,758	3,087	Czech Republic	3,321	3,497	3,661	3,472	3,595
Rest of Arabia	3,151	3,294	3,359	3,256	3,265	Denmark	3,334	3,344	3,433	3,319	3,365
Saudi Arabia	2,936	3,055	3,128	3,020	3,045	Finland	3,168	3,184	3,269	3,162	3,208
Sudan	2,329	2,463	2,705	2,427	2,625	France	3,481	3,477	3,552	3,445	3,474
Syria	3,125	3,490	3,887	3,450	3,779	Germany	3,420	3,446	3,540	3,416	3,465
Tajikistan	2,154	2,416	2,577	2,388	2,497	Greece	3,544	3,584	3,730	3,554	3,653
Tunisia	3,287	3,612	3,879	3,572	3,769	Hungary	3,208	3,322	3,485	3,299	3,423
Turkey	3,596	3,661	3,698	3,620	3,596	Iceland	3,254	3,304	3,416	3,279	3,352
Turkmenistan	3,016	3,448	3,468	3,413	3,371	Ireland	3,399	3,444	3,536	3,412	3,458
Uzbekistan	2,563	2,849	3,024	2,820	2,935	Italy	3,507	3,537	3,668	3,504	3,580
Yemen	2,084	2,101	2,315	2,080	2,256	Netherlands	3,193	3,208	3,303	3,180	3,234
EAST ASIA & PACIFIC	2,906	2,994	3,076	2,962	3,004	Norway	3,411	3,404	3,437	3,376	3,364
Australia	3,133	3,213	3,291	3,183	3,221	Other Balkans	2,898	3,044	3,163	3,020	3,106
Japan	2,770	2,787	2,842	2,757	2,773	Poland	3,441	3,615	3,734	3,588	3,667
New Zealand	3,107	3,164	3,305	3,135	3,236	Portugal	3,543	3,595	3,734	3,559	3,646
South Korea (ROK)	3,139	3,347	3,429	3,310	3,346	Romania	3,447	3,640	3,765	3,608	3,686
EAST & SOUTHERN AFRICA	2,228	2,441	2,697	2,367	2,553	Slovakia	2,890	3,027	3,144	3,005	3,085
Botswana	2,173	2,435	2,581	2,372	2,453	Slovenia	3,159	3,234	3,348	3,206	3,283
Burundi	1,965	2,206	2,647	2,156	2,531	Spain	3,242	3,227	3,292	3,197	3,221
Djibouti	2,227	2,372	2,564	2,330	2,456	Sweden	3,092	3,139	3,217	3,115	3,154
Eritrea	1,742	1,849	2,211	1,819	2,123	Switzerland	3,371	3,411	3,506	3,383	3,435
Ethiopia	2,067	2,307	2,614	2,267	2,533	United Kingdom	3,372	3,433	3,543	3,401	3,465
Kenya	2,133	2,395	2,708	2,300	2,523	**EURplus (EUR+FSU)**	3,326	3,427	3,526	3,396	3,450
Lesotho	2,596	2,732	2,882	2,598	2,649	**FORMER SOVIET UNION**	3,205	3,429	3,526	3,396	3,445
Madagascar	1,996	2,118	2,371	2,046	2,214	Belarus	3,088	3,206	3,216	3,175	3,150
Malawi	2,260	2,368	2,523	2,232	2,292	Moldova	2,690	3,045	3,289	3,020	3,220

TABLE 1B Per capita food consumption (kcal per capita per day)

Region	Without CC 2010	Without CC 2030	Without CC 2050	With CC 2030	With CC 2050
Russia	3,227	3,450	3,532	3,417	3,451
Ukraine	3,201	3,434	3,581	3,400	3,499
LATIN AMERICA & CARIBBEAN	2,878	3,036	3,184	2,985	3,080
Argentina	3,171	3,327	3,426	3,297	3,354
Belize	2,760	2,820	3,009	2,779	2,920
Bolivia	2,133	2,406	2,687	2,360	2,590
Brazil	3,143	3,336	3,492	3,292	3,398
Chile	2,911	3,129	3,245	3,086	3,152
Colombia	2,645	2,804	2,957	2,759	2,868
Costa Rica	2,850	3,047	3,202	3,006	3,115
Cuba	2,913	3,124	3,339	3,071	3,227
Dominican Republic	2,411	2,616	2,790	2,577	2,707
Ecuador	2,330	2,502	2,711	2,468	2,637
El Salvador	2,561	2,657	2,822	2,596	2,706
Guatemala	2,375	2,508	2,794	2,447	2,675
Guyanas	2,386	2,604	2,768	2,566	2,681
Haiti	1,858	2,087	2,382	2,051	2,303
Honduras	2,502	2,672	2,976	2,614	2,863
Jamaica	2,721	2,859	3,258	2,823	3,161
Mexico	3,040	3,133	3,240	3,053	3,095
Nicaragua	2,269	2,488	2,800	2,431	2,685
Other Caribbean	2,778	3,017	3,261	2,983	3,172
Panama	2,427	2,721	2,831	2,676	2,738
Paraguay	2,667	2,826	3,052	2,772	2,945
Peru	2,472	2,752	2,886	2,700	2,781
Uruguay	2,939	3,138	3,274	3,095	3,177
Venezuela	2,536	2,626	2,763	2,579	2,669
NORTH AMERICA	3,713	3,724	3,733	3,685	3,647
Canada	3,465	3,495	3,581	3,457	3,493
Greenland	2,707	2,707	2,707	2,707	2,707
United States	3,741	3,750	3,751	3,711	3,665
SOUTH ASIA	2,365	2,680	2,976	2,634	2,863
Bangladesh	2,426	2,714	2,911	2,653	2,781
Bhutan	2,166	2,455	2,536	2,363	2,369
India	2,354	2,697	2,997	2,651	2,883
Nepal	2,425	2,695	3,186	2,625	3,028
Pakistan	2,379	2,540	2,862	2,514	2,787
Sri Lanka	2,396	2,719	2,909	2,658	2,775
SOUTHEAST ASIA	2,875	3,335	3,463	3,279	3,340
Cambodia	2,348	2,515	2,614	2,463	2,508
China, PR	3,047	3,616	3,745	3,562	3,624
Fiji	2,819	3,107	3,752	3,060	3,615
Indonesia	2,540	2,988	3,279	2,904	3,101
Lao PDR	2,267	2,417	2,483	2,364	2,378
Malaysia	2,838	3,173	3,462	3,143	3,383
Mongolia	2,353	3,275	3,516	3,235	3,408
Myanmar	2,168	2,468	2,586	2,415	2,479
North Korea (DPR)	2,147	2,094	2,067	2,052	1,988
Other Indian Ocean	1,558	1,798	2,016	1,772	1,947
Other Pacific Ocean	2,184	2,431	2,751	2,398	2,665
Other Southeast Asia	2,990	3,013	2,967	2,967	2,872
Papua New Guinea	2,298	2,789	3,151	2,750	3,063
Philippines	2,503	2,641	2,777	2,602	2,690
Solomon Islands	2,487	2,770	3,088	2,733	2,998
Thailand	2,742	3,012	3,183	2,974	3,103
Timor l'Este	2,085	2,437	2,643	2,386	2,539
Vanuatu	2,824	3,094	3,597	3,039	3,467
Viet Nam	2,512	2,709	2,827	2,652	2,708
WEST & CENTRAL AFRICA	2,479	2,730	2,996	2,656	2,846
Angola	2,464	2,662	2,678	2,586	2,529
Benin	2,449	2,565	2,926	2,470	2,741
Burkina Faso	2,523	2,693	2,896	2,621	2,765
Cameroon	2,280	2,534	2,759	2,456	2,605
Central African Rep.	2,003	2,322	2,815	2,255	2,661
Chad	1,998	2,088	2,377	2,039	2,265
Congo, Rep. of	2,176	2,496	2,746	2,448	2,628
DRC	1,948	2,398	3,003	2,330	2,853
Equatorial Guinea	2,707	2,671	2,782	2,601	2,635
Gabon	2,695	2,847	3,050	2,773	2,886
Gambia	2,467	2,676	2,936	2,612	2,794
Ghana	2,763	3,134	3,249	3,044	3,071
Guinea	2,399	2,995	3,280	2,919	3,122
Guinea-Bissau	2,214	2,362	2,586	2,287	2,433
Côte d"Ivoire	2,734	3,133	3,256	3,048	3,091
Liberia	2,231	2,504	2,715	2,444	2,585
Mali	2,583	2,660	2,864	2,598	2,741
Mauritania	2,622	2,949	3,302	2,898	3,160
Niger	2,425	2,598	2,937	2,548	2,831
Nigeria	2,751	2,942	3,135	2,862	2,978
Other Atlantic	1,895	2,265	2,618	2,231	2,535
Senegal	2,229	2,384	2,590	2,305	2,426
Sierra Leone	2,112	2,310	2,431	2,250	2,307
Togo	2,240	2,480	2,669	2,382	2,484
WORLD	2,796	3,034	3,192	2,983	3,078

TABLE 1C Hunger (millions of people at risk)

Region	Without CC 2010	Without CC 2030	Without CC 2050	With CC 2030	With CC 2050
CENTRAL & WEST ASIA & NORTH AFRICA	54.6	60.4	55.2	63.9	63.7
Afghanistan	7.0	9.4	7.9	10.1	10.4
Algeria	1.9	1.9	1.9	2.0	2.2
Armenia	0.2	0.1	0.1	0.1	0.1
Azerbaijan	0.2	0.3	0.3	0.3	0.3
Cyprus*	0.0	0.0	0.0	0.0	0.0
Egypt	1.6	2.2	2.5	2.2	2.5
Georgia	0.4	0.3	0.2	0.3	0.2
Iran	4.7	5.2	4.4	5.7	5.3
Iraq	7.8	7.5	8.5	7.9	9.6
Israel	0.1	0.1	0.1	0.1	0.1
Jordan	0.2	0.2	0.3	0.2	0.3
Kazakhstan	0.6	0.4	0.5	0.4	0.5
Kyrgyzstan	0.4	0.2	0.2	0.2	0.2
Lebanon	0.1	0.1	0.1	0.1	0.1
Libya	0.2	0.2	0.2	0.2	0.2
Morocco	1.7	1.9	2.0	1.9	2.0
Rest of Arabia	0.2	0.3	0.4	0.3	0.4
Saudi Arabia	1.3	1.4	1.5	1.5	1.8
Sudan	11.4	12.8	9.2	13.8	11.1
Syria	0.8	0.7	0.8	0.7	0.8
Tajikistan	2.9	2.1	1.5	2.2	1.8
Tunisia	0.3	0.2	0.3	0.2	0.3
Turkey	1.8	2.2	2.4	2.2	2.4
Turkmenistan	0.1	0.1	0.2	0.1	0.2
Uzbekistan	2.4	0.8	0.8	0.9	0.8
Yemen	6.1	9.9	8.9	10.3	10.1
EAST ASIA & PACIFIC	3.2	2.7	2.0	3.1	2.7
Australia	0.3	0.3	0.3	0.3	0.3
Japan	2.3	2.0	1.2	2.3	1.9
New Zealand	0.1	0.1	0.0	0.1	0.1
South Korea	0.6	0.4	0.4	0.4	0.4
EAST & SOUTHERN AFRICA	115.9	118.4	91.3	133.9	119.5
Botswana	0.6	0.4	0.3	0.5	0.5
Burundi	3.3	3.2	0.9	3.5	1.5
Djibouti	0.2	0.2	0.2	0.2	0.2
Eritrea	3.3	4.6	3.5	4.8	4.1
Ethiopia	32.7	32.3	22.5	34.7	26.5
Kenya	10.2	8.9	5.0	10.8	8.3
Lesotho	0.2	0.2	0.1	0.2	0.2
Madagascar	6.7	8.8	7.5	9.9	10.1
Malawi	3.3	4.7	5.3	6.0	8.3
Mozambique	7.6	6.7	5.8	7.7	7.5

Region	Without CC 2010	Without CC 2030	Without CC 2050	With CC 2030	With CC 2050
Namibia	0.8	0.5	0.2	0.6	0.4
Rwanda	4.0	2.9	2.4	3.3	2.7
Somalia	5.3	5.3	1.3	5.5	1.7
South Africa	1.9	1.5	1.6	1.5	1.6
Swaziland	0.3	0.3	0.1	0.4	0.2
Tanzania, United Rep.	15.6	17.8	17.8	20.5	23.3
Uganda	8.5	10.4	11.3	11.8	14.0
Zambia	6.8	6.4	4.2	7.7	6.4
Zimbabwe	4.5	3.3	1.4	3.9	2.1
EUROPE	7.3	7.0	6.1	7.2	6.9
Albania*	0.0	0.0	0.0	0.0	0.0
Austria	0.1	0.1	0.1	0.1	0.1
Baltic States	0.1	0.1	0.0	0.1	0.0
Belgium-Luxembourg	0.1	0.1	0.1	0.1	0.1
Bulgaria	0.1	0.1	0.1	0.1	0.1
Croatia	0.1	0.0	0.0	0.0	0.0
Czech Republic	0.1	0.1	0.1	0.1	0.1
Denmark	0.1	0.1	0.1	0.1	0.1
Finland	0.1	0.1	0.1	0.1	0.1
France	0.9	1.0	0.6	1.0	1.1
Germany	1.2	1.1	1.0	1.1	1.1
Greece	0.1	0.1	0.1	0.1	0.1
Hungary	0.2	0.1	0.1	0.1	0.1
Iceland*	0.0	0.0	0.0	0.0	0.0
Ireland	0.1	0.1	0.1	0.1	0.1
Italy	0.6	0.6	0.5	0.6	0.5
Netherlands	0.2	0.3	0.2	0.3	0.2
Norway	0.1	0.1	0.1	0.1	0.1
Other Balkans	0.5	0.3	0.2	0.3	0.2
Poland	0.5	0.3	0.3	0.3	0.3
Portugal	0.1	0.1	0.1	0.1	0.1
Romania	0.3	0.2	0.1	0.2	0.1
Slovakia*	0.0	0.0	0.0	0.0	0.0
Slovenia*	0.0	0.0	0.0	0.0	0.0
Spain	0.7	0.8	0.8	0.9	0.9
Sweden	0.1	0.1	0.1	0.2	0.2
Switzerland	0.1	0.1	0.1	0.1	0.1
United Kingdom	0.9	1.0	1.0	1.0	1.0
EURplus (EUR+FSU)	9.9	8.7	7.7	9.0	8.5
FORMER SOVIET UNION	2.5	1.7	1.5	1.7	1.6
Belarus	0.1	0.1	0.1	0.1	0.1
Moldova*	0.0	0.0	0.0	0.0	0.0
Russia	1.8	1.2	1.1	1.2	1.2

TABLE 1C Hunger (millions of people at risk)

Region	Without CC 2010	Without CC 2030	Without CC 2050	With CC 2030	With CC 2050
Ukraine	0.6	0.4	0.3	0.4	0.3
LATIN AMERICA & CARIBBEAN	39.5	32.1	24.0	35.8	28.8
Argentina	0.7	0.6	0.7	0.6	0.7
Belize	0.0	0.0	0.0	0.0	0.0
Bolivia	2.7	1.8	0.8	2.0	1.2
Brazil	3.7	3.2	3.1	3.2	3.1
Chile	0.4	0.3	0.3	0.3	0.3
Colombia	5.0	3.9	2.7	4.5	3.6
Costa Rica	0.2	0.2	0.2	0.2	0.2
Cuba	0.5	0.3	0.1	0.3	0.2
Dominican Republic	1.8	1.3	0.9	1.5	1.2
Ecuador	2.2	1.7	0.9	1.9	1.2
El Salvador	0.7	0.4	0.4	0.6	0.4
Guatemala	2.1	2.2	1.3	2.5	1.7
Guyanas	0.2	0.1	0.1	0.1	0.1
Haiti	5.0	4.4	3.1	4.6	3.5
Honduras	1.1	1.0	0.7	1.1	0.8
Jamaica	0.2	0.1	0.1	0.1	0.1
Mexico	5.3	5.4	5.3	6.1	6.2
Nicaragua	1.2	1.0	0.6	1.1	0.7
Other Caribbean*	0.1	0.1	0.0	0.1	0.1
Panama	0.5	0.3	0.2	0.3	0.3
Paraguay	0.8	0.6	0.4	0.7	0.4
Peru	3.6	2.0	1.4	2.3	1.8
Uruguay*	0.0	0.0	0.0	0.0	0.0
Venezuela	1.4	1.3	0.7	1.6	1.2
NORTH AMERICA	3.0	3.6	3.7	3.6	4.0
Canada	0.5	0.6	0.4	0.6	0.7
Greenland*	0.0	0.0	0.0	0.0	0.0
United States	2.6	3.0	3.3	3.0	3.3
SOUTH ASIA	261.5	128.9	79.8	151.6	86.6
Bangladesh	26.0	11.3	6.9	14.8	8.7
Bhutan	0.1	0.0	0.0	0.1	0.1
India	189.7	73.9	45.0	90.6	44.9
Nepal	2.7	2.0	0.8	2.4	1.5
Pakistan	37.6	38.0	24.4	39.9	28.0
Sri Lanka	5.4	3.5	2.7	3.9	3.3
SOUTHEAST ASIA	268.0	109.1	92.4	116.7	105.5
Cambodia	2.4	1.9	1.5	2.2	2.0
China, PR	173.4	45.0	41.2	45.0	41.2
Fiji	0.0	0.0	0.0	0.0	0.0
Indonesia	32.4	12.9	7.2	15.5	11.2
Lao, PDR	1.5	1.4	1.3	1.6	1.7
Malaysia	0.9	0.8	0.9	0.8	0.9
Mongolia	0.7	0.2	0.1	0.2	0.1
Myanmar	10.5	6.5	4.9	7.2	6.1
North Korea (DPR)	10.2	11.9	11.9	12.7	13.4
Other Indian Ocean	3.0	2.8	2.4	2.9	2.6
Other Pacific Ocean	0.3	0.2	0.1	0.2	0.1
Other Southeast Asia	0.2	0.2	0.3	0.3	0.3
Papua New Guinea	0.9	0.2	0.3	0.2	0.3
Philippines	12.1	12.2	11.0	13.2	13.1
Solomon Islands*	0.0	0.0	0.0	0.0	0.0
Thailand	6.2	3.1	1.8	3.5	2.3
Timor l'Este	0.3	0.2	0.2	0.2	0.2
Vanuatu*	0.0	0.0	0.0	0.0	0.0
Viet Nam	12.9	9.5	7.2	10.9	9.7
WEST & CENTRAL AFRICA	82.5	64.1	50.2	75.7	59.5
Angola	4.0	4.6	6.2	5.2	8.1
Benin	1.2	1.5	0.8	1.8	1.2
Burkina Faso	3.5	4.6	5.3	5.1	6.0
Cameroon	2.8	2.2	1.3	2.6	2.1
Central African Rep.	1.5	1.2	0.5	1.4	0.7
Chad	4.8	6.8	5.5	7.3	6.9
Congo, Rep. of	1.3	0.9	0.6	1.1	0.8
DRC	37.6	20.2	6.7	25.0	6.7
Equatorial Guinea	0.2	0.3	0.3	0.3	0.4
Gabon	0.1	0.1	0.1	0.1	0.1
Gambia	0.1	0.1	0.0	0.1	0.1
Ghana	1.6	1.3	0.9	1.4	1.8
Guinea	2.0	1.1	0.4	1.2	0.4
Guinea-Bissau	0.4	0.4	0.3	0.5	0.4
Côte d'Ivoire	2.8	0.7	0.9	0.7	0.8
Liberia	1.4	1.7	1.8	1.9	2.2
Mali	0.5	0.7	0.6	0.8	0.8
Mauritania	0.3	0.2	0.1	0.2	0.1
Niger	1.8	2.3	2.1	2.6	2.4
Nigeria	9.7	8.5	11.6	10.7	11.5
Other Atlantic	0.1	0.1	0.0	0.1	0.0
Senegal	1.8	1.9	1.3	2.3	2.2
Sierra Leone	1.7	1.9	2.0	2.1	2.4
Togo	1.3	1.1	0.9	1.4	1.3
WORLD	838.1	528.0	406.2	593.3	478.7

Note: * 0.0 indicates a number less than 0.1.

TABLE 2A Total production (million metric tons)

Commodity	Region	Without CC			With CC	
		2010	2030	2050	2030	2050
MEATS includes pork, beef, poultry, sheep, and goats	Central & West Asia & North Africa	15	27	43	27	43
	East Asia & Pacific	11	13	16	13	15
	East & Southern Africa	6	10	15	10	15
	Europe	44	52	56	52	55
	EURplus (EUR+FSU)	52	61	65	61	64
	Former Soviet Union	8	9	9	9	9
	Latin America & Caribbean	44	67	85	66	84
	North America	45	61	73	60	72
	South Asia	10	19	30	19	29
	Southeast Asia	88	117	120	116	119
	West & Central Africa	4	7	13	7	13
	World	274	381	460	380	455
CEREALS includes barley, maize, millet, rice, sorghum, wheat, and aggregated other cereals	Central & West Asia & North Africa	153	203	232	197	224
	East Asia & Pacific	43	55	65	53	60
	East & Southern Africa	52	83	112	84	114
	Europe	311	319	334	315	339
	EURplus (EUR+FSU)	439	490	540	499	573
	Former Soviet Union	129	172	206	184	234
	Latin America & Caribbean	164	245	322	236	294
	North America	436	572	711	478	511
	South Asia	274	379	448	357	408
	Southeast Asia	537	628	676	633	691
	West & Central Africa	56	91	128	87	116
	World	2,155	2,746	3,235	2,621	2,990
FRUITS & VEGETABLES includes bananas, plantains, aggregated temperate fruits, aggregated tropical fruits, and aggregated vegetables	Central & West Asia & North Africa	179	292	416	290	412
	East Asia & Pacific	41	55	70	57	75
	East & Southern Africa	46	85	142	79	125
	Europe	156	207	255	195	226
	EURplus (EUR+FSU)	196	259	315	244	280
	Former Soviet Union	41	52	60	49	54
	Latin America & Caribbean	164	236	299	225	273
	North America	91	114	147	114	149
	South Asia	156	315	463	298	413
	Southeast Asia	669	886	1048	904	1,091
	West & Central Africa	50	92	145	86	128
	World	1,592	2,334	3,044	2,297	2,945

TABLE 2A Total production (million metric tons)

Commodity	Region	Without CC 2010	Without CC 2030	Without CC 2050	With CC 2030	With CC 2050
OILS — includes groundnuts, rapeseed, soybean, sunflower, and aggregated other oilseeds	Central & West Asia & North Africa	10	14	17	14	16
	East Asia & Pacific	2	2	2	2	2
	East & Southern Africa	5	7	9	7	9
	Europe	40	53	60	52	58
	EURplus (EUR+FSU)	54	71	82	70	80
	Former Soviet Union	14	18	21	18	22
	Latin America & Caribbean	126	185	216	181	207
	North America	110	139	155	134	144
	South Asia	41	52	57	50	52
	Southeast Asia	288	500	680	499	679
	West & Central Africa	48	84	106	81	99
	World	685	1,055	1,325	1,039	1,288
PULSES — includes beans, chickpeas, cowpeas, lentils, pigeonpeas, and aggregated other pulses	Central & West Asia & North Africa	4	6	8	5	6
	East Asia & Pacific	2	3	3	3	3
	East & Southern Africa	5	8	10	8	10
	Europe	5	7	9	7	8
	EURplus (EUR+FSU)	8	11	14	11	13
	Former Soviet Union	3	4	5	4	5
	Latin America & Caribbean	7	11	16	11	15
	North America	7	10	12	11	15
	South Asia	16	21	24	20	23
	Southeast Asia	11	14	16	13	16
	West & Central Africa	6	11	18	11	17
	World	66	94	121	92	118
ROOTS & TUBERS — includes cassava, potato, sweet potato, yams, and aggregated other roots and tubers	Central & West Asia & North Africa	31	41	50	42	50
	East Asia & Pacific	7	8	9	8	9
	East & Southern Africa	53	82	112	75	96
	Europe	68	77	82	72	69
	EURplus (EUR+FSU)	142	155	152	123	98
	Former Soviet Union	74	78	70	52	29
	Latin America & Caribbean	60	82	97	83	99
	North America	26	29	33	27	28
	South Asia	50	74	102	79	119
	Southeast Asia	241	268	254	269	251
	West & Central Africa	170	266	377	257	353
	World	780	1,006	1,185	963	1,103

TABLE 2B Per capita food consumption (kg per capita per year)

Commodity	Region	2010	Without CC 2030	2050	With CC 2030	2050
MEATS includes pork, beef, poultry, sheep, and goats	Central & West Asia & North Africa	130	166	202	166	201
	East Asia & Pacific	240	303	349	302	346
	East & Southern Africa	86	112	156	112	154
	Europe	383	389	409	386	404
	EURplus (EUR+FSU)	342	358	379	356	375
	Former Soviet Union	234	269	287	268	284
	Latin America & Caribbean	292	330	359	327	355
	North America	479	484	489	482	485
	South Asia	25	43	69	43	69
	Southeast Asia	371	474	493	472	490
	West & Central Africa	51	81	133	80	132
	World	225	255	268	254	266
CEREALS includes barley, maize, millet, rice, sorghum, wheat, and aggregated other cereals	Central & West Asia & North Africa	1,596	1,595	1,578	1,573	1,516
	East Asia & Pacific	1,027	997	973	984	937
	East & Southern Africa	1,102	1,174	1,236	1,122	1,137
	Europe	934	959	991	948	951
	EURplus (EUR+FSU)	993	1,013	1,036	1,001	997
	Former Soviet Union	1,150	1,168	1,175	1,158	1,137
	Latin America & Caribbean	1,074	1,085	1,085	1,054	1,023
	North America	812	812	808	798	768
	South Asia	1,360	1,374	1,398	1,345	1,324
	Southeast Asia	1,436	1,467	1,467	1,434	1,387
	West & Central Africa	1,023	1,100	1,135	1,058	1,054
	World	1,267	1,290	1,299	1,261	1,229
FRUITS & VEGETABLES includes bananas, plantains, aggregated temperate fruits, aggregated tropical fruits, and aggregated vegetables	Central & West Asia & North Africa	224	239	251	235	243
	East Asia & Pacific	177	190	196	188	190
	East & Southern Africa	121	157	206	153	196
	Europe	212	222	232	219	226
	EURplus (EUR+FSU)	197	216	229	213	222
	Former Soviet Union	156	197	219	194	212
	Latin America & Caribbean	169	196	218	192	210
	North America	202	224	228	220	220
	South Asia	106	191	336	187	324
	Southeast Asia	237	294	301	289	291
	West & Central Africa	182	244	314	237	297
	World	182	229	279	224	269

TABLE 2B Per capita food consumption (kg per capita per year)

Commodity	Region	2010	Without CC 2030	Without CC 2050	With CC 2030	With CC 2050
OILS (includes groundnuts, rapeseed, soybean, sunflower, and aggregated other oilseeds)	Central & West Asia & North Africa	33	37	40	36	37
	East Asia & Pacific	98	103	106	98	96
	East & Southern Africa	37	44	51	42	46
	Europe	25	26	27	25	25
	EURplus (EUR+FSU)	22	23	24	22	22
	Former Soviet Union	14	14	14	14	14
	Latin America & Caribbean	38	38	36	36	33
	North America	94	95	94	91	88
	South Asia	27	35	34	33	31
	Southeast Asia	117	154	145	150	138
	West & Central Africa	89	100	109	95	99
	World	64	75	72	73	68
PULSES (includes beans, chickpeas, cowpeas, lentils, pigeonpeas, and aggregated other pulses)	Central & West Asia & North Africa	67	76	84	76	84
	East Asia & Pacific	15	16	17	16	16
	East & Southern Africa	126	154	189	152	185
	Europe	29	31	32	31	32
	EURplus (EUR+FSU)	26	28	29	28	29
	Former Soviet Union	17	19	20	19	20
	Latin America & Caribbean	106	117	129	117	127
	North America	43	45	46	45	45
	South Asia	91	103	114	102	113
	Southeast Asia	19	23	26	23	26
	West & Central Africa	110	131	157	129	152
	World	61	75	90	74	89
ROOTS & TUBERS (includes cassava, potato, sweet potato, yams, and aggregated other roots and tubers)	Central & West Asia & North Africa	77	78	78	73	71
	East Asia & Pacific	63	64	65	62	62
	East & Southern Africa	311	337	347	329	330
	Europe	137	134	133	130	127
	EURplus (EUR+FSU)	167	162	159	156	151
	Former Soviet Union	247	243	240	233	226
	Latin America & Caribbean	115	110	105	107	100
	North America	104	102	101	97	93
	South Asia	54	70	75	65	67
	Southeast Asia	156	162	156	157	148
	West & Central Africa	540	543	543	531	521
	World	150	165	174	159	165

TABLE 2C Net trade (million metric tons)

Commodity	Region	Without CC 2010	Without CC 2030	Without CC 2050	With CC 2030	With CC 2050
MEATS includes pork, beef, poultry, sheep, and goats	Central & West Asia & North Africa	-2,455	-2,929	227	-2,923	132
	East Asia & Pacific	-853	-1,284	-470	-1,361	-717
	East & Southern Africa	-260	-1,285	-5,036	-1,237	-4,844
	Europe	1,182	5,639	6,240	5,891	6,439
	EURplus (EUR+FSU)	-1,329	2,987	3,302	3,297	3,569
	Former Soviet Union	-2,511	-2,652	-2,938	-2,594	-2,870
	Latin America & Caribbean	7,163	16,669	25,855	16,726	25,970
	North America	4,365	12,479	18,655	12,250	17,826
	South Asia	284	-2,598	-10,436	-2,493	-10,023
	Southeast Asia	-6,313	-21,119	-22,750	-21,387	-22,810
	West & Central Africa	-600	-2,920	-9,348	-2,873	-9,104
CEREALS includes barley, maize, millet, rice, sorghum, wheat, and aggregated other cereals	Central & West Asia & North Africa	-68,611	-109,674	-163,128	-111,688	-157,219
	East Asia & Pacific	-27,468	-28,455	-23,209	-30,746	-29,349
	East & Southern Africa	-12,236	-24,160	-44,395	-18,282	-28,982
	Europe	9,411	-24,334	-22,378	-27,767	-19,644
	EURplus (EUR+FSU)	34,805	43,893	85,225	55,403	121,954
	Former Soviet Union	25,394	68,227	107,603	83,170	141,598
	Latin America & Caribbean	-23,416	-18,355	-5,823	-18,128	-64,180
	North America	124,251	208,287	318,137	151,034	192,239
	South Asia	-3,196	-2,414	-41,025	-16,531	-56,842
	Southeast Asia	-7,168	-32,946	-53,501	24,113	89,961
	West & Central Africa	-16,961	-36,175	-72,282	-35,175	-67,583
FRUITS & VEGETABLES includes banana, plantain, aggregated temperate fruits, aggregated tropical fruits, and aggregated vegetables	Central & West Asia & North Africa	2,622	40,435	105,069	42,443	110,119
	East Asia & Pacific	-6,227	3,042	18,760	5,664	24,871
	East & Southern Africa	1,753	1,032	-2,793	-2,912	-13,151
	Europe	-6,447	28,465	63,956	18,250	40,291
	EURplus (EUR+FSU)	-9,432	28,853	69,385	16,478	40,895
	Former Soviet Union	-2,985	388	5,428	-1,773	604
	Latin America & Caribbean	46,336	76,306	108,327	67,357	88,710
	North America	2,862	-1,683	15,429	37	22,152
	South Asia	-29,416	-126,356	-462,403	-134,965	-479,818
	Southeast Asia	-8,874	-16,915	167,382	14,309	235,812
	West & Central Africa	376	-4,713	-19,157	-8,412	-29,591

TABLE 2C Net trade (million metric tons)

Commodity	Region	2010	Without CC 2030	Without CC 2050	With CC 2030	With CC 2050
OILS includes groundnuts, rapeseed, soybean, sunflower, and aggregated other oilseeds	Central & West Asia & North Africa	-7,544	-9,130	-10,690	-8,583	-9,483
	East Asia & Pacific	-8,660	-9,231	-9,638	-8,770	-8,688
	East & Southern Africa	-150	-559	-1,597	-442	-1,158
	Europe	-16,816	-15,778	-15,406	-14,715	-13,369
	EURplus (EUR+FSU)	-16,136	-14,070	-12,730	-12,806	-10,330
	Former Soviet Union	680	1,709	2,676	1,909	3,039
	Latin America & Caribbean	27,240	46,345	56,619	43,482	49,592
	North America	31,502	37,319	40,843	34,571	35,589
	South Asia	469	-4,478	-9,659	-4,719	-9,869
	Southeast Asia	-27,189	-45,844	-50,617	-42,493	-43,569
	West & Central Africa	467	-353	-2,531	-241	-2,083
PULSES includes beans, chickpeas, cowpeas, lentils, pigeonpeas, and aggregated other pulses	Central & West Asia & North Africa	-1,241	-2,124	-3,038	-3,034	-5,173
	East Asia & Pacific	559	653	784	601	585
	East & Southern Africa	-808	-3,388	-7,917	-3,238	-7,464
	Europe	-1,238	477	2,009	352	1,761
	EURplus (EUR+FSU)	-856	1,635	4,167	1,544	4,096
	Former Soviet Union	382	1,158	2,158	1,192	2,335
	Latin America & Caribbean	-684	1,150	4,434	667	3,064
	North America	3,849	6,041	8,024	7,313	11,363
	South Asia	-2,861	-6,095	-10,021	-6,210	-10,480
	Southeast Asia	1,818	2,165	4,106	2,493	4,786
	West & Central Africa	223	-36	-539	-135	-777
ROOTS & TUBERS includes cassava, potato, sweet potato, yams, and aggregated other roots and tubers	Central & West Asia & North Africa	-764	-467	697	2,789	5,480
	East Asia & Pacific	-2,954	-2,289	-1,525	-1,878	-1,062
	East & Southern Africa	-3,202	-8,713	-14,075	-13,064	-23,305
	Europe	-1,585	7,546	12,840	5,153	4,408
	EURplus (EUR+FSU)	6,458	23,872	26,350	-1,351	-19,052
	Former Soviet Union	8,042	16,325	13,510	-6,505	-23,459
	Latin America & Caribbean	188	11,529	20,352	16,174	29,784
	North America	-444	-1,358	-1,258	-1,901	-3,198
	South Asia	-6,127	-23,917	-30,065	-12,219	1,702
	Southeast Asia	4,310	3,011	13,391	15,474	28,083
	West & Central Africa	2,536	-1,670	-13,868	-4,024	-18,432

TABLE 1 NOTE

World and regional figures include other regions and countries not reported separately. Aggregate food production is an index, by weight, of cereals, meats, fruits and vegetables, oilseeds, pulses, and roots and tubers (which are reported separately in Table 2). Per capita food consumption is a projection of daily dietary energy supply. Estimates of the number of people at risk of hunger are based on a quadratic specification of the relationship between national-level calorie supply and the share of population that is undernourished as defined by the FAO. Values reported for 2010 are calibrated model results. Projections for 2030 and 2050 assume changes in population and income as reflected in the IPCC's Shared Socioeconomic Pathway 2. Climate change impacts are simulated using the IPCC's Representative Concentration Pathway 8.5 and the HadGEM general circulation model. Further documentation is available at www.ifpri.org/program/impact-model.

TABLE 2 NOTE

World and regional figures include other regions and countries not reported separately. Total production is aggregated across irrigated and rainfed systems at the national level and aligned with years as reported in FAOSTAT. Per capita food consumption is based on food availability at the national level. Net trade includes negative and positive numbers indicating that a region is a net importer or exporter, respectively, and balances to zero at the global level. Cereals include barley, maize, millet, rice, sorghum, wheat, and aggregated other cereals. Meats include beef, pork, poultry, and sheep and goats. Fruits and vegetables include banana, plantain, aggregated temperate fruits, aggregated tropical fruits, and aggregated vegetables. Oilseeds include groundnuts, rapeseed, soybean, sunflower, and aggregated other oilseeds. Pulses include beans, chickpeas, cowpeas, lentils, pigeonpeas, and aggregated other pulses. Roots and tubers include cassava, potato, sweet potato, yams, and aggregated other roots and tubers. Values reported for 2010 are calibrated model results. Projections for 2030 and 2050 assume changes in population and income as reflected in the IPCC's Shared Socioeconomic Pathway 2. Climate change impacts are simulated using the IPCC's Representative Concentration Pathway 8.5 and the HadGEM general circulation model. Further documentation is available at www.ifpri.org/program/impact-model.

Notes

CHAPTER 1

1. IPCC (Intergovernmental Panel on Climate Change), *Climate Change 2022: Impacts, Adaptation, and Vulnerability*, Contribution of Working Group II to the Sixth Assessment Report of the Intergovernmental Panel on Climate Change, eds. H.O. Pörtner, D.C. Roberts, M. Tignor, et al. (Cambridge: Cambridge University Press, 2022).

2. IPCC, "Summary for Policymakers, in *Climate Change and Land: An IPCC Special Report on Climate Change, Desertification, Land Degradation, Sustainable Land Management, Food Security, and Greenhouse Gas Fluxes in Terrestrial Ecosystems,* eds. P.R. Shukla, J. Skea, E. Calvo Buendia, et. al. (Geneva: IPCC, 2019).

3. IFPRI, Foresight Modeling with IFPRI's Impact Model, https://www.ifpri.org/project/ifpri-impact-model.

4. FAO (Food and Agriculture Organization of the United Nations), IFAD (International Fund for Agricultural Development), UNICEF, WFP (World Food Programme), and WHO (World Health Organization), *The State of Food Security and Nutrition in the World 2021: Transforming Food Systems for Food Security, Improved Nutrition and Affordable Healthy Diets for All* (Rome: FAO, 2021).

5. FSIN (Food Security Information Network) and Global Network Against Food Crises, *Global Report on Food Crises* (Rome: FSIN, 2021).

6. IPCC, *Climate Change 2022: Impacts, Adaptation, and Vulnerability*, Contribution of Working Group II to the Sixth Assessment Report of the Intergovernmental Panel on Climate Change, ed. H.O. Pörtner, et al.(Cambridge: Cambridge University Press, 2022).

7. M. Crippa, D. Guizzardi, E. Solazzo, et al., *GHG Emissions of All World Countries: 2021 Report* (Luxembourg: European Union, 2021).

8. C. Arndt, D. Arent, F. Hartley, B. Merven, and A.H. Mondal "Faster than You Think: Renewable Energy and Developing Countries," *Annual Review of Resource Economics* 11 (2019): 149–168.

9. F.N. Tubiello, K. Karl, A. Flammini, et al., "Pre- and Post-production Processes along Supply Chains Increasingly Dominate GHG emissions from Agri-food Systems Globally and in Most Countries," *Earth Systems Science Data Discussion* [preprint 2021].

10. IPCC, "Summary for Policymakers, in *Climate Change and Land: An IPCC Special Report on Climate Change, Desertification, Land Degradation, Sustainable Land Management, Food Security, and Greenhouse Gas Fluxes in Terrestrial Ecosystems*, eds. P.R. Shukla, et al. (Geneva: 2019); F.N. Tubiello, K. Karl, A. Flammini, et al., "Pre- and Post-production Processes along Supply Chains Increasingly Dominate GHG Emissions from Agri-food Systems Globally and in Most Countries," *Earth Systems Science Data Discussion* [preprint 2021].

11. IPCC, *Climate Change 2022: Impacts, Adaptation, and Vulnerability*, Contribution of Working Group II to the Sixth Assessment Report of the Intergovernmental Panel on Climate Change, ed. H.O. Pörtner, et al. (Cambridge: Cambridge University Press, 2022).

12. F.N. Tubiello, G. Conchedda, N. Wanner, S. Federici, S. Rossi, and G. Grassi, "Carbon Emissions and Removals from Forests: New Estimates, 1990-2020," *Earth System Science Data* 13, 4 (2021a): 1681-1691.

13. IPCC, *Climate Change 2022: Impacts, Adaptation, and Vulnerability*, Contribution of Working Group II to the Sixth Assessment Report of the Intergovernmental Panel on Climate Change, ed. H.O. Pörtner, et al. (Cambridge: Cambridge University Press, 2022).

14. W.F. Lamb, T. Wiedmann, J. Pongratz, et al., "A Review of Trends and Drivers of Greenhouse Gas Emissions by Sector from 1990 to 2018," *Environmental Research Letters* 16, 7 (2021): 073005.

15. P. Potapov, S. Turubanova, M.C. Hansen, et al., "Global Maps of Cropland Extent and Change Show Accelerated Cropland Expansion in the Twenty-First Century," *Nature Food* 3 (2022):19–28.

16. EPA (U.S. Environmental Protection Agency), "Inventory of US Greenhouse Gas Emissions and Sinks: 1990-2020," (EPA, Washington, DC, 2022).

17. IPCC, "Summary for Policymakers, in *Climate Change and Land: An IPCC Special Report on Climate Change, Desertification, Land Degradation, Sustainable Land Management, Food Security, and Greenhouse Gas Fluxes in Terrestrial Ecosystems*, ed. P.R. Shukla, et al. (Geneva: 2019); W. Willet, J. Rockström, B. Loken, et al., "Food in the Anthropocene: The EAT-Lancet Commission on Healthy Diets from Sustainable Food Systems," *Lancet* 393, 10170 (2019): 447-492.

18. See also, D. Gollin, C. Hansen, and A. Wingender, "Two Blades of Grass: Agricultural Innovation, Productivity, and Economic Growth," *Journal of Political Economy* 129, 8 (2021); J. Alston, P. Pardey, and X. Rao, "Payoffs to a Half Century of CGIAR Research," *American Journal of Agricultural Economics* 104, 2 (2022): 502-529.

19. See also, M. Gautam, D. Laborde, A. Mamun, W. Martin, V. Piñeiro, and R. Vos, *Repurposing Agricultural Policies and Support: Options to Transform Agriculture and Food Systems to Better Serve the Health of People, Economies, and the Planet* (Washington, DC: World Bank and IFPRI, 2022).

20. M. Gautam, D. Laborde, A. Mamun, W. Martin, V. Piñeiro, and R. Vos, *Repurposing Agricultural Policies and Support: Options to Transform Agriculture and Food Systems for Better Health of People, Economies and the Planet*, Technical Report (Washington, DC: World Bank and IFPRI, 2022).

21. A. Afshin, P.J. Sur, K.A. Fay, et al., "Health Effects of Dietary Risks in 195 Countries, 1990-2017: A Systematic Analysis for the Global Burden of Disease Study 2017," *Lancet* 393, 10184 (2019): 1958-1972.

22. T. Garnett, "What Is a Sustainable Healthy Diet?" Discussion paper (Food Climate Research Network, Oxford, UK, 2014).

23. F.N. Tubiello, K. Karl, A. Flammini, et al., "Pre- and Post-production Processes along Supply Chains Increasingly Dominate GHG emissions from Agri-food Systems Globally and in Most Countries," *Earth Systems Science Data Discussion* [preprint 2021].

24. M. Gautam, D. Laborde, A. Mamun, W. Martin, V. Piñeiro, and R. Vos, *Repurposing Agricultural Policies and Support: Options to Transform Agriculture and Food Systems for Better Health of People, Economies and the Planet*, Technical Report (Washington, DC: World Bank and IFPRI, 2022).

BOX 1

a. A. Kalibata. "The Food Systems Summit: A New Deal for People, Planet and Prosperity," (UN, 2021).

b. Ministry of Foreign Affairs of Japan, "Tokyo Compact on Global Nutrition for Growth," Tokyo Nutrition for Growth Summit 2021.

c. United Nations, "Secretary-General's Chair Summary and Statement of Action on the UN Food Systems Summit" (UNFSS, 2021).

d. UN Climate Change Conference UK 2021, Glasgow Leaders' Declaration on Forests and Land Use (Nov. 2, 2021).

CHAPTER 2

1. Some countries also impose taxes on the agriculture sector (mostly on exports), which imply "negative" support. During 2018-2020, such taxes amounted to US$104 billion per year globally (see Box 1).

2. M. Gautam, D. Laborde, A. Mamun, W. Martin, V. Piñeiro, and R. Vos, *Repurposing Agricultural Policies and Support: Options to Transform Agriculture and Food Systems for Better Health of People, Economies and the Planet*, Technical Report (Washington, DC: World Bank and IFPRI, 2022).

3. OECD (Organisation for Economic Co-Operation and Development), *Agricultural Policy Monitoring and Evaluation 2021* (Paris: 2021).

4. T. Searchinger, C. Malins, P. Dumas, D. Baldok, J. Glauber, T. Jayne, J. Huang, and P. Marenya, *Revising Public Agricultural Support to Mitigate Climate Change* (Washington, DC: World Bank, 2020); D. Laborde, A. Mamun, W. Martin, V. Piñeiro, and R. Vos, "Agricultural Subsidies and Global Greenhouse Gas Emissions," *Nature Communications* 12 (2021): 2601; M. Gautam, D. Laborde, A. Mamun, W. Martin, V. Piñeiro, and R. Vos, *Repurposing Agricultural Policies and Support: Options to Transform Agriculture and Food Systems for Better Health of People, Economies and the Planet*, Technical Report (Washington, DC: World Bank and IFPRI, 2022).

5. https://foodsystems.community/game-changing-propositions-solution-clusters/repurposing-public-support-to-food-and-agriculture-2/

6. D. Laborde, A. Mamun, W. Martin, V. Piñeiro, and R. Vos, "Modeling the Impacts of Agricultural Support Policies on Emissions from Agriculture," IFPRI Discussion Paper, No. 1954 (Washington, DC: IFPRI, 2020); D. Laborde, A. Mamun, W. Martin, V. Piñeiro, and R. Vos, "Agricultural Subsidies and Global Greenhouse Gas Emissions," *Nature Communications* 12 (2021): 2601; FAO (Food and Agriculture Organization), UNDP (United Nations Development Programme), and UNEP (United Nations Environment Programme), *A Multi-Billion-Dollar Opportunity. Repurposing Agricultural Support to Transform Food Systems* (Rome: FAO, 2021); M. Gautam, D. Laborde, A. Mamun, W. Martin, V. Piñeiro, and R. Vos, *Repurposing Agricultural Policies and Support: Options to Transform Agriculture and Food Systems for Better Health of People, Economies and the Planet*, Technical Report (Washington, DC: World Bank and IFPRI, 2022).

7. M. Gautam, D. Laborde, A. Mamun, W. Martin, V. Piñeiro, and R. Vos, *Repurposing Agricultural Policies and Support: Options to Transform Agriculture and Food Systems for Better Health of People, Economies and the Planet*, Technical Report (Washington, DC: World Bank and IFPRI, 2022).

8. P.A. Hall, "The Role of Interests, Institutions, and Ideas in the Political Economy of Industrialized Nations," in *Comparative Politics: Rationality, Culture and Structure*, eds. M. Lichbach and A. Zuckerman (Cambridge University Press: 1997), 174-207.

9. OECD, *Agricultural Policy Monitoring and Evaluation 2021* (Paris: 2021).

10. J. Swinnen, *The Political Economy of Agricultural and Food Policies* (London, New York: Palgrave/MacMillan, 2018).

11. G. Pursell, A. Gulatie, and K. Gupta, "India," in *Distortions to Agricultural Incemtives in Asia*, eds. K. Anderson and W. Martin (Washington, DC: World Bank, 2009).

12. C. Gouel, M. Gautam, and W. Martin, "Managing Food Price Volatility in a Large Open Country: The Case of Wheat in India," *Oxford Economic Papers* 68, 3 (2016): 811-835.

13. S. Hussain, "Farm Laws: Potential for Positive Outcomes," *International Growth Centre Blog*, October 15, 2020.

14. European Commission, *How the Future CAP Will Contribute to the EU Green Deal* (Brussels: 2020); European Commission, *Farm to Fork Strategy: For Fair, Healthy and Environmentally Friendly Food System* (Brussels: 2020).

15. V. Bjornlund, H. Bjornlund, and A.F. van Rooyen, "Exploring the Factors Causing the Poor Performances of Most Irrigation: Schemes in Post-Independence Sub-Saharan Africa," *International Journal of Water Resources Development* 36, Supp. 1 (2020): S54-101.

16. T. Mogues, "What Determines Public Expenditure Allocations? A Review of Theories and Implications for Agricultural Public Investments," IFPRI Discussion Paper 1216 (IFPRI, Washington, DC, 2013).

17. J. Swinnen, *The Political Economy of Agricultural and Food Policies* (London, New York: Palgrave/MacMillan, 2018).

18. G-J. Stads and S. Doumbia, "Côte d'Ivoire: Recent Developments in Agricultural Research," ASTI Country Notes 7645 (Washington, DC: IFPRI, 2010).

19. D. Resnick, S. Haggblade, S. Babu, S. Hendricks, and D. Mathers, "The Kaleidoscope Model of Policy Change: Applications to Food Security Policy in Zambia," *World Development* 109 (2018): 101-120.

BOX 1

a. OECD, *Agricultural Policy Monitoring and Evaluation 2021* (Paris: 2021).

BOX 2

a. India, Ministry of Agriculture, *Agricultural Statistics at a Glance* (New Delhi: Ministry of Agriculture, Department of Agriculture and Co-operation, 2020).

b. P. Kumar, S. Kumar, and L. Joshi, "Problem of Residue Management Due to Rice Wheat Crop Rotation in Punjab," in *Socioeconomic and Environmental Implications of Agricultural Residue Burning* (New Delhi: Springer, 2015).

c. B. Ramaswami, *Agricultural Subsidies*, Report submitted to the Finance Commission (New Delhi: Government of India, 2019).

d. K. Bozhinova, J. Satre, and S. Rajagopal, "An Unlikely Success: Demonstrations against Farm Laws in India," Armed Conflict Location and Event Data Project (ACLED, 2021).

BOX 3

a. J.K. Huang and G.L. Yang, "Understanding Recent Challenges and New Food Policy in China," *Global Food Security* 12 (2017): 119-126.

b. Y. Chai, P.G. Pardey, C. Chang-Kang, J.K. Huang, K. Lee, and W.L. Dong, "Passing the Food and Agricultural R&D Buck? The United States and China," *Food Policy* 86 (2019): 101729.

BOX 4

a. A. Swinbank, *The Interactions between the EU's External Action and the Common Agricultural Policy*, Report for the European Parliament's Committee on Agriculture and Rural Development (Strasbourg, France: European Parliament, 2016).

b. J. Swinnen, *The Perfect Storm: The Political Economy of the Fischler Reforms of the Common Agricultural Policy* (Brussels: Centre for European Policy, 2008).

c. OECD, *Evaluation of Agricultural Policy Reforms in the European Union: The Common Agricultural Policy* 2014-20 (Paris: 2017).

d J. Swinnen, *The Imperfect Storm: The Political Economy of the 2013 Reforms of the Common Agricultural Policy* (Brussels: Centre for European Policy, 2014).

e European Commission, *How the Future CAP Will Contribute to the EU Green Deal* (Brussels: 2020); European Commission, *Farm to Fork Strategy: For Fair, Healthy and Environmentally Friendly Food System* (Brussels: 2020).

f M. Gautam, D. Laborde, A. Mamun, W. Martin, V. Piñeiro, and R. Vos, *Repurposing Agricultural Policies and Support: Options to Transform Agriculture and Food Systems for Better Health of People, Economies and the Planet*, Technical Report (Washington, DC: World Bank and IFPRI, 2022).

BOX 5

a R. Lawrence, "How Good Politics Results in Bad Policy: The Case of Biofuels Mandates," CID Working Paper 200 (Center for International Development, Cambridge, MA, 2010).

b W. Tyner, "The US Ethanol and Biofuels Boom: Its Origins, Current Status, and Future Prospects," *Bioscience* 58, 7 (2008): 646-653.

c B. Wright, "Global Biofuels: Key to the Puzzle of Grain Market Behavior," *Journal of Economic Perspectives* 28, 1 (2014): 73-98.

d CRS (Congressional Research Service), *The Renewable Fuel Standard (RFS): An Overview*, CRS Report R43325 (Washington, DC: 2022).

e T. Searchinger, R. Heimlich, R.A. Houghton, et al., "Use of U.S. Croplands for Biofuels Increases Greenhouse Gases through Emissions from Land-Use Change," *Science* 319, 5867 (2008): 1238-1240; EPA, *Biofuels and the Environment: Second Triennial Report to Congress*, EPA/600/R-18/195F (Washington, DC: 2018).

f T. Lark, N. Hendricks, A. Smith, N. Pates, S. Spawn-Lee, M. Bougie, E. Booth, C. Kucharik, and H. Gibbs, "Environmental Outcomes of the US Renewable Fuel Standard," *PNAS* 119, 9 (2022): e210108419.

g R. Lawrence, "How Good Politics Results in Bad Policy," CID Working Paper 200 (Center for International Development, Cambridge, MA, 2010).

CHAPTER 3

1 United Nations, *Transforming Our World: The 2030 Agenda for Sustainable Development*, Doc. A/RES/70/1 (New York: 2015).

2 E. Díaz-Bonilla and J. Hepburn, *Trade, Food Security and the 2030 Agenda* (Geneva: ICTSD, 2016).

3 U.L.C. Baldos and T. Hertel, "The Role of International Trade in Managing Food Security Risks from Climate Change," *Food Security* 7, 2 (2015): 275-290.

4 W.B. Traill, *Transnational Corporations, Food Systems and Their Impacts on Diets in Developing Countries*, Trade Policy Technical Note No. 17 (Rome: FAO, 2017); A. Thow, E. Reeve, T. Naseri, T. Martyn, and C. Bollars, "Food Supply, Nutrition, and Trade Policy: Reversal of an Import Ban on Turkey Tails," *Bulletin of the World Health Organization* 95, 10 (2017): 723-725.

5 A. Thow, B. Swinburn, S. Colagiuri, M. Diligolevu, C. Quested, P. Vivili, and S. Leeder, "Trade and Food Policy: Case Studies from Three Pacific Island Countries," *Food Policy* 35, 6 (2010): 556-564.

6 W. Martin and J. Glauber, "Trade Policy and Food Security," in *COVID-19 and Trade Policy: Why Turning Inward Won't Work*, eds. R. Baldwin and S. Evenett (London: CEPR Press, 2020), 89-102.

7 G. Hufbauer, "Divergent Climate Change Policies among Countries Could Spark a Trade War: The WTO Should Step In," *PIIE Trade and Investment Policy Watch Blog* (August 30, 2021).

8 R.D. Smith, K. Lee, and N. Drager, "Trade and Health: An Agenda for Action," *Lancet* 373, 9665 (2009): 768-773.

9 J. Schmidhuber and S. Meyer, "Has the Treadmill Changed Direction? WTO Negotiations in the Light of a Potential New Global Agricultural Market Environment," in *Tackling Agriculture in the Post-Bali Context: A Collection of Short Essays*, eds. R. Meléndez-Ortiz, C. Bellmann, and J. Hepburn (Geneva: ICRSD, 2014).

10 WTO (World Trade Organization), *World Trade Statistical Review 2021* (Geneva: 2021).

11 V.H. Smith and J.W. Glauber, "Trade, Policy and Food Security," *Agricultural Economics* 51, 1 (2020): 159-171.

12 K.A. Ingersent and A. Rayner, *Agricultural Policy in Western Europe and the United States* (Cheltenham, England: Edward Elgar, 1999).

13 V.H. Smith and J.W. Glauber, "Trade, Policy and Food Security," *Agricultural Economics* 51, 1 (2020): 159-171.

14 UNCTAD (UN Conference on Trade and Development), Statistics database, accessed 2021. https://unctad.org/statistics

15 T. Josling, S. Tangermann, and T. Warley, *Agriculture in the GATT* (London: MacMillan Press, 1996).

16 WTO, *World Trade Statistical Review 2021* (Geneva: 2021).

17 J-C. Bureau, H. Guimbard, and S. Jean, "Agricultural Trade Liberalization in the 21st Century: Has It Done Its Business?" *Journal of Agricultural Economics* 70, 1 (2018): 3-25.

18 K. Anderson, M. Ivanic, and W.J. Martin, "Food Price Spikes, Price Insulation, and Poverty," in *The Economics of Food Price Volatility* (Chicago: University of Chicago Press, 2014), 311-339.

19 V.H. Smith and J.W. Glauber, "Trade, Policy and Food Security," *Agricultural Economics* 51, 1 (2020): 159-171.

20 FAO (Food and Agriculture Organization of the United Nations), UNDP (UN Development Programme), and UNEP (UN Environmental Programme), *A Multi-Billion-Dollar Opportunity: Repurposing Agricultural Support to Transform Food Systems* (Rome: FAO; 2021).

21 J. Swinnen, "Economics and Politics of Food Standards, Trade, and Development," *Agricultural Economics* 47, S1 (2016): 7-19.

22 This discussion is without prejudice to the effectiveness of the tax. The effectiveness of consumption taxes in shaping health outcomes has been hotly debated; see J.M. Fletcher, D. Frisvold, and N. Tefft, "Can Soft Drink Taxes Reduce Population Weight?" *Contemporary Economic Policy* 28, 1 (2010): 23-35; D. Just and G. Gabrielyan, "Food and Consumer Behavior: Why the Details Matter," *Agricultural Economics* 47, S1 (2016): 73-83.

23 G. Hufbauer, "Divergent Climate Change Policies among Countries Could Spark a Trade War: The WTO Should Step In," *PIIE Trade and Investment Policy Watch Blog* (August 30, 2021).

24 J. Bhagwati, "US Trade Policy: The Infatuation with Free Trade Agreements," in *The Dangerous Drift Towards Free Trade Agreements*, eds. J. Bhagwati and A. Krueger (Washington, DC: AEI Press, 1995).

25 European Commission, *Impact Assessment: Minimising the Risk of Deforestation and Forest Degradation Associated with Products Placed on the EU Market* (Brussels: 2021).

26 OECD (Organisation for Economic Co-operation and Development), *Agricultural Policy Monitoring and Evaluation 2021: Addressing the Challenges Facing Food Systems* (Paris: 2021).

27 M. Gautam, D. Laborde, A. Mamun, W. Martin, V. Piñeiro, and R. Vos, *Repurposing Agricultural Policies and Support : Options to Transform Agriculture and Food Systems to Better Serve the Health of People, Economies, and the Planet* (Washington, DC: World Bank, 2022); FAO, UNDP, and UNEP, A Multi-Billion-Dollar Opportunity: Repurposing Agricultural Support to Transform Food Systems (Rome: FAO, 2021). Some of those results are discussed in more detail in Chapter 2.

28 FAO, UNDP, and UNEP, *A Multi-Billion-Dollar Opportunity: Repurposing Agricultural Support to Transform Food Systems* (Rome: FAO, 2021).

29 G. Anania, "Export Restrictions and Food Security," in *Tackling Agriculture in the Post-Bali Context: A Collection of Short Essays*, eds R. Melendez-Ortiz, C. Bellmann, and J. Hepburn (Geneva: ICTSD, 2014); J.W. Glauber, D. Laborde Debucquet, W. Martin, and R. Vos, "COVID-19: Trade Restrictions Are Worst Possible Response to Safeguard Food Security," in *COVID-19 and Global Food Security*, eds. J. Swinnen and J. McDermott (Washington, DC: IFPRI, 2020), 66-68.

30 K. Anderson, M. Ivanic, and W.J. Martin, "Food Price Spikes, Price Insulation, and Poverty," in *The Economics of Food Price Volatility* (Chicago: University of Chicago Press, 2014), 311-339.

31 J.W. Glauber, "Negotiating Agricultural Trade in a New Policy Environment," IFPRI Discussion Paper 01831 (IFPRI, Washington, DC, 2019).

32 E. Díaz-Bonilla and J. Hepburn, *Trade, Food Security and the 2030 Agenda* (Geneva: ICTSD, 2016); G. Anania, "Export Restrictions and Food Security," in *Tackling Agriculture in the Post-Bali Context: A Collection of Short Essays*, eds. R. Melendez-Ortiz, C. Bellmann, and J. Hepburn (Geneva: ICTSD, 2014).

33 G20, "Action Plan on Food Price Volatility and Agriculture," Ministerial Declaration from the Meeting of G20 Agriculture Ministers, Paris, June 22-23, 2011.

34 J.W. Glauber, "Negotiating Agricultural Trade in a New Policy Environment," IFPRI Discussion Paper 01831 (IFPRI, Washington, DC, 2019).

35 G.M. Nelson, A. Palazzo, I. Gray, et al., *Food Security, Farming, and Climate Change to 2050: Scenarios, Results, Policy Options* (Washington, DC: IFPRI, 2010); U.L.C. Baldos and T. Hertel, "The Role of International Trade in Managing Food Security Risks from Climate Change," *Food Security* 7, 2 (2015): 275-290.

BOX 1

a W. Martin, "Border Carbon Adjustments: Should Production or Consumption be Taxed?" Working Paper EUI RSC, 2022/02 (San Domenico di Fiesole, Italy, 2022).

CHAPTER 4

1 N. Alexandratos and J. Bruinsma, "World Agriculture towards 2030/2050: The 2012 Revision," ESA Working Paper 12-03 (FAO [Food and Agriculture Organization of the United Nations, Rome], 2012); T. Sulser, K.D. Wiebe, S. Dunston et al., *Climate Change and Hunger: Estimating Costs of Adaptation in the Agrifood System*, Food Policy Report (Washington, DC: IFPRI, 2021).

2 M. Herrero, P.K. Thornton, D. Mason-D'Croz, et al., "Innovation Can Accelerate the Transition towards a Sustainable Food System," *Nature Food* 1, 5 (2020): 266-272.

3 D. Mason-D'Croz, T. Sulser, K. Wiebe, et al., "Agricultural Investments and Hunger in Africa Modeling Potential Contributions to SDG2-Zero Hunger," *World Development* 116 (2019): 38-53; M. Rosegrant, T. Sulser, D. Mason-D'Croz, et al., *Quantitative Foresight Modeling to Inform the CGIAR Research Portfolio*, Project Report for USAID (Washington, DC: IFPRI, 2017); IFPRI, *Estimating the Global Investment Gap in Research and Innovation for Sustainable Agriculture Intensification in the Global South* (Colombo: CoSAI, 2021).

4 D. Mason-D'Croz, T. Sulser, K. Wiebe, et al., "Agricultural Investments and Hunger in Africa Modeling Potential Contributions to SDG2-Zero Hunger," *World Development* 116 (2019): 38-53.

5 J.M. Alston, P.G. Pardey, and X. Rao, *Payoff to Investing in CGIAR Research* (Arlington, VA: SoAR, 2020).

6 T. Mogues, S. Fan, and S. Benin, "Public Investments in and for Agriculture," *European Journal of Development Research* 27, 3 (2015): 337-352; E. Díaz-Bonilla, D. Orden, and A. Kwieciński, "Enabling Environment for Agricultural Growth and Competitiveness," OECD Food, Agriculture and Fisheries Papers 67 (OECD, Paris, 2014).

7 S. Benin, L. McBride, and T. Mogues, "Why Do African Countries Underinvest in Agricultural R&D?" in *Agricultural Research in Africa: Investing in Future Harvests*, eds J. Lynam, N. Beintema, J. Roseboom, and O. Badiane (Washington, DC: IFPRI, 2016), 109-138.

8 R. Benfica and A. Nin-Pratt, "Understanding the Impacts of R&D Investments on Poverty and Undernourishment: A Causal Mediation Approach," IFPRI Discussion Paper 2048 (IFPRI, Washington, DC, 2021).

9 P.L. Pingali, "Green Revolution: Impacts, Limits, and the Path Ahead," *PNAS* 109, 31 (2012): 12302-12308.

10 K. Fuglie, "The Growing Role of the Private Sector in Agricultural Research and Development World-Wide," *Global Food Security* 10 (2016): 29-38.

11 K. Fuglie, "The Growing Role of the Private Sector in Agricultural Research and Development World-Wide," *Global Food Security* 10 (2016): 29-38.

12 FAO, IFAD (International Fund for Agricultural Development), UNICEF, WFP (World Food Program), and WHO (World Health Organization), *The State of Food Security and Nutrition in the World: Transforming Food Systems for Food Security, Improved Nutrition, and Affordable Healthy Diets for All* (Rome: FAO, 2021).

13 W. Willett, J. Rockstrom, B. Loken, et al., "Food in the Anthropocene: The EAT-Lancet Commission on Healthy Diets from Sustainable Food Systems," *Lancet* 393, 10170 (2019):447-492

14 M. Ivanic and W. Martin, "Sectoral Productivity Growth and Poverty Reduction: National and Global Impacts," *World Development* 109 (2018): 429-439; E. Ligon and E. Sadoulet, "Estimating the Relative Benefits of Agricultural Growth on the Distribution of Expenditures," *World Development* 109 (2018): 417-428.

15 T. Sulser, et al., *Climate Change and Hunger: Estimating Costs of Adaptation in the Agrifood System*, Food Policy Report (Washington, DC: IFPRI, 2021).

16 A. Steensland and M. Zeigler, "Productivity in Agriculture for a Sustainable Future," in *The Innovation Revolution in Agriculture*, ed. H. Campos (Cham, Switzerland: Springer, 2021).

17 K. Wiebe, H. Lotze-Campen, R. Sands, et al., "Climate Change Impacts on Agriculture in 2050 under a Range of Plausible Socioeconomic and Emissions Scenarios," *Environmental Research Letters* 10, 8 (2015): 085010; A. Ortiz-Bobea, T. Ault, C. Carrillo, R. Chambers, and D. Lobell, "Anthropogenic Climate Change Has Slowed Global Agricultural Productivity Growth," *Nature Climate Change* 11 (2021): 306-312; T. Sulser, K.D. Wiebe, S. Dunston et al., *Climate Change and Hunger* (Washington, DC: IFPRI, 2021).

18 J.M. Alston, P.G. Pardey, and X. Rao, *Payoff to Investing in CGIAR Research* (Arlington, VA: SoAR, 2020).

19. T. Mogues, B. Yu, S. Fan, and L. Mcbride, "The Impacts of Public Investment in and for Agriculture: Synthesis of the Existing Evidence," ESA Working Paper No. 12-07 (FAO, 2012).

20. "Attainable levels" of national agricultural R&D investment are based on four variables: the size of a country's economy, its income level, the level of diversification of its agricultural production, and the availability of relevant technology spillovers from other countries. This weighted measurement allows for the identification of potential investment gaps and the quantification of the additional investment needed to close those gaps based on comparisons with countries of similar status. For more detail on the calculation of the attainable investment targets and R&D investment gaps, see A. Nin-Pratt, "Agricultural R&D Investment Intensity: A Misleading Conventional Measure and a New Intensity Index," *Agricultural Economics* 52 (2021): 317-328.

21. Small research systems are defined as those that invest less than $10 million per year (in 2011 PPP prices); medium-sized research systems invest between $10 million and $60 million per year (in 2011 PPP prices), see R. Benfica and A. Nin-Pratt, "Understanding the Impacts of R&D Investments on Poverty and Undernourishment: A Causal Mediation Approach," IFPRI Discussion Paper 2048 (Washington, DC: IFPRI, 2021).

22. M. Springmann, et al., "Options for Keeping the Food System within Environmental Limits," *Nature* 562 (2018): 519-525.

23. X. Xu, P. Sharma, S. Shu, T-S. Lin, P. Ciais, F.N. Tubiello, P. Smith, N. Campbell, and A.K. Jain, "Global Greenhouse Gas Emissions from Animal-based Foods are Twice Those of Plant-based Foods," *Nature Food* 2 (2021): 724-732.

24. Dalberg Asia, "Innovation Investment Study," (Colombo: CoSAI, 2021).

25. S.S. Myers, M.R. Smith, S. Guth, et al., "Climate Change and Global Food Systems: Potential Impacts on Food Security and Undernutrition," *Annual Review of Public Health* 20, 38 (2017): 259-277; M. Springmann, M. Clark, D. Mason-D'Croz, et al., "Options for Keeping the Food System within Environmental Limits," *Nature* 562 (2018): 519-525.

26. R. Benfica, J. Chambers, J. Koo, A. Nin-Pratt, J. Falck-Zepeda, G-J. Stads, and C. Arndt, *Food System Innovations and Digital Technologies to Foster Productivity Growth and Rural Transformation*, Food Systems Summit Brief (Bonn: IFPRI/United Nations Food Systems Summit, 2021).

27. A. Loboguerrero, B. Campbell, P. Cooper, J. Hansen, T. Rosenstock, and E. Wollenberg, "Food and Earth Systems: Priorities for Climate Change Adaptation and Mitigation for Agriculture and Food Systems," *Sustainability* 11, 5 (2019): 1372.

28. P.K. Thornton and M. Herrero. 2010. "Potential for Reduced Methane and Carbon Dioxide Emissions from Livestock and Pasture Management in the Tropics," *PNAS* 107 (2010): 19667-19672; P. Kurukulasuriya and S. Rosenthal, "Climate Change and Agriculture: A Review of Impacts and Adaptations," Climate Change Series Paper No. 91 (World Bank Group, Washington DC, 2003).

29. M.W. Rosegrant, J. Koo, N. Cenacchi, et al., *Food Security in a World of Natural Resource Scarcity: The Role of Agricultural Technologies* (Washington, DC: IFPRI, 2014).

30. IFPRI, *Estimating the Global Investment Gap in Research and Innovation for Sustainable Agriculture Intensification in the Global South* (Colombo: CoSAI, 2021).

31. E. Wollenberg, M. Richards, P. Smith, et al., "Reducing Emissions from Agriculture to Meet the 2°C target," *Global Change Biology* 22 (2016): 3859-3864.

32. UNEP (United Nations Environment Programme), *Food Waste Index Report 2021* (Nairobi: 2021).

33. J.M. Antle and S.M. Capalbo, "Adaptation of Agricultural and Food Systems to Climate Change: An Economic and Policy Perspective," *Applied Economic Perspectives and Policy* 32, 3 (2010): 386-416.

34. T. Reardon, R. Echeverria, J. Berdegué, B. Minten, S. Liverpool-Tasie, D. Tschirley, and D. Zilberman, "Rapid Transformation of Food Systems in Developing Regions: Highlighting the Role of Agricultural Research & Innovations," *Agricultural Systems* 172 (2019): 47-59.

35. T. Sulser, et al., *Climate Change and Hunger*, Food Policy Report (Washington, DC: IFPRI, 2021).

36. T. Sulser, et al., *Climate Change and Hunger*, Food Policy Report (Washington, DC: IFPRI, 2021).

37. G. Nelson, J. Bogard, K. Lividini, et al., "Income Growth and Climate Change Effects on Global Nutrition Security to Mid-Century," *Nature Sustainability* 1 (2018): 773-781.

38. C. Raymond, R. Horton, J. Zscheischler, et al., "Understanding and Managing Connected Extreme Events," *Nature Climate Change* 10 (2020): 611-621.

39. M.W. Rosegrant, J. Koo, N. Cenacchi, et al., *Food Security in a World of Natural Resource Scarcity* (Washington, DC: IFPRI, 2014).

40. ASTI, "CGIAR Data Tool | ASTI" (database), accessed Oct. 2021. https://www.asti.cgiar.org/cgiar-data

41. J. von Braun, K. Afsana, L.O. Fresco, and M. Hassan, "Food Systems: Seven Priorities to End Hunger and Protect the Planet," *Nature* 597, 7874 (2021): 28-30.

42. This prioritization exercise can be informed with the use of analytical tools such as the Rural Investment and Policy Analysis (RIAPA) economy-wide model that allows for the adequate prioritization of agricultural value chains and the identification of the policies and investments that maximize sustainable development objectives. See R. Benfica, B. Cunguara, and J. Thurlow, "Linking Agricultural Investments to Growth and Poverty: An Economywide Approach Applied to Mozambique," *Agricultural Systems* 172 (2019): 91-100; J. Thurlow, A. Sahoo, and R. Benfica, "Rural Investment and Policy Analysis (RIAPA): An Economy-wide Model for Prioritizing Value-Chain Investments in Developing Countries," (IFPRI, 2020 unpublished); and R. Benfica, "Agricultural R&D Investment and Policy Development Goals in sub-Saharan Africa: Assessing Prioritization of Value Chains in Senegal," IFPRI Discussion Paper 2102 (IFPRI, Washington, DC, 2022).

CHAPTER 5

1. See for instance, E. Díaz-Bonilla (*Financing SGD2 and Ending Hunger*, Food Systems Summit Brief [Bonn: United Nations Food Systems Summit, 2021]), which discusses the financing options for achieving SDG2, with an emphasis on zero hunger.

2. Article 2, Paragraph 1 (c). This is the third objective immediately after the two key objectives on mitigation and adaptation and resilience.

3. The UNFCCC defined "climate finance" as "local, national or transnational financing – drawn from public, private and alternative sources of financing – that seeks to support mitigation and adaptation actions that will address climate change."

4. IFPRI, *Estimating the Global Investment Gap in Research and Innovation for Sustainable Agriculture Intensification in the Global South* (Colombo: CoSAI, 2021).

5. R. Echeverría, "How to Feed the World Without Starving the Planet Is a $15 Billion Question," *Inter-Press Service*, August 23, 2021.

6. E. Díaz-Bonilla, J. Swinnen, and R. Vos, "Financing the Transformation to Healthy, Sustainable, and Equitable Food Systems," in *Global Food Policy Report 2021: Transforming Food Systems after COVID-19* (Washington, DC: IFPRI, 2021), 20–23.

7. UNFCCC SCF (Standing Committee on Finance), *Biennial Assessment and Overview of Climate Finance Flows*, Technical Report (New York: UNFCCC, 2021).

8. R. Macquarie, B. Naran, P. Rosane, M. Solomon, C. Wetherbee, and B. Buchner, *Updated View of the Global Landscape of Climate Finance 2019* (London: Climate Policy Initiative, 2020).

9. The data show financial flows that have been applied to a mitigation or adaptation activity, directly in the case of households and corporations (out of their own incomes/savings); indirectly in the case of development flows, banking systems and capital markets; and directly or indirectly by governments. The approach tries to avoid double counting. There is an important difference between the classification in Díaz-Bonilla, Swinnen, and Vos ("Financing the Transformation to Healthy, Sustainable, and Equitable Food Systems") and the Climate Policy Initiative (CPI) methodology: in Figure 1, consumer food expenditures are the cash flow of producers and operators in food systems, while in Table 1 expenditures by households and corporations are unrelated to each other. Both are expenditures on mitigation and adaptation activities and equipment (such as electric vehicles, solar panels, etc.) bought from third parties. We thank Baysa Naran from CPI for generously providing us with the data on AFOLU.

10. UNFCCC SCF, *Biennial Assessment and Overview of Climate Finance Flows*; R. Macquarie, et al., *Updated View of the Global Landscape of Climate Finance 2019* (London: Climate Policy Initiative, 2020).

11. E. Díaz-Bonilla, *Financing SGD2 and Ending Hunger*, Food Systems Summit Brief (Bonn: UNFSS, 2021).

12. R. Macquarie et al., *Updated View of the Global Landscape of Climate Finance 2019* (London: Climate Policy Initiative, 2020).

13. These include, among others, the Global Environment Facility (GEF), the Green Climate Fund (GCF), the Special Climate Change Fund (SCCF), the Least Developed Countries Fund (LDCF), and the Adaptation Fund (AF) (see C. Watson and L. Schalatek, *The Global Climate Finance Architecture*, Climate Finance Fundamentals No 2 (Washington, DC: Heinrich Böll Stiftung; London: Overseas Development Institute, 2021).

14. IFPRI, *Estimating the Global Investment Gap in Research and Innovation for Sustainable Agriculture Intensification in the Global South*; T. Sulser, K.D. Wiebe, S. Dunston, et al., *Climate Change and Hunger: Estimating Costs of Adaptation in the Agrifood System*, Food Policy Report (Washington, DC: IFPRI, 2021); D. Laborde, M. Parent, and C. Smaller, *Ending Hunger, Increasing Incomes, and Protecting the Climate: What Would it Cost Donors?* (Washington, DC: IFPRI; Winnipeg, Canada: IISD; Ithaca, New York: Cornell University, 2020); ZEF (Center for Development Research) and FAO, *Investment Costs and Policy Action Opportunities for Reaching a World without Hunger (SDG 2)*, Joint Report (Bonn: ZEF; Rome: FAO, 2020); (FOLU, 2019; UNEP, 2021)

15. IEA (International Energy Agency), *Net Zero by 2050: A Roadmap for the Global Energy Sector* (Paris: 2021).

16. E. Díaz-Bonilla, *Financing SGD2 and Ending Hunger*, Food Systems Summit Brief (Bonn: UNFSS, 2021).

17. E. Díaz-Bonilla, *Financing SGD2 and Ending Hunger*.

18. https://unfccc.int/news/end-of-coal-in-sight-at-cop26

19. UNFCCC SFC, *Biennial Assessment and Overview of Climate Finance Flows*.

20. CFLI (Climate Finance Leadership Initiative), EDFI (Association of European Development Finance Institutions) and the GIF (Global Infrastructure Facility), *Unlocking Private Climate Finance in Emerging Markets: Private Sector Considerations for Policymakers* (New York: 2021).

21. For a brief discussion of different experiences, see O. Wongpiyabovorn, A. Plastina, and S. Lence, "Futures Market for Ag Carbon Offsets under Mandatory and Voluntary Emission Targets," *Agricultural Policy Review* (Fall 2021).

22. If the focus is on other SDGs as well, such as those related to poverty and hunger, then social expenditures should also be expanded, the operations better focused, and the design of the programs improved (see the discussion in E. Díaz-Bonilla, *Financing SGD2 and Ending Hunger*; R. Echeverría, "How to Feed the World Without Starving the Planet is a $15 Billion Question").

23. These are 2019 current values, updated from E. Díaz-Bonilla, *Financing SGD2 and Ending Hunger* (2021). Consumers also have further expenditures and investments in energy and appliances related to the conservation and preparation of food.

24. The main options currently discussed do not seem to have multiplier effects in mobilizing further public or private money. Therefore, Díaz-Bonilla (*Financing SGD2 and Ending Hunger*; "Using the New IMF Special Drawing Rights for Larger Purposes: Guaranteeing 'Pandemic Recovery Bonds,'" *IFPRI Blog*, October 22, 2021) and J. von Braun and E. Díaz-Bonilla ("Letter: Perpetual Bonds Can Help States Fight Hunger," *Financial Times*, June 6, 2021) have suggested an alternative: using a percentage to SDRs to set up a fund to guarantee the issuing of "zero hunger bonds" or "pandemic recovery bonds" as perpetual bonds. This would have a larger multiplier effect (between 4 and almost 7 times the value of the guarantee fund), mobilizing further financial resources for humanitarian and environmental objectives.

25. Dalberg Asia, *Funding Agricultural Innovation for the Global South: Does It Promote Sustainable Agricultural Intensification?* (Colombo: CoSAI, 2021).

26. J. von Braun, K. Afsana, L. Fresco and M. Hassan, "Food Systems: Seven Priorities to End Hunger and Protect the Planet," *Nature* 597 (2021): 28–30.

27. E. Díaz-Bonilla, *Financing SGD2 and Ending Hunger* (UNFSS, 2021).

28. Extrapolating from I. Parry, S. Black, and N. Vernon, "Still Not Getting Energy Prices Right: A Global and Country Update of Fossil Fuel Subsidies," IMF Working Paper WP/21/236 (IMF, Washington, DC, 2021).

29. Parry, Black, and Vernon ("Still Not Getting Energy Prices Right: A Global and Country Update of Fossil Fuel Subsidies,") estimate the implicit subsidies of mispricing, the externalities of fossil fuels to be close to $5.1 trillion. If all the explicit and implicit subsidies were eliminated, then the revenues generated in 121 developing countries in 2025 would be about $3 trillion.

30. E. Díaz-Bonilla, *Financing SGD2 and Ending Hunger* (UNFSS, 2021).

31. R. Macquarie et al., *Updated View of the Global Landscape of Climate Finance 2019*.

32. SIFMA (Securities Industry and Financial Markets Association), *Capital Markets Fact Book* (New York: 2021).

33. R. Macquarie et al., *Updated View of the Global Landscape of Climate Finance 2019*.

34. A report by Rainforest Action Network (RAN), BankTrack, Indigenous Environmental Network (IEN), Oil Change International (OCI), Reclaim Finance, and the Sierra Club estimates that some US$3.8 trillion was invested by 60 major banks in supporting fossil fuel operations (*Banking on Climate Chaos*, Fossil Fuel Finance Report 2021).

35. There has been some confusion about the attention-grabbing number of $130 trillion that refers to the assets of the financial institutions involved. Assets are investment flows that have been already committed and allocated; to that extent, they cannot finance new investments. Even a process

of divestiture of financial instruments from climate-negative investments does not change much, considering that those financial instruments need to be bought by someone else (balancing the financial flows), while the physical asset (say a coal-fired energy plant) remains the same. And if the financial institution writes off that investment, it is basically giving a subsidy to the offending investment. What the institutions were pledging, as mentioned in the text, was to mobilize $100 trillion in new finance (a flow) until 2050 (see a discussion in E. Díaz-Bonilla, "Clarifying Some of the Confusion about Climate Finance in COP26," *IFPRI Blog*, Dec. 13, 2021).

36 See the proposal in E. Díaz-Bonilla, A.M. Loboguerrero, L. Verchot, E. Viglizzo, and A. Mirzabaev, *Financing "A Sustainable Food Future,"* T20 Argentina: Food Security and Sustainable Development Task Force Brief (Washington, DC: IFPRI, 2018); also A. Apampa, C. Clubb, B.E. Cosgrove et al., "Scaling Up Critical Finance for Sustainable Food Systems through Blended Finance," Discussion Paper (CGIAR Research Program on Climate Change, Agriculture and Food Security, Wageningen, Netherlands, 2021).

37 Using central bank discounts to finance credit lines for agriculture was the norm in many developing countries until the policy changes of the 1980s and 1990s. A current example of that approach is India; see a longer analysis in E. Díaz-Bonilla, *Macroeconomics, Agriculture, and Food Security: A Guide to Policy Analysis in Developing Countries* (Washington, DC: IFPRI, 2015).

38 See a discussion of the evolution of approaches toward public development banks in E. Díaz-Bonilla, *Macroeconomics, Agriculture, and Food Security*.

39 See the "Finance in Common Summit" organized in November 2020, as a joint initiative of the International Development Finance Club (IDFC), the World Federation of Development Finance Institutions (WFDFI), SAFIN, IFAD, and the Government of France.

40 A discussion of a variety of instruments can be found in E. Díaz-Bonilla, *Macroeconomics, Agriculture, and Food Security*.

41 As part of the UNFSS, countries submitted outlines for their plans to transform their food systems ("national pathways"). As of this writing some 111 countries have presented written outlines. They are mostly short qualitative statements that will have to be significantly expanded with quantitative objectives, instruments, organization and institutional aspects, costs, and financing, to become operational guide.

42 CFLI, EDFI, and GIF, *Unlocking Private Climate Finance in Emerging Markets: Private Sector Considerations for Policymakers*.

BOX 1

a Network for Greening the Financial System, *Adapting Central Bank Operations to a Hotter World: Reviewing Some Options* (Paris: 2021).

BOX 2

a UNFSS Finance Lever Group, *Food Finance Architecture: Financing a Healthy, Equitable and Sustainable Food System* (UNFSS, World Bank, Food and Land Use Coalition, and IFPRI 2021).

CHAPTER 6

1 K. Beegle, A. Coudouel, and E. Monsalve, *Realizing the Full Potential of Social Safety Nets in Africa*, Africa Development Forum (Washington, DC: World Bank, 2018).

2 M. Hidrobo, J. Hoddinott, N. Kumar, and M. Olivier, "Social Protection, Food Security, and Asset Formation," *World Development* 101 (2018): 88-103.

3 F. Bastagli, J. Hagen-Zanker, L. Harman, V. Barca, G. Sturge, and T. Schmidt with L. Pellerano, *Cash Transfers: What Does the Evidence Say? A Rigorous Review of Program Impact and of the Role of Design and Implementation Features* (London: Overseas Development Institute, 2016).

4 C.B. Barrett and M.A. Constas, "Toward a Theory of Resilience for International Development Applications," *PNAS* 111, 40 (2014): 14625-14630.

5 R. Godfrey Wood, "Is There a Role for Cash Transfers in Climate Change Adaptation?" *IDS Bulletin* 46, 2 (2011): 79-85.

6 K. Beegle, et al., *Realizing the Full Potential of Social Safety Nets in Africa*, Africa Development Forum (Washington, DC: World Bank, 2018).

7 T. Bowen, C. del Ninno, C. Andrews, et al., *Adaptive Social Protection: Building Resilience to Shocks*, International Development in Focus (Washington, DC: World Bank, 2020).

8 M. Hidrobo, et al., "Social Protection, Food Security, and Asset Formation," *World Development* 101 (2018): 88-103.

9 A. Fiszbein, R. Kanbur, and R. Yemtsov, "Social Protection and Poverty Reduction: Global Patterns and Some Targets," *World Development* 61 (2014): 167-177.

10 F. Bastagli, et al., *Cash Transfers: What Does the Evidence Say? A Rigorous Review of Programme Impact and of the Role of Design and Implementation Features* (London: Overseas Development Institute, 2016).

11 D.K. Evans, C. Gale and K. Kosec, "The Educational Impacts of Cash Transfers for Children with Multiple Indicators of Vulnerability," CGD Working Paper 563 (Center for Global Development, Washington, DC, 2021).

12 L. Phadera, H. Michelson, A. Winter-Nelson, and P. Goldsmith, "Do Asset Transfers Build Household Resilience?" *Journal of Development Economics* 138 (2019): 205-227.

13 D.O. Gilligan and J. Hoddinott, "Is There Persistence in the Impact of Emergency Food Aid? Evidence on Consumption, Food Security and Assets in Rural Ethiopia," *American Journal of Agricultural Economics* 89, 2 (2007): 225-242.

14 K. Ambler, A. de Brauw, and S. Godlonton, "Cash Transfers and Management Advice for Agriculture: Evidence from Senegal," *World Bank Economic Review* 34, 3 (2020): 597-617.

15 P.B. Siegel, J. Gatsinzi and A. Kettlewell, "Adaptive Social Protection in Rwanda: 'Climate-Proofing' the Vision 2020 Umurenge Programme," *IDS Bulletin* 42, 6 (2011): 71-78.

16 N. Rao, E.T. Lawson, W.N. Raditloaneng, D. Solomon, and M.N. Angula, "Gendered Vulnerabilities to Climate Change: Insights from the Semi-arid Regions of Africa and Asia," *Climate and Development* 11, 1 (2017): 14-26.

17 H.N. Adam, "Mainstreaming Adaptation in India: The Mahatma Gandhi National Rural Employment Guarantee Act and Climate Change," *Climate and Development* 7, 2 (2015): 142-152.

18 J.D. Tenzing, 2019. "Integrating Social Protection and Climate Change Adaptation: A Review," *WIREs Climate Change* 11, 2 (2019): e626.

19 K. Ambler, A. de Brauw, and S. Godlonton, "Agriculture Support Services in Malawi: Direct Effects, Complementarities, and Time Dynamics," IFPRI Discussion Paper 1725 (IFPRI, Washington, DC, 2018).

20 K. Hirvonen, E. Machado, A. Simons and V. Taraz, "More Than a Safety Net: Ethiopia's Flagship Public Works Program Increases Tree Cover," SSRN pre-print, available online March 25, 2021.

21. N. Jensen, M. Ikegami, and A. Mude, "Integrating Social Protection Strategies for Improved Impact: A Comparative Evaluation of Cash Transfers and Index Insurance in Kenya," *Geneva Papers on Risk and Insurance-Issues and Practice* 42 (2017): 675-707.

22. A. Amare, B. Simane, J. Nyangaga, A. Defisa, D. Hamza, and B. Gurmessa, "Index-based Livestock Insurance to Manage Climate Risks in Borena Zone of Southern Oromia, Ethiopia," *Climate Risk Management* 25 (2019): 100191.

23. G. Bryan, S. Chowdhury, and A.M. Mobarak, "Underinvestment in a Profitable Technology: The Case of Seasonal Migration in Bangladesh," *Econometrica* 82, 5 (2014): 1671-1748.

24. World Bank, "The World Bank in Social Protection: Overview," last updated April 10, 2020.

25. T. Bowen, C. del Ninno, C. Andrews, et al., *Adaptive Social Protection: Building Resilience to Shocks*, International Development in Focus (Washington, DC: World Bank, 2020).

26. U. Gentilini, M. Almenfi, J. Blomquist, et al., "Social Protection and Jobs Responses to COVID-19: A Real-Time Review of Country Measures" (Washington, DC: World Bank, updated May 14, 2021).

27. E. Knippenberg and J. Hoddinott, "Shocks, Social Protection, and Resilience: Evidence from Ethiopia," Ethiopia Strategy Support Program Working Paper 109 (IFPRI, Washington, DC, 2017).

28. H. Alderman, U. Gentilini, and R. Yemtsov, *The 1.5 Billion People Question: Food, Vouchers or Cash Transfers?* (Washington, DC: World Bank, 2018).

29. S. Devereux and J. Cuesta, "Urban-Sensitive Social Protection: How Universalised Social Protection Can Reduce Urban Vulnerabilities Post COVID-19," *Progress in Development Studies* 21, 4 (2021): 340-360.

BOX 1

a. C. Andrews, A. de Montesquiou, I. Arévalo Sánchez, P. Vasudeva Dutta, B. Varghese Paul, S. Samaranayake, J. Heisey, T. Clay, and S. Chaudhary, *The State of Economic Inclusion Report 2021: The Potential to Scale* (Washington, DC: World Bank, 2021).

b. Urban Climate Change Research Network, *The Future We Don't Want: How Climate Change Could Impact the World's Greatest Cities*, Technical Report (New York: 2018).

CHAPTER 7

1. J.C. Milder, L.E. Buck, F. DeClerck, and S.J. Scherr, "Landscape Approaches to Achieving Food Production, Natural Resource Conservation, and the Millennium Development Goals," in *Integrating Ecology and Poverty Reduction*, eds. J.C. Ingram, F. DeClerck, and C. Rumbaitis del Rio (New York: Springer, 2012), 77-108; P.A. Minang, M. van Noordwijk, et al. (eds.), *Climate-Smart Landscapes: Multifunctionality in Practice* (Nairobi: World Agroforestry Centre [ICRAF], 2015).

2. C. Mbow, C. Neely, and P. Dobie, "How Can an Integrated Landscape Approach Contribute to the Implementation of the Sustainable Development Goals (SDGs) and Advance Climate-Smart Objectives?" in *Climate-Smart Landscapes*, eds. P.A. Minang, et al. (Nairobi: World Agroforestry Centre, 2015), 103-118; J. Sayer, "Forward" in *Climate-Smart Landscapes*, eds. P.A. Minang, et al. (Nairobi: World Agroforestry Centre, 2015), xviii-xix.

3. J.C. Milder, L.E. Buck, F. DeClerck, and S.J. Scherr, "Landscape Approaches to Achieving Food Production, Natural Resource Conservation, and the Millennium Development Goals," in *Integrating Ecology and Poverty Reduction*, eds. J.C. Ingram, F. DeClerck, and C. Rumbaitis del Rio (New York: Springer, 2012), 77-108; P.A. Minang, M. van Noordwijk, O.E. Freeman, et al. (eds.) *Climate-Smart Landscapes* (Nairobi: World Agroforestry Centre, 2015).

4. R. Meinzen-Dick, Q. Bernier, and E. Haglund, "The Six "Ins" of Climate-Smart Agriculture: Inclusive Institutions for Information, Innovation, Investment, and Insurance," CAPRi Working Paper No. 114 (IFPRI, Washington, DC, 2013).

5. J. Sayer, "Forward," in *Climate-Smart Landscapes*, eds. P.A. Minang, et al. (Nairobi: World Agroforestry Centre, 2015), xviii-xix.

6. T.D. Nielsen, "From REDD+ Forests to Green Landscapes? Analyzing the Emerging Integrated Landscape Approach Discourse in the UNFCCC," *Forest Policy and Economics* 73 (2016): 177-184.

7. J. Sayer, "Forward," in *Climate-Smart Landscapes*, eds. P.A. Minang, et al. (Nairobi: World Agroforestry Centre, 2015), xviii-xix.

8. C. Mbow, C. Neely, and P. Dobie, "How Can an Integrated Landscape Approach Contribute to the Implementation of the SDGs and Advance Climate-Smart Objectives?" in *Climate-Smart Landscapes*, eds. P.A. Minang, et al. (Nairobi: World Agroforestry Centre, 2015), 103-118; J. Reed, A. Ickowitz, C. Chervier, et al., "Integrated Landscape Approaches in the Tropics: A Brief Stock-Take," *Land Use Policy* 99 (2020): 104822.

9. J. Sayer, T. Sunderland, J. Ghazoul, et al., "Ten Principles for a Landscape Approach to Reconciling Agriculture, Conservation, and Other Competing Land Uses," *PNAS* 110, 21 (2013): 8349-8356; J. Reed, et al., "Integrated Landscape Approaches to Managing Social and Environmental Issues in the Tropics: Learning from the Past to Guide the Future," *Global Change Biology* 22, 7 (2016): 2540-2554; J. Reed, et al., "Integrated Landscape Approaches in the Tropics: A Brief Stock-Take," *Land Use Policy* 99 (2020): 104822.

10. J. Sayer, et al., "Ten Principles for a Landscape Approach to Reconciling Agriculture, Conservation, and Other Competing Land Uses," *PNAS* 110, 21 (2013): 8349-8356.

11. C. Mbow, C. Neely, and P. Dobie, "How Can an Integrated Landscape Approach Contribute to the Implementation of the SDGs and Advance Climate-Smart Objectives?" in *Climate-Smart Landscapes*, eds. P.A. Minang, et al. (Nairobi: World Agroforestry Centre, 2015), 103-118.

12. P.D. Aligica and V. Tarko, "Polycentricity: From Polanyi to Ostrom, and Beyond," *Governance* 25, 2 (2012): 237-262; A. Thiel, W. Blomquist, and D. Garrick (eds.), *Governing Complexity: Analyzing and Applying Polycentricity* (Cambridge: Cambridge University Press, 2019).

13. M. Lazdinis, P. Angelstam, and H. Pülzl, "Towards Sustainable Forest Management in the European Union through Polycentric Forest Governance and an Integrated Landscape Approach," *Landscape Ecology* 34, 7 (2019): 1737-1749.

14. R. Meinzen-Dick, R. Chaturvedi, S. Kandikuppa, K. Rao, J.P. Rao, B. Bruns, and H. ElDidi, "Securing the Commons in India: Mapping Polycentric Governance," *International Journal of the Commons* 15, 1 (2021): 218-235.

15. M.A. Hajer, "Policy without Polity? Policy Analysis and the Institutional Void," *Policy Sciences* 36, 2 (2003): 175-195.

16. C. Mbow, C. Neely, and P. Dobie, "How Can an Integrated Landscape Approach Contribute to the Implementation of the SDGs and Advance Climate-Smart Objectives?" in *Climate-Smart Landscapes*, eds. P.A. Minang, et al. (Nairobi: World Agroforestry Centre, 2015), 103-118.

17. M. Lazdinis, P. Angelstam, and H. Pülzl, "Towards Sustainable Forest Management in the European Union through Polycentric Forest Governance and an Integrated Landscape Approach," *Landscape Ecology* 34, 7 (2019): 1737-1749; J. Sayer, "Forward," in *Climate-Smart Landscapes*,

eds. P.A. Minang, et al. (Nairobi: World Agroforestry Centre, 2015), xviii–xix.

18 O.E. Freeman, L.A. Duguma, and P.A. Minang, "Operationalizing the Integrated Landscape Approach in Practice," *Ecology and Society* 20, 1 (2015): 24; J. Sayer, "Forward," in *Climate-Smart Landscapes*, eds. P.A. Minang, et al. (Nairobi: World Agroforestry Centre, 2015), xviii–xix.

19 F. Engels, A. Wentland, and S.M. Pfotenhauer, "Testing Future Societies? Developing a Framework for Test Beds and Living Labs as Instruments of Innovation Governance," *Research Policy* 48, 9 (2019): 103826.

20 Y. Voytenko, K. McCormick, J. Evans, and G. Schliwa, "Urban Living for Sustainability and Low Carbon Cities in Europe: Towards a Research Agenda," *Journal of Cleaner Production* 123 (2016): 45-54.

21 European Network of Living Labs (ENoLL), accessed July 18, 2021, https://enoll.org/about-us/what-are-living-labs/; S.M. Pfotenhauer, "Co-producing Emirati Science and Society at Masdar Institute of Science and Technology," in *Accelerating Science and Technology Development in the Middle East: Unleashing the Potential of Near Ties*, eds. A. Siddiqi and L. Diaz-Anadon (Berlin: Gerlach Press, 2017).

22 P. Angelstam, G. Barnes, M. Elbakidze, et al., "Collaborative Learning to Unlock Investments for Functional Ecological Infrastructure: Bridging Barriers in Social-Ecological Systems in South Africa," *Ecosystem Services* 27, Part B (2017): 291-304; P. Angelstam, M. Elbakidze, A. Lawrence, et al., "Barriers and Bridges for Landscape Stewardship and Knowledge Production to Sustain Functional Green Infrastructures," in *Ecosystem Services from Forest Landscapes*, eds. A. Pereira, U. Peterson, G. Pastur, and L. Iverson (Berlin: Springer, 2018), 127-167.

23 J. Reed, et al., "Integrated Landscape Approaches in the Tropics: A Brief Stock-Take," *Land Use Policy* 99 (2020): 104822.

24 L. Denier, S. Scherr, S. Shames, P. Chatterton, L. Hovani, and N. Stam, *The Little Sustainable Landscapes Book: Achieving Sustainable Development Through Integrated Landscape Management* (Oxford: Global Canopy Programme, 2015); K. Kusters, L. Buck, M. de Graaf, P. Minang, C. van Oosten, and R. Zagt, "Participatory Planning, Monitoring and Evaluation of Multistakeholder Platforms in Integrated Landscape Initiatives," *Environmental Management* 62, 1 (2018): 170-181.

25 J. Reed, et al. "Integrated Landscape Approaches in the Tropics: A Brief Stock-Take," *Land Use Policy* 99 (2020): 104822.

26 M. Markopoulos, *Collaboration and Multi-Stakeholder Dialogue: A Review of the Literature*, Version 1.1 (Gland, Switzerland: IUCN, 2012).

27 R. Meinzen-Dick, et al., "Securing the Commons in India: Mapping Polycentric Governance," *International Journal of the Commons* 15, 1 (2021): 218-235.

28 M. Acosta, E.L. Ampaire, P. Muchunguzi, et al., "The Role of Multi-Stakeholder Platforms for Creating an Enabling Climate Change Policy Environment in East Africa," in *The Climate-Smart Agriculture Papers*, eds. T. Rosenstock, A. Nowak, and E. Girvetz (Cham, Switzerland: Springer, 2019), 267-276.

29 B.D. Ratner, A.M. Larson, J.P. Sarmiento Barletti, et al., "Multi-Stakeholder Platforms for Natural Resource Governance: Lessons from Eight Landscape-Level Cases," *Ecology and Society* 27, 2 (2022): 2.

30 A.K. Hart, P. McMichael, J.C. Milder, and S.J. Scherr, "Multi-Functional Landscapes from the Grassroots? The Role of Rural Producer Movements," *Agriculture and Human Values* 33, 2 (2016): 305-322.

31 N. Faysse, "Troubles on the Way: An Analysis of the Challenges Faced by Multi-Stakeholder Platforms," *Natural Resources Forum* 30, 3 (2006): 219-229; B.D. Ratner, et al., "Multi-Stakeholder Platforms for Natural Resource Governance: Lessons from Eight Landscape-Level Cases," *Ecology and Society*, 27, 2 (2022): 2.

32 L. d'Armengol, M.P. Castillo, I. Ruiz-Mallén, and E. Corbera, "A Systematic Review of Co-Managed Small-Scale Fisheries: Social Diversity and Adaptive Management Improve Outcomes." *Global Environmental Change* 52 (2018): 212-225.

33 B.D. Ratner, et al., "Multi-Stakeholder Platforms for Natural Resource Governance: Lessons from Eight Landscape-Level Cases," *Ecology and Society*, 27, 2 (2022): 2.

34 P. Mudliar and T. Koontz, "The Muting and Unmuting of Caste across Inter-Linked Action Arenas: Inequality and Collective Action in a Community-Based Watershed Group," *International Journal of the Commons* 12, 1 (2018): 225-248.

35 M. di Gregorio, K. Hagedorn, M. Kirk, B. Korf, N. McCarthy, and R. Meinzen-Dick, "Property Rights, Collective Action, and Poverty: The Role of Institutions for Poverty Reduction," CAPRi Working Paper No. 81 (IFPRI, Washington, DC, 2008); N. Faysse, "Troubles on the Way: An Analysis of the Challenges Faced by Multi-Stakeholder Platforms," *Natural Resources Forum* 30, 3 (2006): 219-229.

36 K. Kusters, L. Buck, M. de Graaf, P. Minang, C. van Oosten, and R. Zagt, "Participatory Planning, Monitoring and Evaluation of Multistakeholder Platforms in Integrated Landscape Initiatives," *Environmental Management* 62, 1 (2018): 170-181.

37 B.D. Ratner, et al., "Multi-Stakeholder Platforms for Natural Resource Governance: Lessons from Eight Landscape-Level Cases," *Ecology and Society*, 27, 2 (2022): 2.

38 K. Deininger, D.A. Ali, S. Holden, and J. Zevenbergen, "Rural Land Certification in Ethiopia: Process, Initial Impact, and Implications for Other African Countries," *World Development* 36, 10 (2008): 1786-1812; D.A. Ali, K. Deininger, and M. Goldstein, "Environmental and Gender Impacts of Land Tenure Regularization in Africa: Pilot Evidence from Rwanda," *Journal of Development Economics* 110 (2014): 262-275.

39 D. Higgins, T. Balint, H. Liversage, and P. Winters, "Investigating the Impacts of Increased Rural Land Tenure Security-A Systematic Review of the Evidence," *Journal of Rural Studies* 61 (2018): 34-62.

40 C. Doss and R. Meinzen-Dick, "Land Tenure Security for Women: A Conceptual Framework," *Land Use Policy* 99 (2020): 105080.

BOX 1

a P.O. Waeber, A. Fellay, R. Carmenta, et al., "A Mixing Board to Break Down the Complexities of Integrated Landscape Approaches," in *Forest Values and Landscape Approaches to Protect Them* (Montpellier, France: European Conference for Tropical Ecology, 2022).

b S.J. Scherr, S. Shames, and R. Friedman, *Defining Integrated Landscape Management for Policy Makers*, Ecoagriculture Policy Focus No. 10 (Oakton, VA: Ecoagriculture Partners, 2013); J. Reed, J. van Vianen, E.L. Deakin, J. Barlow, and T. Sunderland, "Integrated Landscape Approaches to Managing Social and Environmental Issues in the Tropics: Learning from the Past to Guide the Future," *Global Change Biology* 22, 7 (2016): 2540-2554; J. Erbaugh and A. Agrawal, "Clarifying the Landscape Approach: A Letter to the Editor on 'Integrated Landscape Approaches to Managing Social and Environmental Issues in the Tropics,'" *Global Change Biology* 23, 11 (2017): 4453-4454.

c S.J. Scherr and J.A. McNeely, "Biodiversity Conservation and Agricultural Sustainability: Towards a New Paradigm of 'Ecoagriculture' Landscapes," *Philosophical Transactions of the Royal Society B* 363, 1491 (2007): 477-494; J. Reed, L. Deakin, and T. Sunder"What Are

'Integrated Landscape-Approaches' and How Effectively Have They Been Implemented in the Tropics: A Systematic Map Protocol," *Environmental Evidence* 4 (2015): 2; J. Reed, et al., "Integrated Landscape Approaches to Managing Social and Environmental Issues in the Tropics: Learning from the Past to Guide the Future," *Global Change Biology* 22, 7 (2016): 2540-2554.

BOX 2

a J. Warner, P. Wester, and A. Bolding, "Going with the Flow: River Basins as the Natural Units for Water Management?" *Water Policy* 10, S2 (2008): 121-138.

CHAPTER 8

1 Global Panel on Agriculture and Food Systems for Nutrition, *Future Food Systems: For People, Our Planet, and Prosperity* (London: 2020).

2 W. Willett, J. Rockström, B. Loken, et al, "Food in the Anthropocene: The EAT-Lancet Commission on Healthy Diets from Sustainable Food Systems," *Lancet* 393, 10170 (2019): 447-492; IPCC (Intergovernmental Panel on Climate Change), "Food Security," in *Climate Change and Land: An IPCC Special Report on Climate Change, Desertification, Land Degradation, Sustainable Land Management, Food Security and Greenhouse Gas Fluxes in Terrestrial Ecosystems*, eds. P.R. Shukla, J. Skea, E. Calvo Buendia, et al. (Geneva: 2019), 437-550; J.A. Foley, N. Ramankutty, K.A. Brauman, et al., "Solutions for a Cultivated Planet," *Nature* 478 (2011): 337-342.

3 W. Willett, e al., "Food in the Anthropocene: the EAT-Lancet Commission on Healthy Diets from Sustainable Food Systems," *Lancet* 393, 10170 (2019): 447-492; M. Crippa, E. Solazzo, D. Guizzardi, F. Monforti-Ferrario, F.N. Tubiello, and A. Leip, "Food Systems Are Responsible for a Third of Global Anthropogenic GHG Emissions," *Nature Food* 2 (2021): 198-209; R.D. Semba, S. Askari, M.W. Bloem, and K. Kraemer, "The Potential Impact of Climate Change on the Micronutrient-Rich Food Supply," *Advances in Nutrition* 13 1 (2022): 80-100.

4 S.S. Myers, M.R. Smith, S. Guth, C.D. Golden, B. Vaitla, N.D. Mueller, A.D. Dangour, and P. Huybers, "Climate Change and Global Food Systems: Potential Impacts on Food Security and Undernutrition," *Annual Review of Public Health* 38 (2017): 259-277; T. Wheeler and J. von Braun, "Climate Change Impacts on Global Food Security," *Science* 34, 6145 (2013): 508-513; J. Fanzo, C. Davis, R. McLaren, and J. Choufani, "The Effect of Climate Change across Food Systems: Implications for Nutrition Outcomes," *Global Food Security* 18 (2018): 12-19; B.A. Swinburn, V.I. Kraak, S. Allender, et al., "The Global Syndemic of Obesity, Undernutrition, and Climate Change: The Lancet Commission Report," *Lancet* 393, 10173 (2019): 791-846.

5 J. Fanzo, et al., "The Effect of Climate Change across Food Systems: Implications for Nutrition Outcomes," *Global Food Security* 18 (2018): 12-19; M.W. Cooper, M.E. Brown, S. Hochrainer-Stigler, G. Pflug, I. McCallum, S. Fritz, J. Silva, and A. Zvoleff, "Mapping the Effects of Drought on Child Stunting," *PNAS* 116, 35 (2019): 17219-17224; R.K. Phalkey, C. Aranda-Jan, S. Marx, B. Höfle, and R. Sauerborn, "Systematic Review of Current Efforts to Quantify the Impacts of Climate Change on Undernutrition," *PNAS* 112, 33 (2015): E4522-E4529; M.T. Niles, B.F. Emery, S. Wiltshire, M.E. Brown, B. Fisher, and T.H. Ricketts, "Climate Impacts Associated with Reduced Diet Diversity in Children across Nineteen Countries," *Environmental Research Letters* 16 (2021): 015010.

6 L. Salm, N. Nisbett, L. Cramer, S. Gillespie, and P. Thornton, "How Climate Change Interacts with Inequity to Affect Nutrition," *Wiley Interdisciplinary Review Climate Change* 12, 2 (2021): e696; B.S. Levy and J.A. Patz, "Climate Change, Human Rights, and Social Justice," *Annals of Global Health* 81, 2 (2015): 310-322.

7 M. Springmann, M. Clark, D. Mason-D'Croz, et al., "Options for Keeping the Food System within Environmental Limits," *Nature* 562 (2018): 519-525; B. Swinburn, "Power Dynamics in 21st-Century Food System," *Nutrients* 11, 10 (2019): 2544; J. Macdiarmid, A.D. Jones, J. Ranganathan, M. Herrero, and J. Fanzo, "The Role of Healthy Diets in Environmentally Sustainable Food Systems," *Food and Nutrition Bulletin* 41, S2 (2020): 31S-58S.

8 Ultra-processing refers to the processing of industrial ingredients derived from foods, including moulding, hydrogenation, and hydrolysis. Ultra-processed foods often include additives such as preservatives, sweeteners, sensory enhancers, colorants, flavours, and processing aids.

9 T. Reardon, D. Tschirley, L.S.O Liverpool-Tasie, et al., "The Processed Food Revolution in African Food Systems and the Double Burden of Malnutrition," *Global Food Security* 28 (2021): 100466; B.M. Popkin and T. Reardon, "Obesity and the Food System Transformation in Latin America," *Obesity Reviews* 19, 8 (2018): 1028-1064; B.M. Popkin, C. Corvalan, and L.M. Grummer-Strawn, "Dynamics of the Double Burden of Malnutrition and the Changing Nutrition Reality," *Lancet* 395, 10217 (2020): 65-74; P. Baker, P. Machado, T. Santos, et al., "Ultra-Processed Foods and the Nutrition Transition: Global, Regional and National Trends, Food Systems Transformations and Political Economy Drivers," *Obesity Reviews* 21, 12 (2020): e13126.

10 G. Wu, J. Fanzo, D.D. Miller, P. Pingali, M. Post, J.L. Steiner, and A.E. Thalacker-Mercer, "Production and Supply of High-Quality Food Protein for Human Consumption: Sustainability, Challenges, and Innovations," *Annals of the New York Academy of Sciences* 1321, 1 (2014): 1-19.

11 P. Baker, P. Machado, T. Santos, et al., "Ultra-Processed Foods and the Nutrition Transition: Global, Regional and National Trends, Food Systems Transformations and Political Economy Drivers," *Obesity Reviews* 21, 12 (2020): e13126; I.D. Brouwer, M.J. van Liere, A. de Brauw, et al., "Reverse Thinking: Taking a Healthy Diet Perspective towards Food Systems Transformations," *Food Security* 13 (2021): 1497-1523; M.A. Clark, M. Springmann, J. Hill, and D. Tilman, "Multiple Health and Environmental Impacts of Foods," *PNAS* 116, 46 (2019): 23357-23362.

12 FAO (Food and Agricultural Organization of the United Nations), FAOStat, (database), accessed Sept. 2021.

13 M.T. Herrero, D. Mason-D'Croz, P.K. Thornton, et al., *Livestock and Sustainable Food Systems-Status, Trends, and Priority Actions*, UN Food Systems Summit 2021 (Geneva: United Nations, 2021).

14 A. Clonan, K.E. Roberts, and M. Holdsworth, "Socioeconomic and Demographic Drivers of Red and Processed Meat Consumption: Implications for Health and Environmental Sustainability," *Proceedings of the Nutrition Society* 75 (2016): 367-373.

15 D. Mason-D'Croz, J.R. Bogard, T.B. Sulser, N. Cenacchi, S. Dunston, M. Herrero, and K. Wiebe, "Gaps between Fruit and Vegetable Production, Demand, and Recommended Consumption at Global and National Levels: An Integrated Modelling Study," *Lancet Planetary Health* 3, 7 (2019): e318-e329.

16 W. Bell, K. Lividini, and W.A. Masters, "Global Dietary Convergence from 1970 to 2010 Altered Inequality in Agriculture, Nutrition and Health," *Nature Food* 2 (2021): 156-165; V. Miller, P. Webb, R. Micha, D. Mozaffarian, and Global Dietary Database, "Defining Diet Quality: A Synthesis of Dietary Quality Metrics and Their Validity for the Double Burden of Malnutrition," *Lancet Planet Health* 4, 8 (2020): e352-e370.

17 P. Baker, P. Machado, T. Santos, et al., "Ultra-Processed Foods and the Nutrition Transition: Global, Regional and National Trends, Food Systems Transformations and Political Economy Drivers," *Obesity Reviews* 21,

12 (2020): e13126; P. Baker, A. Kay, and H. Walls, "Trade and Investment Liberalization and Asia's Noncommunicable Disease Epidemic: A Synthesis of Data and Existing Literature," *Globalalization and Health* 10 (2014): 66; B.J. Rolls, P.M. Cunningham, and H.E. Diktas, "Properties of Ultraprocessed Foods That Can Drive Excess Intake," *Nutrition Today* 55, 3 (2020): 109–115; K. Buse, S. Tanaka, and S. Hawkes, "Healthy People and Healthy Profits? Elaborating a Conceptual Framework for Governing the Commercial Determinants of Non-Communicable Diseases and Identifying Options for Reducing Risk Exposure," *Globalization and Health* 13 (2017): 34.

18 R. Parajuli, G. Thoma, and M.D. Matlock, "Environmental Sustainability of Fruit and Vegetable Production Supply Chains in the Face of Climate Change: A Review," *Science of the Total Environment* 650, 2 (2019): 2863–2879.

19 M. Springmann, M. Clark, D. Mason-D'Croz, et al., "Options for Keeping the Food System within Environmental Limits," *Nature* 562 (2018): 519–525; M.A. Clark, M. Springmann, J. Hill, and D. Tilman, "Multiple Health and Environmental Impacts of Foods," *PNAS* 116, 46 (2019): 23357–23362.

20 S. Clune, E. Crossin, and K. Verghese, "Systematic Review of Greenhouse Gas Emissions for Different Fresh Food Categories," *Journal of Cleaner Production* 140, 2 (2017): 766–783.

21 F.N. Tubiello, C. Rosenzweig, G. Conchedda, et al., "Greenhouse Gas Emissions from Food Systems: Building the Evidence Base," *Environmental Research Letters* 16 (2021): 065007.

22 P. Seferidi, D. Scrinis, I. Huybrechts, J. Woods, P. Vineis, and C. Millett, "The Neglected Environmental Impacts of Ultra-Processed Foods," *Lancet Planetary Health* 4, 10 (2020): e437–e438.

23 K. Anastasiou, P. Baker, M. Hadjikakou, G.A. Hendrie, and M. Lawrence, "A Review of the Environmental Impacts of Ultra-processed Foods: What Are the Implications for Sustainable Food Systems?" (forthcoming).

24 U. Jaroenkietkajorn and S.H. Gheewala, "Understanding the Impacts on Land Use Through Ghg-Water-Land-Biodiversity Nexus: The Case of Oil Palm Plantations in Thailand," *Science of the Total Environment* 800 (2021): 149425; J. Yates, M. Deeney, H.B. Rolker, H. White, S. Kalamatianou, and S. Kadiyala, "A Systematic Scoping Review of Environmental, Food Security and Health Impacts of Food System Plastics," *Nature Food* 2 (2021): 80–87.

25 J.T. de Silva, J.M.F. Garzillo, F. Rauber, et al., "Greenhouse Gas Emissions, Water Footprint, and Ecological Footprint of Food Purchases According to Their Degree of Processing in Brazilian Metropolitan Areas: A Time-Series Study from 1987 to 2018," *Lancet Planetary Health* 5, 11 (2021): e775–e785.

26 W. Willett, et al., "Food in the Anthropocene: the EAT-Lancet Commission on Healthy Diets from Sustainable Food Systems," *Lancet* 393, 10170 (2019): 447–492.

27 T. Beal, F. Ortenzi, and J. Fanzo, "Uncertainties in the Micronutrient Adequacy fo the EAT-*Lancet* Planetary Health Diet" (forthcoming).

28 L.A. Moreno, R. Meyer, S.M. Donovan, O. Goulet, J. Haines, F.J. Kok, and P van't Veer, "Perspective: Striking a Balance Between Planetary and Human Health: Is There a Path Forward?" *Advances in Nutrition* (2021): nmab139.

29 K. Hirvonen, Y. Bai, D. Headey, and W.A. Masters, "Affordability of the EAT-Lancet Reference Diet: a Global Analysis," *Lancet Global Health* 8, 1 (2020): e59–e66.

30 D.D Headey and H.H. Alderman, "The Relative Caloric Prices of Healthy and Unhealthy Foods Differ Systematically across Income Levels and Continents," *Journal of Nutrition* 149, 11 (2019): 2020–2033.

31 T-A. Kenny, M. Fillion, J. MacLean, S.D. Wesche, and H.M. Chan, "Calories Are Cheap, Nutrients Are Expensive-The Challenge of Healthy Living in Arctic Communities," *Food Policy* 80 (2018): 39–54; K.B. Amolegbe, J. Upton, E. Bageant, and S. Blom, "Food Price Volatility and Household Food Security: Evidence from Nigeria," *Food Policy* 102 (2021): 102061; S. Charlebois, M. McCormick, and M. Juhasz, "Meat Consumption and Higher Prices: Discrete Determinants Affecting Meat Reduction or Avoidance amidst Retail Price Volatility," *British Food Journal* 118, 9 (2016): 2251–2270.

32 K. Wiebe, H. Lotze-Campen, R. Sands, et al., "Climate Change Impacts on Agriculture in 2050 under a Range of Plausible Socioeconomic and Emissions Scenarios," *Environmental Research Letters* 10 (2015): 085010.

33 T. Reardon, et al., "The Processed Food Revolution in African Food Systems and the Double Burden of Malnutrition," *Global Food Security* 28 (2021): 100466; C.M. Sauer, T. Reardon, D. Tschirley et al., "Consumption of Processed Food and Food Away from Home in Big Cities, Small Towns, and Rural Areas of Tanzania," *Agricultural Economics* 52, 5 (2021): 749–770.

34 I.D. Brouwer, M.J. van Liere, A. de Brauw et al., ""Reverse Thinking: Taking a Healthy Diet Perspective towards Food Systems Transformations," *Food Security* 13 (2021): 1497–1523.

35 Income elasticities of demand measure the responsiveness of demand for a particular good to changes in consumer income. Price elasticities of demand measure the responsiveness of demand for a particular good to changes in its price.

36 L. Colen, P.C. Melo, Y. Abdul-Salam, D. Roberts, S. Mary, Y. Gomez, and S. Paloma, "Income Elsaticities for Food, Calories, and Nutrients across Africa: A Meta-Analysis," *Food Policy* 77 (2018): 116–132.

37 "Edutainment" is a form of entertainment that includes an educational component. For example, *C'est la Vie* is a TV series produced in Senegal that includes messaging on women's empowerment, sexual abuse, and use of contraceptives, among other topics.

38 I.D. Brouwer, et al., "Reverse Thinking: Taking a Healthy Diet Perspective towards Food Systems Transformations," *Food Security* 13 (2021): 1497–1523.

39 P. Menon, P.. Nguyen, K.K. Saha, et al., "Impacts on Breastfeeding Practices of At-Scale Strategies that Combine Intensive Interpersonal Counseling, Mass Media, and Community Mobilization: Results of Cluster-Randomized Program Evaluations in Bangladesh and Viet Nam," *PLoS Medicine* 13 (2018): e1002159; S.K. Kim, S. Park, J. Oh, J. Kim, and S. Ahn, "Interventions Promoting Exclusive Breastfeeding up to Six Months after Birth: A Systematic Review and Meta-analysis of Randomized Controlled Trials," *International Journal of Nursing Studies* 80 (2018): 94–105.

40 C.G. Victora, R. Bahl, A.J.D. Barros, et al., "Breastfeeding in the 21st Century: Epidemiology, Mechanism, and Lifelong Effect," *Lancet* 387, 10017 (2020): 475–490.

41 J.P. Smith, "A Commentary on the Carbon Footprint of Milk Formula: Harms to Planetary Health and Policy Implications," *International Breastfeeding Journal* 14 (2019): 49.

42 M.T. Ruel, H. Alderman, and Maternal and Child Nutrition Study Group, "Nutrition-Sensitive Interventions and Programmes: How Can They Help to Accelerate Progress in Improving Maternal and Child Nutrition?" *Lancet* 382, 9891 (2013) 536–551; C. Hawkes, M.T. Ruel, L. Salm, B. Sinclair, and F. Branca, "Double-Duty Actions: Seizing Programme and Policy Opportunities to Address Malnutrition in All Its Forms," *Lancet* 395, 10218 (2020): 142–155; B.M Popkin, S. Barquera, C. Corvalan, K.J. Hofman, C. Monteiro, S.W. Ng, E.C. Swart, and L.S. Taillie, "Towards Unified and Impactful Policies to Reduce Ultra-Processed Food Consumption and Promote Healthier Eating," *Lancet Diabetes & Endocrinology* 9, 7 (2021): 462–470.

43 A. Herforth, M. Arimond, C. Álvarez-Sánchez, J. Coates, K. Christianson, and E. Muehlhoff, "A Global Review of Food-Based Dietary guidelines," *Advances in Nutrition* 10, 4 (2019): nmy130.

44. L.A. Moreno, et al., "Perspective: Striking a Balance Between Planetary and Human Health: Is There a Path Forward?" *Advances in Nutrition* (2021): nmab139; C.W. Binns, M.K. Lee, B. Maycock, L.E. Torheim, K. Nanishi, and D.T.T. Duong, "Climate Change, Food Supply, and Dietary Guidelines," *Annual Review of Public Health* 42 (2021): 233-255.

45. M.L. Niebylski, K.A. Redburn, T. Duhaney, and N.R. Campbell, "Healthy Food Subsidies and Unhealthy Food Taxation: A Systematic Review of the Evidence," *Nutrition* 31, 6 (2015): 787-795; A.M. Thow, S. Downs, and S. Jan, "A Systematic Review of the Effectiveness of Food Taxes and Subsidies to Improve Diets: Understanding the Recent Evidence," *Nutrition Reviews* 72, 9 (2014): 551-565; J.C. Caro, P. Valizadeh, A. Correa, A. Silva, and S.W. Ng, "Combined Fiscal Policies to Promote Healthier Diets: Effects on Purchases and Consumer Welfare," *PLoS One* 15 (2020): e0226731.

46. M.L. Niebylski, K.A. Redburn, T. Duhaney, and N.R. Campbell, "Healthy Food Subsidies and Unhealthy Food Taxation: A Systematic Review of the Evidence," *Nutrition* 31, 6 (2015): 787-795.

47. B.M Popkin, S. Barquera, C. Corvalan, et al., "Towards Unified and Impactful Policies to Reduce Ultra-Processed Food Consumption and Promote Healthier Eating," *Lancet Diabetes & Endocrinology* 9, 7 (2021): 462-470.

48. World Bank, *Taxes on Sugar-Sweetened Beverages: Summary of International Evidence and Experiences* (Washington, DC: 2020).

49. G. Sacks, J. Kwon, and K. Backholer, "Do Taxes on Unhealthy Foods and Beverages Influence Food Purchases?" *Current Nutrition Reports* 10 (2021): 179-187.

50. M. Springmann, D. Mason-D'Croz, S. Robinson, K. Wiebe, H.C.J. Godfray, M. Rayner, and P. Scarborough, "Health-Motivated Taxes on Red and Processed Meat: A Modelling Study on Optimal Tax Levels and Associated Health Impacts," *PLoS One* 13 (2018): e0204139.

51. M. Springmann, D. Mason-D'Croz, S. Robinson, et al., "Mitigation Potential and Global Health Impacts from Emissions Pricing of Food Commodities," *Nature Climate Change* 7 (2017): 69-74.

52. C. Latka, M. Kuiper, S. Frank, et al., "Paying the Price for Environmentally Sustainable and Healthy EU Diets," *Global Food Security* 28 (2021): 100437.

53. I.D. Brouwer, et al., "Reverse Thinking: Taking a Healthy Diet Perspective towards Food Systems Transformations," *Food Security* 13 (2021): 1497-1523.

54. M. Reyes, M.L. Garmendia, S. Olivares, C. Aqueveque, I. Zacarías, and C. Corvalán, "Development of the Chilean Front-of-Package Food Warning Label," *BMC Public Health* 19 (2019): 906.

55. B.M Popkin, et al., "Towards Unified and Impactful Policies to Reduce Ultra-Processed Food Consumption and Promote Healthier Eating," *Lancet Diabetes & Endocrinology* 9, 7 (2021): 462-470.

56. S. Shangguan, A. Afshin, M. Shulkin, et al., "A Meta-Analysis of Food Labeling Effects on Consumer Diet Behaviors and Industry Practices," *American Journal of Preventive Medicine* 56, 2 (2019): 300-314; R. An, Y. Shi, J. Shen, T. Bullard, G. Liu, Q. Yang, N. Chen, and L. Cao, "Effect of Front-of-Package Nutrition Labeling on Food Purchases: A Systematic Review," *Public Health* 191 (2021): 59-67.

57. S. Shangguan, et al., "A Meta-Analysis of Food Labeling Effects on Consumer Diet Behaviors and Industry Practices," *American Journal of Preventive Medicine* 56, 2 (2019): 300-314.

58. A. Bastounis, J. Buckell, J. Hartmann-Boyce, et al., "The Impact of Environmental Sustainability Labels on Willingness-to-Pay for Foods: A Systematic Review and Meta-Analysis of Discrete Choice Experiments," *Nutrients* 13, 8 (2013): 2677; A. Rondoni and S. Grasso, "Consumers Behaviour Towards Carbon Footprint Labels on Food: A Review of the Literature and Discussion of Industry Implications," *Journal of Cleaner Production* 301 (2021): 127031.

59. L.S. Taillie, E. Busey, F. Mediano Stoltze, and F.R. Dillman Carpentier, "Governmental Policies to Reduce Unhealthy Food Marketing to Children," *Nutrition Reviews* 77, 11 (2019): 787-816.

60. S.A. Chambers, R. Freeman, A.S. Anderson, and S. MacGillivray, "Reducing the Volume, Exposure and Negative Impacts of Advertising for Foods High in Fat, Sugar and Salt to Children: A Systematic Review of the Evidence from Statutory and Self-Regulatory Actions and Educational Measures," *Preventive Medicine* 75 (2015): 32-43.

61. R. Kanter, M. Reyes, B. Swinburn, S. Vandevijvere, and C. Corvalán, "The Food Supply Prior to the Implementation of the Chilean Law of Food Labeling and Advertising," *Nutrients* 11, 1 (2018): 52; S. Vandevijvere and L. Vanderlee, "Effect of Formulation, Labelling, and Taxation Policies on the Nutritional Quality of the Food Supply," *Current Nutrition Reports* 8 (2019): 240-249; L.S. Taillie, M. Reyes, M.A. Colchero, B. Popkin, and C. Corvalán, "An Evaluation of Chile's Law of Food Labeling and Advertising on Sugar-Sweetened Beverage Purchases from 2015 to 2017: A Before-and-After Study," *PLoS Medicine* 17, 2 (2020): e1003015.

CHAPTER 9

1. IEA (International Energy Agency), *World Energy Outlook 2021* (Paris: 2021).

2. IPCC (Intergovernmental Panel on Climate Change), "Summary for Policymakers," in *Global Warming of 1.5°C*, ed. V. Masson-Delmotte, P. Zhai, H-O. Pörtner, et al. (Geneva: 2018), 1-24.

3. IEA, *World Energy Outlook 2021* (Paris: 2021).

4. T. Shah, A. Rajan, G.P. Rai, et al., "Solar Pumps and South Asia's Energy-Groundwater Nexus: Exploring Implications and Reimagining Its Future," *Environmental Research Letters* 13 (2018): 115003

5. A. Kishore, P.K. Joshi, and D. Pandey. "Droughts, Distress, and Policies for Drought-Proofing Agriculture in Bihar, India," in *Climate Smart Agriculture in South Asia*, B.Pal, A. Kishore, P.K. Joshi, and N. Tyagi, eds. (Singapore: Springer, 2019).

6. E. Borgstein, D.K. Mekonnen, and K. Wade, *Capturing the Productive Use Dividend: Valuing the Synergies between Rural Electrification and Smallholder Agriculture in Ethiopia*, Insight Brief (Basalt, Colorado: Rocky Mountain Institute, 2020).

7. M. Jeuland, T.R. Fetter, Y. Li, et al., "Is Energy the Golden Thread? A Systematic Review of the Impacts of Modern and Traditional Energy Use in Low- and Middle-Income Countries," *Renewable and Sustainable Energy Reviews* 135 (2021): 110406

8. F. Mamuye, B. Lemma, and T. Woldeamanuel, "Emissions and Fuel Use Performance of Two Improved Stoves and Determinants of Their Adoption in Dodola, Southeastern Ethiopia," *Sustainable Environment Research* 28, 1 (2018): 32-38; J. Fanzo, C. Davis, R. McLaren, and J. Choufani, "The Effect of Climate Change across Food Systems: Implications for Nutrition Outcomes," *Global Food Security* 18 (2018): 12-19.

9. K. Kaygusuz, "Energy for Sustainable Development: A Case of Developing Countries," *Renewable and Sustainable Energy Reviews* 16, 2 (2012): 1116-1126; S. Dutta, A. Kooijman, and E. Cecelski, *Energy Access and Gender: Getting the Right Balance*, (Washington, DC: World Bank, 2017); D. Mekonnen, E. Bryan, T. Alemu, and C. Ringler, "Food versus Fuel: Examining Tradeoffs in the Allocation of Biomass Energy Sources to Domestic and Productive Uses in Ethiopia," *Agricultural Economics* 48, 4 (2017): 425-435.

10 J.A. Burney, S.J. Davis, and D.B. Lobell, "Greenhouse Gas Mitigation by Agricultural Intensification," *PNAS* 107, 26 (2010): 12052-12057.

11 WHO (World Health Organization), *Fuel for Life: Household Energy and Health* (Washington, DC: 2017).

12 M. Amare, C. Arndt, K.A. Abay, and T. Benson, "Urbanization and Child Nutritional Outcomes," *World Bank Economic Review* 34, 1 (2020): 63-74.

13 T. Fujii, A.S. Shonchoy, and S. Xu, "Impact of Electrification on Children's Nutritional Status in Rural Bangladesh," *World Development* 102 (2018): 315-330.

14 A. Rajan, K. Ghosh, and A. Shah, "Carbon Footprint of India's Groundwater Irrigation," *Carbon Management* 11, 3 (2020): 265-280.

15 J.R. Rohr, C.B. Barrett, D.J. Civitello, et al., "Emerging Human Infectious Diseases and the Links to Global Food Production," *Nature Sustainability* 2 (2019): 445-456.

16 A. Closas and E. Rap, "Solar-Based Groundwater Pumping for Irrigation: Sustainability, Policies, and Limitations," *Energy Policy* 104 (2017): 33-37; H. Xie, C. Ringler, and A. Mondal, "Solar or Diesel: A Comparison of Costs Comparison Analysis for Groundwater-Fed Irrigation in Sub-Saharan Africa under Two Energy Solutions," *Earth's Future* 9, 4 (2021): e2020EF001611.

17 H.H. El-Ghetany, W.I.A. Aly, S.E. Baraka, and T.A.A. Ismail, "Experimental Investigation and Performance Evaluation of a Solar Space Heating/Cooling and Ventilation System for a Poultry House in Egypt," *Journal of International Society for Science and Engineering* 2, 4 (2020): 70-76.

18 K. Gebrehiwot, A.H. Mondal, C. Ringler, and A. Getaneh-Gebremeskel, "Optimization and Cost-Benefit Assessment of Hybrid Power Systems for Off-Grid Rural Electrification in Ethiopia," *Energy* 177, 15 (2019): 234-246.

19 C. Arndt, D. Arent, F. Hartley, B. Merven and A.H. Mondal, "Faster than You Think: Renewable Energy and Developing Countries," *Annual Review of Resource Economics* 11 (2019): 149-168.

20 P. Pavelic, M. Magombeyi, P. Schmitter, and I. Jacobs-Mata, *Sustainable Expansion of Groundwater-Based Solar Water Pumping for Smallholder Farmers in sub-Saharan Africa* (Washington, DC: Efficiency for Access Coalition, 2021).

21 IWMI (International Water Management Institute), "Potentional for Solar Photovolatic Based Irrigation," http://sip.africa.iwmi.org.

22 H. Xie, C. Ringler, and A. Mondal, "Solar or Diesel: A Comparison of Costs Comparison Analysis for Groundwater-Fed Irrigation in Sub-Saharan Africa under Two Energy Solutions," *Earth's Future* 9, 4 (2021): e2020EF001611.

23 R. Shirley, Y. Liu, J. Kakande, and M. Kagarura, "Identifying High-Priority Impact Areas for Electricity Service to Farmlands in Uganda through Geospatial Mapping," *Journal of Agriculture and Food Research* 5 (2021): 100172.

24 E. Borgstein, D.K. Mekonnen, and K. Wade, *Capturing the Productive Use Dividend: Valuing the Synergies between Rural Electrification and Smallholder Agriculture in Ethiopia*, Insight Brief (Basalt, Colorado: Rocky Mountain Institute, 2020).

25 P. Pavelic, et al., *Sustainable Expansion of Groundwater-Based Solar Water Pumping for Smallholder Farmers in sub-Saharan Africa* (Washington, DC: Efficiency for Access Coalition, 2021).

26 S. Uhlenbrook, W. Yu, P. Schmitter, and D. Smith, "Optimising the Water We Eat—Rethinking Policy to Enhance Productive and Sustainable Use of Water in Agri-food Systems across Scales," *Lancet Planetary Health* 6, 1 (2022): e59-e65.

27 O. Seidou, C. Ringler, S. Kalcic, L. Ferrini, T.A. Ramani, and A. Guero, "A Semi-Qualitative Approach to the Operationalization of the Food-Environment-Energy-Water (FE2W) Nexus Concept for Infrastructure Planning: A Case Study of the Niger Basin," *Water International* 46, 5 (2021): 744-770.

28 South African Government, Renewable Independent Power Producer Program, www.gov.za

29 M. Otoo, N. Lefore, P. Schmitter, J. Barron, and G. Gebregziabher, *Business Model Scenarios and Suitability: Smallholder Solar Pump-Based Irrigation in Ethiopia. Agricultural Water Management—Making a Business Case for Smallholders*, IWMI Research Report 172 (Colombo: International Water Management Institute, 2018).

30 J.S. Clancy, S. Dutta, N. Mohlakoana, A. Rojas, and M. Matinga, "The Predicament of Women," in *International Energy and Poverty: The Emerging Contours*, eds., L. Guruswamy (London and New York: Routledge Studies in Energy Policy, 2016), 24-38; L. Mehta, T. Oweis, C. Ringler, and S. Varghese, *Water for Food Security, Nutrition and Social Justice* (New York: Routledge, 2019).

31 S. Theis, N. Lefore, R.S. Meinzen-Dick, and E. Bryan, "What Happens After Technology Adoption? Gendered Aspects of Small-Scale Irrigation Technologies in Ethiopia, Ghana, and Tanzania," *Agriculture and Human Values* 35, 3 (2018): 671-684.

32 NRDC (Natural Resources Defense Council) and SEWA (Self Employed Women's Association), *It Takes a Village: Advancing Hariyali Clean Energy Solutions in Rural India* (New York: NRDC; Ahmedabad, India: SEWA, 2021).

BOX 1

a I. Pappis, A. Sahlberg, R. Walle, O. Broad, E. Eludoyin, M. Howells, and W. Usher, "Influence of Electrification Pathways in the Electricity Sector of Ethiopia: Policy Implications Linking Spatial Electrification Analysis and Medium to Long-Term Energy Planning," *Energies* 14, 4 (2021): 1209.

b "List of Countries by Electricity Consumption," Wikipedia, updated April 13, 2022.

c A. Mekonnen and R. Hiremath, "Rural Electrification Challenges and Implementation Strategies through Micro Grid Approch in Ethiopian Context," *International Journal of Advanced Research in Electrical, Electronics and Instrumentation Engineering* 10, 5 (2021): 1375-1385.

BOX 2

a Statistics South Africa, General Household Survey (2019).

b South Africa, Department of Energy, Non-Grid Policy Guidelines for Integrated National Electrification Programme (2018).

CHAPTER 10

1 M. Chui, M. Evers, J. Manyika, A. Zheng, and T. Nisbet, *The Bio Revolution: Innovations Transforming Economies, Societies, and Our Lives* (San Francisco and Hamburg: McKinsey Global Institute, 2020); World Economic Forum, *Bio-Innovation in the Food System Towards a New Chapter in Multistakeholder Collaboration*, Bio-Innovation Dialogue Initiative (Geneva: 2018).

2 C.C.M. van der Wiel, J.G. Schaart, L.A.P. Lotz, and M.J.M. Smulders, "New Traits in Crops Produced by Genome Editing Techniques Based on Deletions," *Plant Biotechnology Reports* 11, 1 (2017): 1-8; K. Chen, Y. Wang, R. Zhang, H. Zhang, and C. Gao, "CRISPR/Cas Genome Editing and Precision Plant Breeding in Agriculture," *Annual Review of Plant Biology* 70, 1 (2019): 667-697; N. Wada, R. Ueta, Y. Osakabe and K. Osakabe, "Precision

Genome Editing in Plants: State-of-the-Art in CRISPR/Cas9-based Genome Engineering," *BMC Plant Biology* 20 (2020): 234.

3   https://www.cimmyt.org/projects/mln-gene-editing-project/

4   N.G. Karavolias, W. Horner, M.N. Abugu, and S.N Evanega, "Application of Gene Editing for Climate Change in Agriculture," *Frontiers in Sustainable Food Systems* 5 (2021): 685801; V. Estes, "Innovation to Tackle Climate Change and Feed a Growing Population: Commercializing Synthetic Biology," *The Digest*, December 1, 2015.

5   K.V. Pixley, J.B. Falck-Zepeda, K.E. Giller, et al. "Genome Editing, Gene Drives, and Synthetic Biology: Will They Contribute to Disease-Resistant Crops, and Who Will Benefit?" *Annual Review Phytopathology* 57 (2019): 165-188; J.R. Lamichhane, M. Barzman, K. Booij et al., "Robust Cropping Systems to Tackle Pests under Climate Change. A Review," *Agronomy for Sustainable Development* 35, 2 (2015): 443-459.

6   X. Ma, M. Mau, and T.F. Sharbel, "Genome Editing for Global Food Security," *Trends in Biotechnology* 36, 2 (2018): 123-127; J. Schiemann, J. Robienski, S. Schleissing, A. Spök, T. Sprink, and R.A. Wilhelm, "Editorial: Plant Genome Editing: Policies and Governance," *Frontiers in Plant Science* 11 (2020): 284.

7   P.C. Ronald and R. W. Adamchak, *Tomorrow's Table: Organic Farming, Genetics, and the Future of Food*, 2nd ed. (New York: Oxford University Press, 2018).

8   D. Jenkins, R. Dobert, A. Atanassova, and C. Pavely, "Impacts of the Regulatory Environment for Gene Editing on Delivering Beneficial Products," *In Vitro Cellular and Developmental Biology–Plant* 57, 4 (2021): 609-626.

9   K. Farhall and L. Rickards, "The 'Gender Agenda' in Agriculture for Development and Its (Lack of) Alignment with Feminist Scholarship," *Frontiers in Sustainable Food Systems* 5 (2021): 573424; C.R. Farnworth, L. Badstue, G J. Williams, A. Tegbaru, and H.I.M. Gaya, "Unequal Partners: Associations between Power, Agency and Benefits among Women and Men Maize Farmers in Nigeria," *Gender, Technology and Development* 24, 3 (2020): 271-296.

10 S. Smyth, W. Kerr, and P.W.B. Phillips, "The Unintended Consequences of Technological Change: Winners and Losers from GM Technologies and the Policy Response in the Organic Food Market," *Sustainability* 7 (2015): 7667-7683.

11 L. Klerkx and D. Rose, "Dealing with the Game-Changing Technologies of Agriculture 4.0: How Do We Manage Diversity and Responsibility in Food System Transition Pathways?" *Global Food Security* 24 (March): 100347; D.C. Rose and J. Chilvers, "Agriculture 4.0: Broadening Responsible Innovation in an Era of Smart Farming," *Frontiers of Sustainable Food Systems* 2 (2-18): 87.

12 National Academies of Sciences, Engineering, and Medicine, *Genetically Engineered Crops: Experiences and Prospects* (Washington, DC: 2016); EASAC (European Academies Science Advisory Council), *EASAC and the New Plant Breeding Techniques* (Brussels: 2018); Leopoldina (German National Academy of Sciences), Union of the German Academies of Sciences and Humanities and the German Research Foundation, *Towards a Scientifically Justified, Differentiated Regulation of Genome Edited Plants in the EU* (Berlin: Leopoldina, 2019).

13 N. Graham, G.B. Patil, D.M. Bubeck, et al., "Plant Genome Editing and the Relevance of Off-Target Changes," *Plant Physiology* 183, 4 (2020): 1453-1471; Leopoldina, et al., *Towards a Scientifically Justified, Differentiated Regulation of Genome Edited Plants in the EU*; European Commission, *New Techniques in Agricultural Biotechnology* (Brussels: 2017); ALLEA (All European Academies), "Genome Editing for Crop Improvement," ALLEA-KVAB Symposium Report Summary (Berlin: October 2020).

14 J. Menz, D. Modrzejewski, F. Hartung, R. Wilhelm, and T. Sprink, "Genome Edited Crops Touch the Market: A View on the Global Development and Regulatory Environment," *Frontiers in Plant Science* 11 (2020): 586027.

15 M. Bagley, *Genome Editing in Latin America: CRISPR Patent and Licensing Policy* (Washington, DC: Inter-American Development Bank, 2021).

16 Fontagro, *FONTAGRO Annual Report 2018-2019* (Washington, DC: Inter-American Development Bank, 2019).

17 L. Cornish, "Who Are the Donors Taking on GMOs?" *DEVEX Inside Development: The GMO Debate* (2018).

18 M. Chui, et al., *The Bio Revolution: Innovations Transforming Economies, Societies, and Our Lives* (San Francisco and Hamburg: McKinsey Global Institute, 2020).

19 S.M. Schmidt, M. Belisle, and W.B. Frommer, "The Evolving Landscape Around Genome Editing in Agriculture," *EMBO Reports* 21 (2020): e50680; J. Entine, M.S.S Felipe, J.H. Groenewald, et al., "Regulatory Approaches for Genome Edited Agricultural Plants in Select Countries and Jurisdictions around the World," *Transgenic Research* 30, 4 (2021): 551-584.

20 O.T. Westengen and T. Winge, "New Perspectives on Farmer-Breeder Collaboration in Plant Breeding," in *Farmers and Plant Breeding: Current Approaches and Perspectives*, eds. O.T. Westengen and T. Winge (Abingdon, VA: Routledge/Bioversity International, 2019); M. Bagley, *Genome Editing in Latin America: CRISPR Patent and Licensing Policy* (Washington, DC: Inter-American Development Bank, 2021).

21 B. Lenaerts, B.C.Y. Collard, and M. Demont, "Review: Improving Global Food Security through Accelerated Plant Breeding," *Plant Science* 287 (2019): 110207.

22 National Academies of Sciences, Engineering, and Medicine, *Genetically Engineered Crops: Experiences and Prospects* (Washington, DC: 2016).

23 N.A. Abdallah, C.S. Prakash, and A.G. McHughen, "Genome Editing for Crop Improvement: Challenges and Opportunities," *GM Crops & Food* 6, 4 (2015): 183-205; J. Falck-Zepeda, J. Yorobe, B. Amir Husin, et al., "Estimates and Implications of the Costs of Compliance with Biosafety Regulations in Developing Countries," *GM Crops & Food* 3, 1 (2012): 52-59.

24 F. Hartung and J. Schiemann, "Precise Plant Breeding Using New Genome Editing Techniques: Opportunities, Safety and Regulation in the EU," *The Plant Journal* 78, 5 (2014): 742-752; D. Jenkins, et al., "Impacts of the Regulatory Environment for Gene Editing on Delivering Beneficial Products," *In Vitro Cellular and Developmental Biology-Plant* 57, 4 (2021): 609-626.

25 A. Gatica-Arias, "The Regulatory Current Status of Plant Breeding Technologies in some Latin American and the Caribbean Countries," *Plant Cell, Tissue and Organ Culture* 141 (2020): 229-242; S.M. Schmidt, M. Belisle, and W.B. Frommer, "The Evolving Landscape Around Genome Editing in Agriculture," *EMBO Reports* 21 (2020): e50680.

26 K.V. Pixley, J.B. Falck-Zepeda, R. Paarlberg. et al., "Genome Edited Crops for Improved Food Security of Smallholder Farmers," *Nature Genetics* (forthcoming).

27 USDA Foreign Agricultural Service, *China: MARA Issues First Ever Gene-Editing Guidelines*, GAIN Report (Washington, DC: 2022).

28 R. Lassoued, S.J. Smyth, P.W.B. Phillips, and H. Hesseln, "Regulatory Uncertainty around New Breeding Techniques," *Frontiers in Plant Science* 9 (2018): 1291.

29 H. De Steur, J. Wesana, D. Blancquaert, D. Van Der Straeten, and X. Gellynck, "The Socioeconomics of Genetically Modified Biofortified Crops: A Systematic Review and Meta-Analysis," *Annals of the New York Academy of Sciences* 1390, 1 (2017): 14-33; A.M. Shew, L. Lanier Nalley, H.A. Snell, R.M. Nayga, and B.L. Dixon, "CRISPR versus GMOs: Public Acceptance

and Valuation," *Global Food Security* 19 (December 2018): 71-80; NRC (National Research Council), "Public Engagement on Genetically Modified Organisms: When Science and Citizens Connect: Workshop Summary" (Washington, DC: National Academies Press, 2015); T. Hiroyuki and G.P. Gruère. "Pressure Group Competition and GMO Regulations in Sub-Saharan Africa: Insights from the Becker Model," *Journal of Agricultural and Food Industrial Organization* 9, 1 (2011): 1-19.

30. K. Ludlow, S. Smyth, and J.B. Falck-Zepeda, "Socio-economic Considerations and Potential Implications for Gene-Edited Crops," *Journal of Regulatory Science* 9, 2 (2021): 1-11.

31. M. Qaim, "Role of New Plant Breeding Technologies for Food Security and Sustainable Agricultural Development," *Applied Economic Perspectives and Policy* 42, 2 (2020): 129-150; J. Wesseler, H. Politiek, and D. Zilberman, "The Economics of Regulating New Plant Breeding Technologies: Implications for the Bioeconomy Illustrated by a Survey Among Dutch Plant Breeders," *Frontiers in Plant Science* 10 (2019): 1597.

32. N. Hillson, M. Caddick, Y. Cai, et al., "Building a Global Alliance of Biofoundries," *Nature Communications* 10, 1 (2019): 2040.

33. K. Fuglie, M. Gautam, A. Goyal, and W.F. Maloney, *Harvesting Prosperity: Technology and Productivity Growth in Agriculture* (Washington, DC: World Bank Group, 2020).

34. Exploring innovative mechanisms to address intellectual property limitations, especially in LMICs, including open-source databases for research inputs and club-type IP mechanisms, will be important (such as Addgene database for sharing materials www.addgene.org and MPEG LA consortia for managing IP https://www.mpegla.com/crispr/).

35. J. Zhou, D. Li, G. Wang, F. Wang., M. Kunjal, D. Joldersma, and Z. Liu, "Application and Future Perspective of CRISPR/Cas9 Genome Editing in Fruit Crops," *Journal of Integrative Plant Biology* 62, 3 (2020): 269-286; Y.C. Kim, Y. Kang, E.Y. Yang, et al., "Applications and Major Achievements of Genome Editing in Vegetable Crops: A Review," *Frontiers in Plant Science* 12 (2021): 688980.

36. A.I. Whelan, P. Gutti, and M.A. Lema, "Gene Editing Regulation and Innovation Economics," *Frontiers in Bioengineering and Biotechnology* 8 (2020): 303.

37. A.I. Whelan, et al., "Gene Editing Regulation and Innovation Economics," *Frontiers in Bioengineering and Biotechnology* 8 (2020): 303.

38. A. Galiè, N. Teufel, A. Webb Girard, et al., "Women's Empowerment, Food Security and Nutrition of Pastoral Communities in Tanzania," *Global Food Security* 23 (2019): 125-134; P. Biermayr-Jenzano, "Food Systems, Obesity, and Gender in Latin America," IFPRI LAC Working Paper 1 (IFPRI, Washington, DC, 2019).

39. P. Biermayr-Jenzano, S. Kassam, B. Dhehibi, and A. Aw-Hassan, "Understanding Gender and Poverty Dimensions of High Value Agricultural Commodity Chains in the Souss Masaa-Draa Region of Southwestern Morocco," in *Agricultural Value Chain*, ed. G. Egilmez (London: IntechOpen, 2014), 25-61; FAO, "Cassava in Latin America and the Caribbean: A Look at the Potential of the Crop to Promote Agricultural Development and Economic Growth" (FAO, Clayuca Corporation, and IICA, 2016).

40. National Academies, *Genetically Engineered Crops: Experiences and Prospects* (Washington, DC: 2016); K.V. Pixley, J.B. Falck-Zepeda, K.E. Giller, et al., "Genome Editing, Gene Drives, and Synthetic Biology: Will They Contribute to Disease-Resistant Crops, and Who Will Benefit?" *Annual Review of Phytopathology* 57 (2019): 165-188.

41. D.J.S. Hamburger, "Normative Criteria and Their Inclusion in a Regulatory Framework for New Plant Varieties Derived from Genome Editing," *Frontiers in Bioengineering and Biotechnology* 6 (2018): 176.

42. J.E. Huesing, D. Andres, M.P. Braverman, et al., "Global Adoption of Genetically Modified (GM) Crops: Challenges for the Public Sector," *Journal Agricultural Food Chemistry* 64 (2016): 394-402.

43. S. Ceccarelli, *People I Have Known* (Cali, Colombia: CIAT-PRGA, 2011); S. Ceccarelli, A. Galie, Y. Mustafa, and S. Grando, "Syria: Participatory Barley Breeding-Farmers' Input Becomes Everyone's Gain," in *The Custodians of Biodiversity*, ed. M. Ruiz and R. Vernooy (New York: Earthscan USA, 2012), 53-66.

44. Y. Song and R. Vernoy, "Seeds of Empowerment: Action Research in the Context of the Feminization of Agriculture in Southwest China," *Gender, Technology & Development* 14, 1 (2010): 25-44; L. Sperling, J.A. Ashby, M.E. Smith, E. Weltzien, and S.M. Guire, "A Framework for Analyzing Participatory Plant Breeding Approaches and Results," *Euphytica* 122 (2001): 439-450.

45. O.C. Ezezika, J. Deadman, and A.S. Daar, "She Came, She Saw, She Sowed: Re-negotiating Gender-Responsive Priorities for Effective Development of Agricultural Biotechnology in sub-Saharan Africa," *Journal of Agricultural and Environmental Ethics* 26 (2012): 461-471.

46. A. Galiè, J. Jiggins, P.C. Struik, S. Grando, and S. Ceccarelli, "Women's Empowerment through Seed Improvement and Seed Governance: Evidence from Participatory Barley Breeding in Pre-War Syria," *NJAS-Wageningen Journal of Life Sciences* 81, 1 (2017): 1-8.

47. E. Díaz-Bonilla, F. Paz, and P. Biermayr-Jenzano, "Nutrition Policies and Interventions for Overweight and Obesity: A Review of Conceptual Frameworks and Classifications," IFPRI LAC Working Paper 6 (IFPRI, Washington, DC, 2020); P. Biermayr-Jenzano, "Obesidad y Género: Enfoque de Género y Salud Alimentaria Nutricional en América Latina," IFPRI LAC Working Paper 8 (IFPRI, Washington, DC, 2020).

48. Z. Tadele, "Challenges of Food Security for Orphan Crops," *Encyclopedia Food Security Sustainability* 1 (2019): 403-408

CHAPTER 11

1. T. Reardon, M.F. Bellemare, and D. Zilberman, "How COVID-19 May Disrupt Food Supply Chains in Developing Countries," Chapter 17, in *COVID-19 and Global Food Security*, J. Swinnen and J. McDermott, eds. (Washington, DC: IFPRI, 2020).

2. M. Burke, E. Miguel, S. Satyanath, J. Dykem, and D. Lobell, "Warming Increases the Risk of War in Africa," *PNAS* 106, 49 (2009): 20670-20674; C. Hendrix and H.-J. Brinkman, "Food Insecurity and Conflict: Causal Linkages and Complex Dynamics," *Stability: International Journal of Security and Development* 2 (2): 26.

3. W. Schlenker and M. Roberts, "Nonlinear Temperature Effects Indicate Severe Damages to U.S. Crop Yields under Climate Change," *PNAS* 106, 37 (2009): 15594-15598.

4. M. Burke, W.M. Davis, and N.S. Diffenbaugh, "Large Potential Reduction in Economic Damages under UN Mitigation Targets," *Nature* 557, 7706 (2018): 549-553.

5. For a detailed description of transaction costs and their implications in value chains, see A. de Brauw and E. Bulte, *African Farmers, Value Chains, and Agricultural Development: An Economic and Institutional Perspective* (London: Palgrave MacMillan, 2021).

6. C.K. Khoury, A.D. Bjorkman, H. Dempewolf, et al., "Increasing Homogeneity in Global Food Supplies and the Implications for Food Security," *PNAS* 111, 11 (2014): 4001-4006.

7   J. Fanzo, R. McLaren, C. Davis, and J. Choufani, "Climate Change and Variability: What are the Risks for Nutrition, Diets, and Food Systems?" IFPRI Discussion Paper 01645 (IFPRI, Washington, DC, 2017).

8   M.C. Tirado, R. Clarke, L.A. Jaykus, A. McQuatters-Gollop, and J.M. Frank, "Climate Change and Food Safety: A Review," *Food Research International* 43, 7 (2010): 1745-1765.

9   FAO (Food and Agriculture Organization of the United Nations), *The State of Food and Agriculture 2019: Moving Forward on Food Loss and Waste Reduction* (Rome: 2019).

10 L. Delgado, M. Schuster, and M. Torero, "On the Origins of Food Loss," *Applied Economic Perspectives and Policy* 43, 2 (2021): 750-780.

11 L. Gatere, V. Hoffmann, M. Murphy, and P. Scollard, *Food Safety in Tomatoes Produced in Laikipia County*, Voices for Change Partnership Policy Brief (Washington, DC: IFPRI, 2020).

12 S.J. Vermeulen, B.M. Campbell, and J.S.I. Ingram, "Climate Change and Food Systems," *Annual Review of Environment and Resources* 37 (2012): 195-222.

13 See, for example, N. Traore, and J. Foltz, "Temperatures, Productivity, and Firm Competitiveness in Developing Countries: Evidence From Africa," paper presented at the Agricultural & Applied Economics Association Annual Meeting (Washington, DC, 2018).

14 L.S.O Liverpool-Tasie and C.M. Parkhi, "Climate Risk and Technology Adoption in the Midstream of Value Chains: Evidence from Nigerian Maize Traders," *Journal of Agricultural Economics* 72, 1 (2021): 158-179.

15 K. Ambler, A. de Brauw, S. Herskowitz and M. Murphy, "Gender and Start-Up Capital for Agri-Food MSMEs in Indonesia and Viet Nam," project note (Washington, DC: IFPRI, 2020).

16 C. Martin-Shields and W. Stojetz, "Food Security and Conflict: Empirical Challenges and Future Opportunities for Research and Policy Making on Food Security and Conflict," FAO Agricultural Development Economics Working Paper 18-04, (Rome: FAO, 2018); M. Humphreys, "Natural Resources, Conflict, and Conflict Resolution: Uncovering the Mechanisms," *Journal of Conflict Resolution* 49, 4 (2016): 508-537; O. Koren and B.E. Bagozzi, "From Global to Local, Food Insecurity is Associated with Contemporary Armed Conflicts," *Food Security* 8, 5 (2016): 999-1010.

17 C.S. Hendrix and H.-J. Brinkman, "Food Insecurity and Conflict Dynamics: Causal Linkages and Complex Feedbacks," *Stability: International Journal of Security & Development* 2, 2 (2013): 1-18.

18 L-E. Cederman, N.B. Weidmann, and K.S. Gleditsch, "Horizontal Inequalities and Ethnonationalist Civil War: A Global Comparison," *American Political Science Review* 105, 3 (2011): 478-495; G. Østby, "Polarization, Horizontal Inequalities and Violent Civil Conflict," *Journal of Peace Research* 45, 2 (2008): 143-162; G. Østby, R. Nordås, and J.K. Rød, "Regional Inequalities and Civil Conflict in Sub-Saharan Africa," *International Studies Quarterly* 53, 2 (2009): 301-324.

19 E.-M. Meemken, C.B. Barrett, H.C. Michelson, M. Qaim, T. Reardon, and J. Sellare, "The Role of Sustainability Standards in Global Agrifood Supply Chains," *Nature Food* 2 (2020): 758-765.

20 E.-M. Meemken, "Do Smallholder Farmers Benefit from Sustainability Standards? A Systematic Review and Meta-Analysis," *Global Food Security* 26 (2020): 100373.

21 B. van Rijsbergen, W. Elbers, R. Ruben, and S.N. Njuguna, "The Ambivalent Impact of Coffee Certification on Farmers' Welfare: A Matched Panel Approach for Cooperatives in Central Kenya," *World Development* 77 (2016): 277-292.

22 The lower estimate is provided by J. Poore and T. Nemecek ("Reducing Food's Environmental Impacts through Producers and Consumers," *Science* 360, 6392 (2018): 987-992), and the higher by a computation based on F. Tubiello, C. Rosenzweig, G. Conchedda, et al. ("Greenhouse Gas Emissions from Food Systems: Building the Evidence Base," *Environmental Research Letters* 16, 6 (2021): 065007).

23 J. Winfree and P. Watson, "The Welfare Economics of 'Buy Local'," *American Journal of Agricultural Economics* 99, 4 (2017): 971-987.

24 For example, N. Magnan, V. Hoffmann, N. Opuku, G. Gajate Garrido, and D.A. Kanyam, "Information, Technology, and Market Rewards: Incentivizing Aflatoxin Control in Ghana," *Journal of Development Economics* 151 (2021): 102620.

25 L. Leavens, J. Bauchet, and J. Ricker-Gilbert, "After the Project Is Over: Measuring Longer-Term Impacts of a Food Safety Intervention in Senegal," *World Development* 141 (2021): 105414.

26 R. Bandyopadhyay, A. Ortega-Beltran, A. Akande, et al., "Biological Control of Aflatoxins in Africa: Current Status and Potential Challenges in the Face of Climate Change," *World Mycotoxin Journal* 9, 5 (2016): 771-789.

27 See for example, M. Sadi, and A. Arabkoohsar, "Techno-Economic Analysis of Off-Grid Solar-Driven Cold Storage Systems for Preventing the Waste of Agricultural Products in Hot and Humid Climates," *Journal of Cleaner Production* 275 (2020): 24143.

28 E. Kusano, "The Cold Chain for Agri-Food Products in ASEAN," ERIA Research Project Report 2018, No. 11. (Jakarta: Economic Research Institute for ASEAN and East Asia, 2019).

29 A. Dresse, A, Fischhendler, J.O. Nielsen, and D. Zikos, "Environmental Peacebuilding: Towards a Theoretical Framework," *Cooperation and Conflict* 54, 1 (2018): 99-119.

30 T. Ide, "Space, Discourse and Environmental Peacebuilding," *Third World Quarterly* 38, 3 (2016): 544-562.

CHAPTER 12

1   S. Ornes, "Core Concept: How Does Climate Change Influence Extreme Weather? Impact Attribution Research Seeks Answers," *PNAS* 115, 33 (2018): 8232-8235.

2   D.N. Mubiru, M. Radeny, F.B. Kyazze, A. Zziwa, J. Lwasa, J. Kinyangi, and C. Mungai, "Climate Trends, Risks and Coping Strategies in Smallholder Farming Systems in Uganda," *Climate Risk Management* 22 (2018): 4-21; B. Traore, M. Corbeels, M.T. van Wijk, M.C. Rufino, and K.E. Giller, "Effects of Climate Variability and Climate Change on Crop Production in Southern Mali," *European Journal of Agronomy* 49 (2013): 115-125; M.T. Wakjira, N. Peleg, D. Anghileri, D. Molnar, T. Alamirew, J. Six, and P. Molnar, "Rainfall Seasonality and Timing: Implications for Cereal Crop Production in Ethiopia," *Agricultural and Forest Meteorology* 310 (2021): 108633.

3   B. Langenbrunner, "Water, Water Not Everywhere," *Nature Climate Change* 11, 8 (2021): 650; UN News, "Madagascar: Severe Drought Could Spur World's First Climate Change Famine," UN News, October 21, 2021.

4   A.B. Smith, *U.S. Billion-Dollar Weather and Climate Disasters, 1980–Present (NCEI Accession 0209268)*, NOAA National Centers for Environmental Information, Dataset (March 17, 2020).

5   C.M. Godde, D. Mason-D'Croz, D.E. Mayberry, P.K. Thornton, and M. Herrero, "Impacts of Climate Change on the Livestock Food Supply Chain: A Review of the Evidence," *Global Food Security* 28 (2021): 100488; J. Woetzel, D. Pinner, H. Samandari, H. Engel, M. Krishnan, C. Kampel, and J. Graabak, *Could Climate Become the Weak Link in Your Supply Chain?* (Washington, DC: McKinsey Global Institute, 2020).

6   GSMA (GSM Association), *Digital Agriculture Maps: 2020 State of the Sector in Low- and Middle-Income Countries* (London: 2020).

7. M. Tsan, S. Totapally, M. Hailu, and B. Addom, *Digitalisation of Africa Agriculture Report: 2018-2019* (Netherlands: Proud Press, 2019).

8. R. Fabregas, M. Kremer, and F. Schilbach, "Realizing the Potential of Digital Development: The Case of Agricultural Advice," *Science* 366, 6471 (2019): eaay3038.

9. G.T. Abate, T. Bernard, S. Makhija, and D.J. Spielman, "Accelerating Technical Change through Video-Mediated Agricultural Extension: Evidence from Ethiopia," IFPRI Discussion Paper 1852 (IFPRI, Washington, DC, 2019).

10. C. Vaughan, J. Hansen, P. Roudier, P. Watkiss, and E. Carr, "Evaluating Agricultural Weather and Climate Services in Africa: Evidence, Methods, and a Learning Agenda," *WIREs Climate Change* 10, 4 (2019): e 586; J. van Etten, K. de Sousa, A. Aguilar, et al., "Crop Variety Management for Climate Adaptation Supported by Citizen Science," *PNAS* 116, 10 (2019): 4194-4199.

11. D-A. An-Vo, S. Mushtaq, K. Reardon-Smith, et al., "Value of Seasonal Forecasting for Sugarcane Farm Irrigation Planning," *European Journal of Agronomy* 104 (2019): 37-48.

12. W. Kropff, D. Jimenez, A. Molero, G. Smith, Z. Mehrabi, et al., CGIAR's Role in Digital Extension Services (Cali, Colombia: Alliance of Bioversity International and CIAT, 2021).

13. B. Kramer, J. Hellin, J. Hansen, A. Rose, and M. Braun, "Building Resilience through Climate Risk Insurance: Insights from Agricultural Research for Development," CCAFS Working Paper No. 287 (CGIAR Research Program on Climate Change, Agriculture and Food Security, Wageningen, Netherlands, 2019).

14. B. Kramer and F. Ceballos, "Enhancing Adaptive Capacity through Climate-Smart Insurance: Theory and Evidence from India," in *10th International Conference of Agricultural Economists* (Vancouver: IFPRI, 2018)

15. E. Benami, Z. Jin, M.R. Carter, A. Ghosh, R.J. Hijmans, A. Hobbs, B. Kenduiywo, and D.B. Lobell, "Uniting Remote Sensing, Crop Modelling and Economics for Agricultural Risk Management," *Nature Reviews Earth & Environment* 2, 2 (2021): 140-159.

16. M.C. Annosi, F. Brunetta, F. Bimbo, and M. Kostoula, "Digitalization within Food Supply Chains to Prevent Food Waste. Drivers, Barriers and Collaboration Practice," *Industrial Marketing Management*, 93 (2021): 208-220; F. Casino, V. Kanakaris, T.K. Dasaklis, S. Moschuris, S. Stachtiaris, M. Pagoni, and N.P. Rachaniotis, "Blockchain-based Food Supply Chain Traceability: A Case Study in the Dairy Sector," *International Journal of Production Research* 59, 19 (2021): 5758-5770; P. Müller and M. Schmid, "Intelligent Packaging in the Food Sector: A Brief Overview," *Foods* 8, 1 (2019): 16; S. Wang, X. Liu, M. Yang, Y. Zhang, K. Xiang, and R. Tang, "Review of Time Temperature Indicators as Quality Monitors in Food Packaging: Review of Time Temperature Indicator," *Packaging Technology and Science* 28, 10 (2015): 839-867.

17. F. Ceballos and B. Kramer, "The Role of Asymmetric Information in Multi-Peril Picture-Based Crop Insurance: Field Experiments in India," IFPRI Discussion Paper 2088 (IFPRI, Washington, DC, 2021).

18. M. Jouanjean, *Digital Opportunities for Trade in the Agriculture and Food Sectors,* OECD Food, Agriculture and Fisheries Papers No. 122 (OECD, 2019); J. von Braun, A. Gulati, and H. Kharas, "Key Policy Actions for Sustainable Land and Water Use to Serve People," *Economics*, 11, 1 (2017): 20170032.

19. ECMWF (European Centre for Medium-Range Weather Forecasts), *Long Range Climate Forecast* (Reading, UK: 2017).

20. C. Zhang, "Bridging the Gap Between Weather and Climate Predictions," *Eos* (June 8, 2020).

21. B. Kramer, R. Rusconi, and J.W. Glauber, "Five Years of Regional Risk Pooling: An Updated Cost-Benefit Analysis of the African Risk Capacity," IFPRI Discussion Paper 1965 (IFPRI, Washington, DC, 2020).

22. D. Perera, O. Seidou, J. Agnihotri, M. Rasmy, V. Smakhtin, P. Coulibaly, and H. Mehmood, *Flood Early Warning Systems: A Review of Benefits, Challenges, and Prospects* (Hamilton, Canada: UN University Institute for Water, Environment and Health, 2019).

23. WMO (World Meteorological Organization), *2019 State of Climate Services* (Geneva: 2019).

24. J.C. Aker and I.M. Mbiti, "Mobile Phones and Economic Development in Africa," *Journal of Economic Perspectives* 24, 3 (2010): 207-232.

25. GSMA, *The State of Mobile Internet Connectivity 2021* (London: 2021).

26. Z. Mehrabi, M.J. McDowell, V. Ricciardi, et al., "The Global Divide in Data-Driven Farming," *Nature Sustainability* 4, 2 (2021): 154-160.

27. M. Tsan, S. Totapally, M. Hailu, and B. Addom, *Digitalisation of Africa Agriculture Report: 2018-2019* (Netherlands: Proud Press, 2019).

28. IEA (International Energy Agency), *Access to Electricity–SDG7: Data and Projections–Analysis* (Paris: 2020).

29. A. Steiner, G. Aguilar, K. Bomba, J.P. Bonilla, and A. Campbell, Actions to Transform Food Systems under Climate Change (Wageningen, Netherlands: CCAFS, 2020).

30. S. Fritz, L. See, J.C.L. Bayas, F. et al., "A Comparison of Global Agricultural Monitoring Systems and Current Gaps," *Agricultural Systems* 168 (2019): 258-272.

31. M. Ng, N. de haan, B. King, and S. Langan, *Promoting Inclusivity and Equity in Information and Communications Technology for Food, Land, and Water, Systems* (Cali: CGIAR Platform for Big Data in Agriculture, 2021).

32. C. Highet, A. Salman, and N. Singh, "The Digital Gender Divide Won't Close by Itself – Here's Why," *CGAP Blog*, December 15, 2020.

33. GSMA, *Connected Women: The Mobile Gender Gap Report 2021* (London: 2021).

34. EQUALS Research Group, *Taking Stock: Data and Evidence on Gender Equality in Digital Access, Skills, and Leadership* (Geneva: EQUALS Global Partnership, 2019).

35. United Nations, "SDG 9 Indicators," 2020. https://unstats.un.org/sdgs/report/2020/goal-09/

36. A. Ferrari, M. Bacco, K. Gaber, et al., "Drivers, Barriers and Impacts of Digitalisation in Rural Areas from the Viewpoint of Experts," *Information and Software Technology* 145 (2022): 106816.

37. ITU (International Telecommunication Union) and UNESCO, *Connecting Africa Through Broadband: A Strategy for Doubling Connectivity by 2021 and Reaching Universal Access by 2030* (Geneva: Broadband Commission for Sustainable Development, 2019).

38. GSMA, *The State of Mobile Internet Connectivity 2021* (London: 2021).

39. ITU and UNESCO, *21st Century Financing Models for Bridging Broadband Connectivity Gaps* (Geneva: Broadband Commission for Sustainable Development, 2021).

40. B. Chakravorti, "How to Close the Digital Divide in the U.S.," *Harvard Business Review*, July 20, 2021; P. Romer, "Taxing Digital Advertising," May 17, 2021 (https://adtax.paulromer.net/).

41. E. Bryan, "Getting 'Gender Right' is Essential for the Sustainability of Food Systems," *Agrilinks Blog,* January 23, 2020.

42. L.J. Brown, Aflatoxins in Food and Feed: Impacts, Risks, and Management Strategies, GCAN Policy Note (Washington, DC: IFPRI, 2018).

43 USAID, "Closing the Gender Digital Divide," updated November 22, 2021; WCC (WomenConnect Challenge), *WomenConnect Fact Sheet: Gram Vaani* (Washington, DC: USAID, 2018); WCC, *WomenConnect Fact Sheet: EPoD India at IFMR* (Washington, DC: USAID, 2018); WCC, *WomenConnect Fact Sheet: Viamo* (Washington, DC: USAID, 2018c).

44 K. Hao, "There's an Easy Way to Make Lending Fairer for Women. Trouble Is, It's Illegal," *MIT Technology Review*, November 15, 2019; IPA (Innovations for Poverty Action), *The Impact of a Women-Specific Credit Scoring Model on Women's Access to Credit in the Dominican Republic* (Washington, DC: 2018).

45 T. Allen and P. Prosperi, "Modeling Sustainable Food Systems," *Environmental Management* 57, 5 (2016): 956-975; T. Hertel, I. Elouafi, M. Tanticharoen, and F. Ewert, "Diversification for Enhanced Food Systems Resilience," *Nature Food* 2, 11 (2021): 832-834.

46 World Bank, *World Development Report 2021: Data for Better Lives* (Washington, DC: 2021).

47 B. Roseth, A. Reyes, and Y.A. Karla, "The Value of Official Statistics: Lessons from Intergovernmental Transfers," IDB Technical Note No. 168 (Washington, DC: Inter-American Development Bank, 2019).

48 M. McMillan, "Understanding African Poverty over the Longue Durée: A Review of Africa's Development in Historical Perspective," *Journal of Economic Literature* 54, 3 (2016): 893-905.

49 J. Koo, A. Mamun, and W. Martin, "From Bad to Worse: Poverty Impacts of Food Availability Responses to Weather Shocks," *Agricultural Economics* 52, 5 (2021): 833-847.

50 J.Fanzo, L. Haddad, R. McLaren, et al., "The Food Systems Dashboard Is a New Tool to Inform Better Food Policy," *Nature Food* 1, 5 (2020): 243-246. See the Dashboard at https://foodsystemsdashboard.org.

51 World Bank, *World Development Report 2021: Data for Better Lives* (Washington, DC: 2021).

52 C. Arndt, A. Dale, P. Hatzenbuehler, J. Kyle, I. Matshe, A. Schlosser, and E. Schmidt, "A Data Revolution for Agricultural Production Statistics in Sub-Saharan Africa," *Ghanaian Journal of Economics* 6, 1 (2018): 54-73.

53 World Bank, *Digital Development Partnership: Annual Review 2020* (Washington, DC: 2020).

54 WMO, *2020 State of Climate Services* (Geneva: 2020); WMO, *State of the Climate in Africa* (Geneva: 2020).

55 R. Chason and R. Ombuor, "A Lack of Weather Data in Africa Is Thwarting Critical Climate Research," *Washington Post*, September 24, 2021.

56 B. King, M. Devare, M. Overduin, et al., *Toward a Digital One CGIAR: Strategic Research on Digital Transformation in Food, Land, and Water Systems in a Climate Crisis* (Cali, Colombia: CIAT, 2021).; S. Melhem and A.H. Jacobsen, *A Global Study on Digital Capabilities for the Digital Economy* (Washington, DC: World Bank Group, 2021).

57 FAO (Food and Agricultural Organization), *Global Outlook on Climate Services in Agriculture* (Rome: 2021).

58 T. Ferdinand, E. Illick-Frank, L. Postema, et al., *A Blueprint for Digital Climate-Informed Advisory Services: Building the Resilience of 300 Million Small-Scale Producers by 2030* (Washington, DC: World Resources Institute, 2021).

59 T. Ferdinand, et al., *A Blueprint for Digital Climate-Informed Advisory Services* (Washington, DC: World Resources Institute, 2021).

60 L.C. Howe, B. MacInnis, J.A. Krosnick, E.M. Markowitz, and R. Socolow, "Acknowledging Uncertainty Impacts Public Acceptance of Climate Scientists' Predictions," *Nature Climate Change* 9, 11 (2019): 863-867; A. Kause, W. Bruine Bruin, S. Domingos, N. Mittal, J. Lowe, and F. Fung, "Communications about Uncertainty in Scientific Climate-Related Findings: A Qualitative Systematic Review," *Environmental Research Letters* 16, 5 (2021): 053005.

61 T. Ferdinand, et al., *A Blueprint for Digital Climate-Informed Advisory Services* (Washington, DC: World Resources Institute, 2021).

62 H. Kreibich, P. Hudson, and B. Merz, "Knowing What to Do Substantially Improves the Effectiveness of Flood Early Warning," *Bulletin of the American Meteorological Society* 102, 7 (2021): E1450-E1463.

63 A.J. Challinor, C. Müller, S. Asseng, et al., "Improving the Use of Crop Models for Risk Assessment and Climate Change Adaptation," *Agricultural Systems* 159 (2018): 296-306.

64 World Bank, *Narok County Flood Early Warning Communication Strategy* (Washington, DC: 2020).

65 M.C. Parlasca, C. Johnen, and M. Qaim, "Use of Mobile Financial Services among Farmers in Africa: Insights from Kenya," *Global Food Security* 32 (2022): 100590.

66 EFI Insight-Finance, *Women in Agriculture Using Digital Financial Services: Lessons Learned from Technical Assistance Support to DigiFarm, Fenix, and myAgro* (Washington, DC: World Bank, 2021).

67 GPFI (Global Partnership for Financial Inclusion), *G20 High-Level Policy Guidelines on Digital Financial Inclusion for Youth, Women and SMEs*, Saudi Arabia G20 Presidency (Washington, DC: World Bank, 2020).

BOX 1

a The dataset is available at https://radiantearth.github.io/stac-browser/#/external/raw.githubusercontent.com/khufkens/EotG_data/main/release_v1/catalog.json.

REGIONAL DEVELOPMENTS

AFRICA

1 T.S. Jayne, F. Meyer, and L.N. Traub, "Africa's Evolving Food Systems: Drivers of Change and the Scope for Influencing Them," IIED Working Paper (International Institute for Economic Development, London, 2014); D. Tschirley, T. Reardon, M. Dolislager, and J. Snyder, "The Rise of a Middle Class in East and Southern Africa: Implications for Food System Transformation," *Journal of International Development* 27, 5 (2015): 628-646; D. Tschirley, S. Haggblade, and T. Reardon, eds., *Africa's Emerging Food System Transformation: Eastern and Southern Africa* (East Lansing, MI: Global Center for Food Systems Innovation, 2014).

2 Yields vary according to farming systems. However, most crops and livestock currently have productivities around one-quarter or less of their potential; see J. Dixon, P. Garrity, J-M. Boffa, T.O. Williams, and T. Amede, *Farming Systems and Food Security in Africa: Priorities for Science and Policy under Global Change* (New York: Routledge, 2019).

3 About 80% to 90% of farms in Africa are smallholdings operating with less than 5 ha; see AGRA (Alliance for a Green Revolution in Africa), *Africa Agricultural Status Report 2016: Progress Towards Agricultural Transformation in Africa* (Nairobi: 2016). About 80% of processed foods in sub-Saharan Africa comes from small and medium enterprises; see T. Reardon, D. Tschirley, L.S.O. Liverpool-Tasie, et al., "The Processed Food Revolution in African Food Systems and the Double Burden of Malnutrition," *Global Food Security* 28 (2021): 100466.

4 A recent review of the literature explores the contribution of women (as well as the trade-offs women make vis-à-vis their access to or control over

resources and decision-making that affect their ability to participate) in various segments of the food system (E. Mkandawire, M. Mentz-Coetzee, M.N. Mangheni, and E. Barusi, "Enhancing the Glopan Food Systems Framework by Integrating Gender: Relevance for Women in African Agriculture," *Sustainability* 13 (2021): 8564).

5   WMO (World Meteorological Organization), *State of Climate in Africa 2019* (Geneva: 2020).

6   Sea-level increase reached 5 mm per year in several oceanic areas surrounding the continent, especially from Madagascar toward and beyond Mauritius, which is more than the average global sea-level rise of 3-4 mm per year. About 56% of the coastlines in Benin, Côte d'Ivoire, Senegal, and Togo are eroding and expected to worsen in the future (WMO, *State of Climate in Africa 2019* [Geneva: 2020]).

7   Africa CDC, "Coronavirus Disease 2019 (COVID-19)," accessed November 11, 2021.

8   J.B. Mwesigwa and K.K. Mwangi, "Desert Locust Invasion in Eastern Africa: Why the Current Invasion Reached Unprecedented Levels," *ICPAC-Medium*, April 8, 2020.

9   H. De Groote, S.C. Kimenju, B. Munyua, et al., "Spread and Impact of Fall Armyworm (*Spodoptera frugiperda*) in Maize Production Areas of Kenya," *Agriculture, Ecosystems & Environment* 292 (2020): 106804; S. Niassy, M.K. Agbodzavu, E. Kimathi, et al., "Bioecology of Fall Armyworm *Spodoptera frugiperda*, Its Management and Potential Patterns of Seasonal Spread in Africa," *PLoS ONE* 16, 6 (2021): e0249042.

10 FAO (Food and Agricultural Organization of the United Nations), *The State of Food Security and Nutrition in the World 2021: Transforming Food Systems for Food Security, Improved Nutrition, and Affordable Healthy Diets for All* (Rome: 2021).

11 FAO, *The Impact of Disasters and Crises on Agriculture and Food Security: 2021* (Rome: 2021).

12 FAO, *The State of Food and Agriculture: Making Agri-food Systems More Resilient to Shocks and Stresses* (Rome: 2021).

13 AUDA (African Union Development Agency), "Malabo Declaration on Accelerated Agricultural Growth and Transformation for Shared Prosperity and Improved Livelihoods," in *Twenty Third Ordinary Session of the AU Assembly* (Malabo: 2014).

14 The indicators used to measure and track these commitments are: 1) policies that enable national or regional response to disasters; 2) budget lines that fund early warning and response systems; and 3) proportion of vulnerable households that are covered by weather-based index insurance or social protection schemes; see AUC (African Union Commission), *Document for Preparing Country Biennial Review Report on Progress Made for Achieving the Malabo Declaration Goals and Targets: Technical Guidelines* (Addis Ababa: 2017).

15 AUC, *Second Biennial Review Report of the African Union Commission on the Implementation of the Malabo Declaration on Accelerated Agricultural Growth and Transformation for Shared Prosperity and Improved Livelihoods* (Addis Ababa: 2020). The report of the third biennial review is expected to be released at the African Union Summit in early 2022.

16 J.E. Kurtz and J.M. Ulimwengu, "Biennial Review 2019: Commitment 6: Enhancing Resilience to Climate Variability," *Africa Agriculture Transformation Scorecard* October 2020 (Washington, DC: IFPRI, 2020).

17 AUC, *First Biennial Review Report of the African Union Commission on the Implementation of the Malabo Declaration* (Addis Ababa: 2018).

18 The two African countries that have not submitted their first NDCs are Libya and Sahrawi (UNFCC, "NDC Registry," accessed November 19, 2021).

19 G. Savvidou, A. Atteridge, K. Omari-Motsumi, and C.H. Trisos, "Quantifying International Public Finance for Climate Change Adaptation in Africa," *Climate Policy* 21, 8 (2021): 1020-1036; AU, *African Union Green Recovery Action Plan 2021-2027* (Addis Ababa: 2021).

20 H. Ritchie, "Who Has Contributed Most to Global CO2 Emissions?" Our World in Data (October 19, 2019).

21 The M&E framework includes both qualitative and quantitative context-appropriate indicators that allow for aggregation and comparison of adaptation and means of implementation, as well as greenhouse gas emissions and actions to reduce them at different levels (UNDP, *Kenya Climate Smart Agriculture Implementation Framework* [Nairobi: 2018]).

22 AfDB (African Development Bank Group), "African NDC Hub," accessed November 19, 2021.

23 O. Badiane, J. Collins, and T. Makombe, "Africa," in *2020 Global Food Policy Report* (Washington, DC: IFPRI, 2020), 66-87.

24 IFC (International Finance Corporation), *Digital Financial Services for Agriculture Handbook* (Washington, DC: 2018).

25 N. Ntukamazina, R.N. Onwonga, R. Sommer, J.C. Rubyogo, C.M. Mukankusi, J. Mburu, and R. Kariuki, "Index-Based Agricultural Insurance Products: Challenges, Opportunities and Prospects for Uptake in Sub-Sahara Africa," *Journal of Agriculture and Rural Development in the Tropics and Subtropics* 118, 2 (2017): 171-185.

26 D. Anghileria, V. Bozzini, P. Molnar, A.J. Jamali, and J. Sheffield, "Comparison of Hydrological and Vegetation Remote Sensing Datasets as Proxies for Rainfed Maize Yield in Malawi," *Agricultural Water Management* 262 (2022): 107375.

27 S. Chantarat, A.G.Mude, C.B.Barrett, and M.R.Carter, "Designing Index-Based Livestock Insurance for Managing Asset Risk in Northern Kenya," *Journal of Risk and Insurance* 80 (2013): 205-237.

28 African Risk Capacity, "Drought,"accessed March 2022. https://www.arc.int/drought

29 ACRE-Africa, "Product Categories," accessed March 2022. https://acreafrica.com/product-categories/

30 F. Cecchi, J. Chegeh, S. D. Aredo, B. Kivuva, B. Kramer, L. Waithaka, and C. Waweru, "Climate-Smart Crop Insurance to Promote Adoption of Stress-Tolerant Seeds: Midterm Findings from a Cluster Randomized Trial," Project Note (IFPRI, Washington, DC, 2021).

31 T.G.A. Hamilton and S. Kelly, "Low Carbon Energy Scenarios for Sub-Saharan Africa: An Input-Output Analysis on the Effects of Universal Energy Access and Economic Growth," *Energy Policy* 105 (2017): 303-319.

32 N. Moore, D. Glandon, J. Tripney, et al., *Systematic Review on the Impact of Access to Electricity on Household Welfare,* Systematic Review Report (Asian Development Bank, 2020).

33 T. Ndwiga, R. Kei, H. Jepngetich, and K. Korrir, "Assessment of Health Effects Related to the Use of Biomass Fuel and Indoor Air Pollution in Kapkokwon Sub-location, Bomet Country, Kenya," *Open Journal of Air Pollution* 3, 3 (2014): 61-69.

34 WMO, *State of Climate in Africa 2019* (Geneva: 2020).

35 F. Kemausuor, M.D. Sedzro, and I. Osei, "Decentralised Energy Systems in Africa: Coordination and Integration of Off-Grid and Grid Power Systems: Review of Planning Tools to Identify Renewable Energy Deployment Options for Rural Electrification in Africa," *Current Sustainable/Renewable Energy Reports* 5 (2018): 214-223.

36 Sustainable Energy for All (SEforALL), *State of the Global Mini-Grids Report 2020* (Vienna: 2020).

37 H. Ritchie, "Who Has Contributed Most to Global CO2 Emissions?" *Our World in Data* (Oct. 1, 2019). Adaptation measures can also be mitigation measures. For example, climate-smart agriculture (including conservation tillage, cover cropping, nutrient management, and agroforesty) can accomplish both by enhancing resilience and reducing emissions, in addition to increasing productivity through sustainable intensification.

38 N. Clay and K. S. Zimmerer, "Who Is Resilient in Africa's Green Revolution? Sustainable Intensification and Climate Smart Agriculture in Rwanda," *Land Use Policy* 97 (2020): 104558.

39 AU, *African Union Green Recovery Action Plan 2021-2027* (Addis Ababa: 2021).

40 A.A. Adenle, J.D. Ford, J. Morton, et al., "Managing Climate Change Risks in Africa: A Global Perspective," *Ecological Economics* 141 (2017): 190-201.

41 A.A. Adenle, et al., "Managing Climate Change Risks in Africa," *Ecological Economics* 141 (2017): 190-201.

42 ERA (previously known as the CSA Compendium) is a meta-dataset and analytical engine with tools to compare the performance of agricultural technologies. See https://era.ccafs.cgiar.org.

43 A.A. Adenle, et al., "Managing Climate Change Risks in Africa," *Ecological Economics* 141 (2017): 190-201.

44 AU, *African Union Green Recovery Action Plan 2021-2027* (Addis Ababa: 2021).

45 K. Omari-Motsumi, M. Barnett, and L. Schalatek, *Broken Connections and Systemic Barriers: Overcoming the Challenge of the 'Missing Middle' in Adaptation Finance* (Rotterdam: Global Center on Adaptation, 2019): A.A. Adenle, et al., "Managing Climate Change Risks in Africa," *Ecological Economics* 141 (2017): 190-201.

MIDDLE EAST AND NORTH AFRICA

1 IMF (International Monetary Fund), *Middle East and Central Asia: Building Forward Better*, Regional Economic Outlook (Washington, DC: 2021); World Bank Group, *Overconfident: How Economic and Health Fault Lines Left the Middle East and North Africa Ill-Prepared to Face COVID-19*, MENA Economic Update (Washington, DC: 2021).

2 K.A. Abay, L. Abdelfattah, C. Breisinger, J. Glauber, and D. Laborde, "The Russia-Ukraine Crisis Poses a Serious Food Security Threat for Egypt," *IFPRI Blog*, March 14, 2022.; S. Kurdi, C. Breisinger, J. Glauber, and D. Laborde, "The Russian Invasion of Ukraine Threatens to Further Exacerbate the Food Security in Yemen," *IFPRI Blog*, March 23, 2022.

3 A.A. Khalil and M.K. Hassanein, "Extreme Weather Events and Negative Impacts on Egyptian Agriculture," *International Journal of Advanced Research* 4, 12 (2016): 1843-1851; "'Dragon Storm' Batters Egypt with Torrential Rains, Killing Nearly 20," *Al-Monitor*, March 13, 2020h; "Can Egypt Be Ready for the Next 'Dragon Storm'?" *Enterprise*, March 18, 2020.

4 N.D. Perez, Y. Kassim, C. Ringler, T.S. Thomas, H. ElDidi, and C. Breisinger, *Climate-Resilience Policies and Investments for Egypt's Agriculture Sector: Sustaining Productivity and Food Security*, Food Policy Report (Washington, DC: IFPRI, 2021).

5 World Bank, *Beyond Scarcity: Water Security in the Middle East and North Africa* (Washington, DC: 2017).

6 World Bank, *Beyond Scarcity* (Washington, DC: 2017).

7 World Bank Group, *MENA Crisis Tracker - 12/6/2021* (Washington, DC: 2021).

8 M. Antonelli and S. Tamea, "Food-Water Security and Virtual Water Trade in the Middle East and North Africa," *International Journal of Water Resources Development* 31, 3 (2015): 326-342; E. Borgomeo, N.A.M. Fawzi, J.W. Hall, et al., "Tackling the Trickle: Ensuring Sustainable Water Management in the Arab Region," *Earth's Future* 8, 5 (2020): e2020EF001495.

9 A. Elmahdi and A. Badawy, "Equity and Egypt's Water Paradigm Shift," *Alternative Policy Solutions* (Cairo: American University in Cairo, 2021).

10 N. Singh and G. Brandolini, *Enhancing Rural Resilience in Yemen: Joint Programme*, Final Evaluation Report (Geneva: UNDP, ILO, FAO and WFP, 2019).

11 K.A. Abay, G.T. Abate, J. Chamberlin, Y. Kassim, and D.J. Spielman, "Digital Tools and Agricultural Market Transformation in Africa: Why Are They Not at Scale Yet, and What Will It Take to Get There?" IFPRI Discussion Paper 2092 (IFPRI, Washington, DC, 2021); A. Elmahdi, M. Nassif, and M. Abi Saab, "Phone App Gives Opportunity to Improve Water Productivity in Lebanon," *IWMI Blog*, April 21, 2020.

12 N.D. Perez, et al., *Climate-Resilience Policies and Investments for Egypt's Agriculture Sector: Sustaining Productivity and Food Security*, Food Policy Report(Washington, DC: IFPRI, 2021).

13 C. Breisinger, A. Mukashov, M. Raouf, and M. Wiebelt, "Energy Subsidy Reform for Growth and Equity in Egypt: The Approach Matters," *Energy Policy* 129 (2019): 661-671; S. Kurdi, M. Mahmoud, K.A. Abay, and C. Breisinger, "Too Much of a Good Thing? Evidence that Fertilizer Subsidies Lead to Overapplication in Egypt," MENA RP Working Paper 27 (IFPRI, Washington, DC, 2020); C. Breisinger, H. ElDidi, H. El-Enbaby, et al., *Egypt's Takaful and Karama Cash Transfer Program: Evaluation of Program Impacts and Recommendations*, IFPRI Policy Brief (Washington, DC: IFPRI, 2018).

14 H.Y. Khudhaire and H. Naji, "Causes of Abandoned Construction Projects: A Case Study in Iraq," *IOP Conference Series: Materials Science and Engineering* 1105 (2021): 012081.

15 N.D. Perez, et al., *Climate-Resilience Policies and Investments for Egypt's Agriculture Sector: Sustaining Productivity and Food Security*, Food Policy Report (Washington, DC: IFPRI, 2021).

CENTRAL ASIA

1 IPCC (Intergovernmental Panel on Climate Change), "Summary for Policymakers," in *Climate Change 2021: The Physical Science Basis*, contribution of Working Group I to the Sixth Assessment Report of the IPCC (Cambridge: Cambridge University Press, 2022); M. Punkari, P. Droogers, W. Immerzeel, N. Korhonen, A. Lutz, and A. Venäläinen, *Climate Change and Sustainable Water Management in Central Asia*, ADB Central and West Asia Working Paper Series No. 5 (Metro Manila, ADB, 2014).

2 USAID (US Agency for International Development, "Central Asia Climate Risk Profile," fact sheet (Washington, DC: April 2018).

3 World Bank and ADB (Asia Development Bank), *Climate Risk Country Profile: Uzbekistan* (Washington, DC: World Bank Group; Metro Manila: ADB, 2021).

4 World Bank and ADB, *Climate Risk Country Profile: Kazakhstan* (Washington, DC: World Bank Group; ADB: Metro Manila, 2021).

5 Two primary types of drought affect the region: meteorological droughts associated with a precipitation deficit and hydrological droughts associated with a deficit in surface and ground water flow. In addition, depending on cropping patterns and land management practices, these droughts may produce agricultural drought. See G. Naumann, L. Alfieri, K. Wyser, et al., "Global Changes in Drought Conditions Under Different Levels of Warming," *Geophysical Research Letters* 45, 7 (2018): 3285-3296.

6 World Bank and ADB, *Climate Risk Country Profile: Kyrgyz Republic* (Washington, DC: World Bank Group; Metro Manila: ADB, 2021); World Bank and ADB, *Climate Risk Country Profile: Tajikistan* (Washington, DC:

World Bank Group; Metro Manila: ADB, 2021); World Bank and ADB, *Climate Risk Country Profile: Turkmenistan* (Washington, DC: World Bank Group; Metro Manila: ADB, 2021).

7   A. Mirzabaev, J. Goedecke, O. Dubovyk, U. Djanibekov, B.L. Quang, and A. Aw-Hassan, "Economics of Land Degradation in Central Asia," in *Economics of Land Degradation and Improvement: A Global Assessment for Sustainable Development*, eds. E. Nkonya, A. Mirzabaev, and J. von Braun (London: Springer Nature, 2016), 261-290.

8   E.B. Barbier and J.P. Hochard, "Land Degradation and Poverty," *Nature Sustainability* 1 (2018): 623-631.

9   A. Mirzabaev, et al., "Economics of Land Degradation in Central Asia," in *Economics of Land Degradation and Improvement*, eds. E. Nkonya, A. Mirzabaev, and J. von Braun (London: Springer Nature, 2016), 261-290.

10 T. Siegfried, T. Bernauer, R. Guiennet, et al., "Will Climate Change Exacerbate Water Stress in Central Asia?" *Climatic Change* 112, 3 (2012): 881-899; R. Gan, Y. Luo, Q. Zuo and L. Sun, "Effects of Projected Climate Change on the Glacier and Runoff Generation in the Naryn River Basin, Central Asia," *Journal of Hydrology*, 523 (2015): 240-251; A.D. Nikanorova, E.V. Milanova, N.M. Dronin, and N.O. Telnova, "Estimation of Water Deficit under Climate Change and Irrigation Conditions in the Fergana Valley of Central Asia," *Arid Ecosystems* 6, 4 (2016): 260-267; Y. Luo, X. Wang, S. Piao, et al., "Contrasting Streamflow Regimes Induced by Melting Glaciers across the Tien Shan-Pamir-North Karakoram," *Scientific Reports* 8 (2018):16470.

11 World Bank and ADB, *Climate Risk Country Profile: Tajikistan* (Washington, DC: World Bank Group).

12 United Nations, *The Millenium Development Goals Report* (New York: 2015).

13 Hydrologists assess water scarcity by looking at the population-water equation. A country experiences water stress when annual water supplies drop below 1,700 m^3 per person. When annual water supplies drop below 1,000 m^3 per person, a country faces water scarcity, and severe water scarcity below 500 m^3 per person (FAO, *The State of Food Security and Nutrition in Central Asia 2017* [Budapest: 2017]).

14 These are alarmingly unsustainable freshwater withdrawal levels. Once a country reaches a withdrawal level above 100 percent, it starts depleting its renewable groundwater resources, relying on nonrenewable fossil groundwater or nonconventional sources, such as desalinated wastewater and agricultural drainage water.

15 Increase in water stress is measured as change in annual demand of water as a share of annual supply of water, assuming that the demand for water stays constant over time, to allow to measure the impact of climate change alone.

16 J. Woetzel, D. Pinner, H. Samandari, H. Engel, et al., , *Climate Risk and Response: Physical Hazards and Socioeconomic Impacts* (New York: McKinsey Global Institute, 2020).

17 FAO, *The State of Food Security and Nutrition in Central Asia 2017* (Budapest: 2017).

18 UNICEF/WHO/World Bank Joint Child Malnutrition Estimates, May 2021, accessed April 7, 2022.

19 T. Thomas, K. Akramov, R. Robertson, V. Nazareth, and J. Ilyasov, "Climate Change, Agriculture and Crop Yields in Central Asia," IFPRI Discussion Paper 2081 (IFPRI, Washington, DC, 2021).

20 P. Khakimov, J. Aliev, T. Thomas, J. Ilyasov, and S. Dunston, "Climate Change Effects on Agriculture and Food Security in Tajikistan," *Silk Road: A Journal of Eurasian Development* 2, 1 (2020): 89-112.

21 FAO, *The State of Food Security and Nutrition in Central Asia 2020* (Budapest: 2020).

22 S. Babu and S. Djalalov, eds., *Policy Reforms and Agricultural Development in Central Asia*, (New York: Springer, 2006)

23 Development partners are implementing a number of technical assistance and loan projects in Central Asia, including the World Bank's Agricultural Modernization project in Uzbekistan and Sustainable Livestock Programs for Sustainable Results project in Kazakhstan, ADB's Climate- and Disaster-Resilient Irrigation and Drainage Modernization project in Tajikistan, USAID's Tajikistan Agriculture and Water Activity and Land Governance projects, and the EU's Budget Support to the Agricultural Sector of the Republic of Uzbekistan.

24 T. Thomas, et al., "Climate Change, Agriculture and Crop Yields in Central Asia," IFPRI Discussion Paper 2081 (IFPRI, Washington, DC, 2021).

25 UN Economic Commission for Europe (UNECE), European Union Water Initiative (EUWI), and Ministry of Agriculture and Melioration of the Kyrgyz Republic (MoAM), *Modern Irrigation Technologies and Possibility of their Application in Kyrgyzstan: National Policy Dialogue on Integrated Water Resources Management in Kyrgyzstan*. Bishkek (Genva: UNECE; Paris: EUWI; Bishkek: MoAM, 2015); K. Djumaboev, T. Yuldashev, B. Holmatov, and Z. Gafurov, "Assessing Water Use, Energy Use, and Carbon Emissions in Lift Irrigated Areas: A Case Study from Karshi Steppe in Uzbekistan," *Irrigation and Drainage* 68, 3 (2019): 409-419.

26 K. Djumaboev, H. Manthrithilake, J. Ayars, T. Yuldashev, B. Akramov, R. Karshiev, and D. Eshmuratov, "Growing Cotton in Karshi Steppe, Uzbekistan: Water Productivity Differences with Three Different Methods of Irrigation," in *Proceedings of 9th International Micro Irrigation Conference, Aurangabad, India* (Bangalore: Ivy League, 2019), 391-397.

27 K. Djumaboev, et al. , "Assessing Water Use, Energy Use, and Carbon Emissions in Lift Irrigated Areas," *Irrigation and Drainage* 68, 3 (2019): 409-419.

28 The evidence-based policy recommendations of the CGIAR Research Program on Water, Land, and Ecosystems made a significant contribution to the policy change and development and adoption of the state program on water-saving technologies. For more details, see https://www.cgiar.org/annual-report/performance-report-2020/water-technologies-shift-energy-policies-in-uzbekistan/.

29 T. Thomas, et al. , "Climate Change, Agriculture and Crop Yields in Central Asia," IFPRI Discussion Paper 2081 (IFPRI, Washington, DC, 2021).

30 L. Goedde, J. Katz, A. Menard, and J. Revellat, "Agriculture's Connected Future: How Technology Can Yield New Growth," *McKinsey & Company: Our Insights*, October 9, 2020.

31 K. Akramov, L. Carrillo, and K. Kosec, "COVID-19, Rural Poverty, and Women's Role in Decision-Making: Evidence from Khatlon Province in Tajikistan," OSF Preprints: https://osf.io/j7vrm/.

32 M. Russell, "Water in Central Asia: An Increasingly Scarce Resource," European Parliamentary Research Service Briefing PE. 625.181 (2018).

SOUTH ASIA

1   M. Kugelman, "Can South Asia Get Serious about Climate Change?" South Asia Brief, *Foreign Policy* (Nov. 4, 2021).

2   For example, compared to 2019-20, the climate budgets for 2020/21 dropped by 20% in India and 34% in Pakistan in nominal terms.

3   Center for Global Development, "Developing Countries Are Responsible for 63 Percent of Current Carbon Emissions" (webpage, undated).

4. IPCC (Intergovernmental Panel on Climate Change), "Summary for Policymakers," in *Climate Change 2021: The Physical Science Basis*, Contribution of Working Group I to the Sixth Assessment Report of the Intergovernmental Panel on Climate Change, ed. V. Masson-Delmotte, P. Zhai, A. Pirani, (Geneva: 2021).

5. IPCC, "Summary for Policymakers," in *Climate Change 2021: The Physical Science Basis*, Contribution of Working Group I to the Sixth Assessment Report of the Intergovernmental Panel on Climate Change, ed. V.Masson-Delmotte, et al. (Geneva: 2021).

6. IPCC, "Summary for Policymakers," *Climate Change 2022: Impacts, Adaptation and Vulnerability Assessment Report 6*, Working Group II (2022, in press).

7. M.J.U. Khan, A.K.M.S. Islam, M.K. Das, K. Mohammed, S.K. Bala, and G.M. Tarekul Islam, "Observed Trends in Climate Extremes over Bangladesh from 1981 to 2010," *Climate Research* 77 (2019): 45-61.

8. R. Krishnan, J. Sanjay, C. Gnanaseelan, M. Mujumdar, A. Kulkarni, and S. Chakraborty, ed., *Assessment of Climate Change over the Indian Region: A Report of the Ministry of Earth Sciences (MoES)*, Government of India (Singapore: Springer Nature, 2021); N. Scovronick, C. Dora, E. Fletcher, A. Haines, and D. Shindell, "Reduce Short-Lived Climate Pollutants for Multiple Benefits," *Lancet* 386, 10006 (2015): e28-e31.

9. M. Almazroui, S. Saeed, F. Saeed, M.N. Islam, and M. Ismail, "Projections of Precipitation and Temperature over the South Asian Countries in CMIP6," *Earth Systems and Environment* 4 (2020): 297-320.

10. R. Krishnan, et al., eds., *Assessment of Climate Change over the Indian Region, Government of India* (Singapore: Springer Nature, 2021).

11. E.S. Im, J.S. Pal, and E.A.B. Eltahir, ""Deadly Heat Waves Projected in the Densely Populated Agricultural Regions of South Asia," *Science Advances* 3, 8 (2017): 1-8; R. Krishnan, et al., eds., *Assessment of Climate Change over the Indian Region: A Report of the Ministry of Earth Sciences (MoES)*, Government of India (Singapore: Springer Nature, 2021).

12. M.J.U. Khan, et al., "Observed Trends in Climate Extremes over Bangladesh from 1981 to 2010," *Climate Research* 77 (2019): 45-61.

13. IPCC, "Summary for Policymakers," in *Climate Change 2021: The Physical Science Basis. Contribution of Working Group I to the Sixth Assessment Report of the Intergovernmental Panel on Climate Change*, eds., V. Masson-Delmotte, et al., (Geneva: 2021); T. Bolch, J.M. Shea, S. Liu, et al., "Status and Change of the Cryosphere in the Extended Hindu Kush Himalaya Region," in *The Hindu Kush Himalaya Assessment*, 209-255 (Cham, Switzlerand: Springer Nature, 2019).

14. M. Almazroui, S. Saeed, F. Saeed, M.N. Islam, and M. Ismail, "Projections of Precipitation and Temperature over the South Asian Countries in CMIP6," *Earth Systems and Environment* 4 (2020): 297-320.; T. Bolch, et al., "Status and Change of the Cryosphere in the Extended Hindu Kush Himalaya Region," in *The Hindu Kush Himalaya Assessment*, 209-255 (Cham, Switzlerand: Springer Nature, 2019); H. Biemans, C. Siderius, A.F. Lutz, et al., "Importance of Snow and Glacier Meltwater for Agriculture on the Indo-Gangetic Plain," *Nature Sustainability* 2 (2019): 594-601.

15. IPCC, "Summary for Policymakers," in *Climate Change 2021: The Physical Science Basis. Contribution of Working Group I to the Sixth Assessment Report of the Intergovernmental Panel on Climate Change*, eds. V.Masson-Delmotte, et al. (Geneva: 2021).

16. A.V. Kulkarni, T.S. Shirsat, A. Kulkarni, H.S. Negi, and I.M. Bahuguna, "State of Himalayan Cryosphere and Implications for Water Security," *Water Security* 14 (2021): 100101.

17. A.V. Kulkarni, et al., "State of Himalayan Cryosphere and Implications for Water Security."

18. M.J.U. Khan, et al., "Observed Trends in Climate Extremes over Bangladesh from 1981 to 2010," *Climate Research* 77 (2019): 45-61.

19. M. Almazroui, S. Saeed, F. Saeed, M.N. Islam, and M. Ismail, "Projections of Precipitation and Temperature over the South Asian Countries in CMIP6," *Earth Systems and Environment* 4 (2020): 297-320.

20. S. Philip, et al., "Attributing the 2017 Bangladesh Floods from Meteorological and Hydrological Perspectives."

21. G. Amarnath, N. Alahacoon, V. Smakhtin, and P. Aggarwal, "Mapping Multiple Climate-Related Hazards in South Asia," IWMI Research Report No. 170 (International Water Management Institute, Colombo, 2017).

22. A.V. Kulkarni, et al., "State of Himalayan Cryosphere and Implications for Water Security," *Water Security* 14 (2021): 100101; V. Mishra, S. Aadhar, and S.S. Mahto, "Anthropogenic Warming and Intraseasonal Summer Monsoon Variability Amplify the Risk of Future Flash Droughts in India," *npj Climate and Atmospheric Science* 4 (2021): 1.

23. K. Ahmed, S. Shahid, and N. Nawaz, "Impacts of Climate Variability and Change on Seasonal Drought Characteristics of Pakistan," *Atmospheric Research* 214 (2018): 364-374.

24. A.V. Kulkarni, et al., "State of Himalayan Cryosphere and Implications for Water Security," *Water Security* 14 (2021): 100101.

25. S. Philip, S. Sparrow, S.F. Kew, et al., "Attributing the 2017 Bangladesh Floods from Meteorological and Hydrological Perspectives," *Hydrology and Earth System Sciences* 23, 3 (2019): 1409-1429; R.H. Rimi, K. Haustein, E.J. Barbour, and M.R. Allen, "Risks of Pre-monsoon Extreme Rainfall Events of Bangladesh: Is Anthropogenic Climate Change Playing a Role?" *Bulletin of the American Meteorological Society* 100, 1 (2019): S61-S65.

26. A. Ortiz-Bobea, T.R. Ault, C.M. Carrillo, R.G. Chambers, and D.B. Lobell, 2021. "Anthropogenic Climate Change Has Slowed Global Agricultural Productivity Growth," *Nature Climate Change* 11 (2021): 306-312.

27. F. Gaupp, J. Hall, S. Hochrainer-Stigler, and S. Dadson, "Changing Risks of Simultaneous Global Breadbasket Failure," *Nature Climate Change* 10 (2020): 54-57.

28. M.K. Hasan and L. Kumar, "Yield Trends and Variabilities Explained by Climatic Change in Coastal and Non-coastal Areas of Bangladesh," *Science of the Total Environment* 795 (2021): 148814.

29. M. Hasan, M. Alauddin, M.A. Rashid Sarker, M. Jakaria, and M. Alamgir, "Climate Sensitivity of Wheat Yield in Bangladesh: Implications for the United Nations Sustainable Development Goals 2 and 6," *Land Use Policy* 87 (2019):104023.

30. S. Ahmad, G. Abbas, M. Ahmed, et al., "Climate Warming and Management Impact on the Change of Phenology of the Rice-Wheat Cropping System in Punjab, Pakistan," *Field Crops Research* 230 (2019): 46-61.

31. I. Jan, M. Ashfaq, and A.A. Chandio, "Impacts of Climate Change on Yield of Cereal Crops in Northern Climatic Region of Pakistan," *Environmental Science and Pollution Research* 28 (2021): 60235-60245.

32. B. Praveen and P. Sharma, "Climate Change and Its Impacts on Indian Agriculture: An Econometric Analysis," *Journal of Public Affairs*, 20, 1 (2020): e1972

33. P.S. Birthal, J. Hazrana, D.S. Negi, and G. Pandey, "Benefits of Irrigation against Heat Stress in Agriculture: Evidence from Wheat Crop in India," *Agricultural Water Management* 255 (2021): 106950.

34. R.K. Srivastava, R.K. Panda, and A. Chakraborty, "Assessment of Climate Change Impact on Maize Yield and Yield Attributes under Different Climate Change Scenarios in Eastern India," *Ecological Indicators* 120 (2021): 106881.

35 B.B. Shrestha, E.D.P. Perera, S. Kudo, et al., "Assessing Flood Disaster Impacts in Agriculture under Climate Change in the River Basins of Southeast Asia," *Natural Hazards* 97, 1 (2019): 157-192.

36 V. Lauria, I. Das, S. Hazra, et al., "Importance of Fisheries for Food Security across Three Climate Change Vulnerable Deltas," *Science of the Total Environment* 640-641 (2018): 1566-1577.

37 S. Dasgupta, M. Huq, M.G. Mustafa, M.I. Sobhan, and D. Wheeler, "The Impact of Aquatic Salinization on Fish Habitats and Poor Communities in a Changing Climate: Evidence from Southwest Coastal Bangladesh," *Ecological Economics* 139 (2017): 128-139; S.K. Dubey, R.K. Trivedi, B.K. Chand, B. Mandal, and S.K. Rout, "Farmers' Perceptions of Climate Change, Impacts on Freshwater Aquaculture and Adaptation Strategies in Climatic Change Hotspots: A Case of the Indian Sundarban Delta," *Environmental Development* 21 (2017): 38-51.

38 N.L. Bindoff, W.W.L. Cheung, J.G Kairo, et al., "Changing Ocean, Marine Ecosystems, and Dependent Communities," in *IPCC Special Report on the Ocean and Cryosphere in a Changing Climate* (Geneva: 2019), 447-588.; K.L. Oremus, J. Bone, C. Costello, et al., "Governance Challenges for Tropical Nations Losing Fish Species Due to Climate Change," *Nature Sustainability* 3 (2020): 277-280.

39 G. Salem, S. Kazama, and S. Shahid, "Groundwater-Dependent Irrigation Costs and Benefits for Adaptation to Global Change," *Mitigation and Adaptation Strategies for Global Change* 23 (2018): 953-979; S. Sekhri, "Wells, Water, and Welfare: The Impact of Access to Groundwater on Rural Poverty and Conflict," *American Economic Journal: Applied Economics* 6, 3 (2014): 76-102.

40 G. Salem, et al., "Groundwater-Dependent Irrigation Costs and Benefits for Adaptation to Global Change; S. Sekhri, "Wells, Water, and Welfare."

41 A. Asoka and V. Mishra, "A Strong Linkage between Seasonal Crop Growth and Groundwater Storage Variability in India," *Journal of Hydrometeorology* 22, 1 (2020): 1-39; M. Jain, R. Fishman, P. Mondal, et al., "Groundwater Depletion Will Reduce Cropping Intensity in India," *Science Advances* 7, 9 (2021): eabd2849.

42 D. Blakeslee, R. Fishman, and V. Srinivasan, "Way Down in the Hole: Adaptation to Long-Term Water Loss in Rural India," *American Economic Review* 110, 1 (2020): 200-224.

43 A. Asoka, Y. Wada, R. Fishman, and V. Mishra, "Strong Linkage Between Precipitation Intensity and Monsoon Season Groundwater Recharge in India," *Geophysical Research Letters* 45, 11 (2018): 5536-5544.

44 HLPE (High Level Panel of Experts), *Impacts of COVID-19 on Food Security and Nutrition: Developing Effective Policy Responses to Address the Hunger and Malnutrition Pandemic* (Rome: Committee on World Food Security, 2020).

45 B. Mahapatra, M. Walia, C.A.R. Rao, B.M.K. Raju, and N. Saggurti, "Vulnerability of Agriculture to Climate Change Increases the Risk of Child Malnutrition: Evidence from a Large-Scale Observational Study in India," *PLoS One* 16, 6 (2021): e0253637.

46 T. Hasegawa, G. Sakurai, S. Fujimori, K. Takahashi, Y. Hijioka, and T. Masui, "Extreme Climate Events Increase Risk of Global Food Insecurity and Adaptation Needs," *Nature Food* 2 (2021): 587-595.

47 G.C. Nelson, M.W. Rosegrant, J. Koo, et al., *Climate Change: Impact on Agriculture and Costs of Adaptation*, Food Policy Report (Washington, DC: IFPRI, 2009).

48 O. Banerjee, M. Mahzab, S. Raihan, and N. Islam, "An Economy-wide Analysis of Climate Change Impacts on Agriculture and Food Security in Bangladesh," *Climate Change Economics* 6, 1 (2015): 1-17.

49 FAO (Food and Agriculture Organization), *The Impact of Disasters and Crises on Agriculture and Food Security* (Rome: 2018).

50 P.S. Ward and S. Makhija, "New Modalities for Managing Drought Risk in Rainfed Agriculture: Evidence from a Discrete Choice Experiment in Odisha, India," *World Development* 107 (2018): 163-175.

51 O. Banerjee, et al. "An Economy-wide Analysis of Climate Change Impacts on Agriculture and Food Security in Bangladesh," *Climate Change Economics* 6, 1 (2015): 1-17.

52 S. Mehvar, T. Filatova, M.H. Sarker, A. Dastgheib, and R. Ranasinghe, "Climate Change-Driven Losses in Ecosystem Services of Coastal Wetlands: A Case Study in the West Coast of Bangladesh," *Ocean & Coastal Management* 169 (2019): 273-283.

53 F. Dolan, J. Lamontagne, R. Link, M. Hejazi, P. Reed, and J. Edmonds, "Evaluating the Economic Impact of Water Scarcity in a Changing World," *Nature Communications* 12 (2021): 1-10.

54 Y. Khan, Q. Bin, and T. Hassan, "The Impact of Climate Changes on Agriculture Export Trade in Pakistan: Evidence from Time-Series Analysis," *Growth and Change* 50 (2019): 1568-1589.

55 H. Pathak, "Greenhouse Gas Emission from Indian Agriculture: Trends, Drivers, and Mitigation Strategies," *Proceedings of the Indian National Science Academy* 81, 5 (2015): 1133-1149; W.F. Lamb, T. Wiedmann, J. Pongratz, et al., "A Review of Trends and Drivers of Greenhouse Gas Emissions by Sector from 1990 to 2018," *Environmental Research Letters* 16 (2021): 073005; T.B. Sapkota, F. Khanam, G.P. Mathivanan, et al., "Quantifying Opportunities for Greenhouse Gas Emissions Mitigation Using Big Data from Smallholder Crop and Livestock Farmers across Bangladesh," *Science of the Total Environment* 786 (2021): 147344.

56 H. Pathak, "Greenhouse Gas Emission from Indian Agriculture: Trends, Drivers, and Mitigation Strategies," *Proceedings of the Indian National Science Academy* 81, 5 (2015): 1133-1149; W.F. Lamb, et al., "A Review of Trends and Drivers of Greenhouse Gas Emissions by Sector from 1990 to 2018," *Environmental Research Letters* 16 (2021): 073005; T.B. Sapkota, et al., "Quantifying Opportunities for Greenhouse Gas Emissions Mitigation Using Big Data from Smallholder Crop and Livestock Farmers across Bangladesh," *Science of the Total Environment* 786 (2021): 147344.

57 L. Rosa, D.D. Chiarelli, M. Sangiorgio, et al., "Potential for Sustainable Irrigation Expansion in a 3 C Warmer Climate," *PNAS* 117, 47 (2020): 29526-29534.

58 A. Mukherji, "Sustainable Groundwater Management in India Needs a Water-Energy-Food Nexus Approach," *Applied Economic Perspectives and Policy*, 44, 1 (2022): 394-410.

59 G.J. Stads and M. Rahija, *Public Agricultural R&D in South Asia: Greater Government Commitment, Yet Underinvestment Persists*, ASTI Synthesis Report (Washington, DC: IFPRI, 2012).

60 Some even argue that the urea subsidy is justified because Indian agriculture is net taxed. However, this argument misses the point that some high-value commodities, which do not use as much urea, face larger negative protection than rice and wheat.

61 A. Kishore, M. Alvi, and T.J. Krupnik, "Development of Balanced Nutrient Management Innovations in South Asia: Perspectives from Bangladesh, India, Nepal, and Sri Lanka," *Global Food Security* 28 (2021): 100464.

62 A. Kishore, K.V. Praveen, and D. Roy, "Direct Cash Transfer System for Fertilisers: Why it Might be Hard to Implement," *Economic and Political Weekly* 48, 52 (2013): 54-63.

63 T. Shah, A. Rajan, G.P. Rai, A. Verma, and N. Durga, "Solar Pumps and South Asia's Energy-Groundwater Nexus: Exploring Implications and

Reimagining Its Future," *Environmental Research Letters* 13, 11 (2018): 115003.

64 M.P. Gulati, S. Priya, and E.W. Bresnyan, *Grow Solar, Save Water, Double Farmer Income* (Washington, DC: World Bank, 2020).

EAST AND SOUTHEAST ASIA

1 ADB (Asian Development Bank), *The Economics of Climate Change in Southeast Asia: A Regional Review* (Manila, Philippines: ADB, 2009); ASEAN Secretariat, *ASEAN State of Climate Change Report* (ASCCR), (Jakarta: ASEAN, 2021).

2 ASEAN Secretariat, *ASEAN State of Climate Change Report* (ASCCR), (Jakarta: ASEAN, 2021).

3 FAO (Food and Agriculture Organization of the United Nations), *Building Sustainable and Resilient Food Systems in Asia and the Pacific* (Rome: 2020).

4 OECD (Organisation for Economic Co-operation and Development), *Building Food Security and Managing Risk: A Focus on Southeast Asia* (Paris: OECD Publishing, 2017).

5 ADB, *Asian Development Outlook (ADO) 2021 Update: Transforming Agriculture in Asia* (Manila: 2021).

6 IPCC (Intergovernmental Panel on Climate Change), *An IPCC Special Report on Climate Change, Desertification, Land Degradation, Sustainable Land Management, Food Security, and Greenhouse Gas Fluxes in Terrestrial Ecosystems* (Geneva: 2020).

7 FAO, FAOSTAT database, 2020.

8 IFAD (International Fund for Agricultural Development), *An Outlook on Asia's Agricultural and Rural Transformation: Prospects and Options for Making It an Inclusive and Sustainable One* (Rome: 2019).

9 R.C. Estoque, M. Ooba, V. Avitabile, et al., "The Future of Southeast Asia's forests," *Nature Communications* 10 (2019): 1829.

10 ADB, *Assessing the Intended Nationally Determined Contributions of ADB Developing Members* (Manila: 2016).

11 https://www.ccacoalition.org/en/resources/asean-regional-guidelines-promoting-climate-smart-agriculture-csa-practices

12 ADB and IFPRI, *Building Climate Resilience in the Agriculture Sector in Asia and the Pacific* (Mandaluyong City, Philippines: ADB, 2009).

13 A.D. Mason and S. Shetty, *A Resurgent East Asia: Navigating a Changing World. World Bank East Asia and Pacific Regional Report* (Washington, DC: World Bank, 2019).

14 W.J. Barbon, B. Punzalan, R. Wassman, et al., "Scaling of Climate-Smart Agriculture via Climate-Smart Villages in Southeast Asia: Insights and Lessons from Vietnam, Laos, Philippines, Cambodia and Myanmar," CCAFS Working Paper No. 376 (CGIAR Research Program on Climate Change, Agriculture and Food Security, Wageningen, the Netherlands, 2021).

15 IPCC, *An IPCC Special Report on Climate Change, Desertification, Land Degradation, Sustainable Land Management, Food Security, and Greenhouse Gas Fluxes in Terrestrial Ecosystems* (Geneva: 2020).

16 Y. Zhang, S. Fan, K. Chen, X. Feng, X. Zhang, Z. Bai, and X. Wang, "Transforming Agrifood Systems to Achieve China's 2060 Carbon Neutrality Goal," in *China and Global Food Policy Report: Rethinking of Agrifood Systems* (Beijing: AGFEP, 2021).

17 ADB, *The Economics of Climate Change in Southeast Asia* (Manila: 2009).

18 ADB, *Financing Sustainable and Resilient Food Systems in Asia and the Pacific* (Mandaluyong City, Philippines: 2021).

19 OECD, *Building Food Security and Managing Risk* (Paris: 2017).

20 OECD, *Agricultural Policies in Philippines* (Paris: 2017).

LATIN AMERICA AND THE CARIBBEAN

1 E. Díaz-Bonilla, "Democracy and Commodity Cycles in Latin America and the Caribbean," *IFPRI Blog*, November 25, 2019.

2 E. Díaz-Bonilla and V. Piñeiro, "Latin America and the Caribbean," in *2021 Global Food Policy Report: Transforming Food Systems after COVID-19* (Washington, DC: IFPRI, 2021), 74–105.

3 E. Díaz-Bonilla, V. Piñeiro, and D. Laborde Debucquet, "Latin America and the Caribbean: Food Systems in Times of the Pandemic, in *Advances in Food Security and Sustainability* vol. 6, ed. M.J. Cohen, 263–288 (Amsterdam: Elsevier, 2021).

4 E. Díaz-Bonilla, "Using the New IMF Special Drawing Rights for Larger Purposes: Guaranteeing 'Pandemic Recovery Bonds,'" *IFPRI Blog*, October 22, 2021.

5 E. Díaz-Bonilla, "América Latina y el Caribe en la economía verde y azul," *Pensamiento Iberoamericano* 7 (2019): 94–105, and the references cited therein.

6 ECLAC (Economic Commission for Latin America and the Caribbean), *Forest Loss in Latin America and the Caribbean from 1990 to 2020: The Statistical Evidence*, ECLAC Statistical Briefings No. 2 (Santiago, Chile: 2021); see also satellite data at http://www.terra-i.org/terra-i.html

7 E. Díaz-Bonilla, "América Latina y el Caribe en la economía verde y azul," *Pensamiento Iberoamericano* 7 (2019): 94–105; M. Morris, A. Sebastian, V.M.E. Perego, et al., *Future Foodscapes: Re-imagining Agriculture in Latin America and the Caribbean* (Washington, DC: World Bank, 2020).

8 C. Navarrete Frias, "Policy Implementation in a Frontier Region: The Case of Deforestation in the Amazon," University of Southampton, Doctoral Thesis, 2020.

9 https://ukcop26.org/glasgow-leaders-declaration-on-forests-and-land-use/

10 C. Navarrete Frias, "Policy Implementation in a Frontier Region," University of Southampton, Doctoral Thesis, 2020.

11 This section draws extensively on R.G. Echeverría, "Innovation for Sustainable, Healthy, and Inclusive Agrifood Systems and Rural Societies in Latin America and the Caribbean: Framework for Action 2021–2025" (Santiago, Chile: FAO, 2021); See also C. Giraldo, K. Camacho, C. Navarro-Racines, D. Martinez-Baron, S.D. Prager, and J. Ramírez-Villegas, "Outcome Harvesting: Assessment of the Transformations Generated by Local Technical Agroclimatic Committees In Latin America," CCAFS Working Paper No.299 (CGIAR Research Program on Climate Change, Agriculture and Food Security, 2020); C.A. Sova, G. Grosjean, T. Baedeker, et al. "Bringing the Concept of Climate-Smart Agriculture to Life: Insights from CSA Country Profiles Across Africa, Asia, and Latin America" (Washington, DC: World Bank and the International Centre for Tropical Agriculture, 2018).

12 FAO, FAOSTAT database.

13 FAO, "Producción pecuaria en América Latina y el Caribe." https://www.fao.org/americas/prioridades/produccion-pecuaria/es/

14 J. Hyland, "A Record Year for Agtech Activity in Latin America?" *Silicon Valley Bank Blog*, May 3, 2017.

Printed in Great Britain
by Amazon